ADVANCES IN LABORATORY AUTOMATION ROBOTICS 1985

Edited by:

Janet R. Strimaitis
and
Gerald L. Hawk

Zymark Corporation
Zymark Center
Hopkinton, MA USA

Zymark Corporation Hopkinton, MA USA

Zymark Corporation

Zymark Center
Hopkinton, MA 01748 USA

ISBN 0-931565-01-4

Telex 174 104

Current Printing (last digit)

10 9 8 7 6 5 4 3 2 1

PRINTED IN THE UNITED STATES OF AMERICA

PREFACE

This volume contains selected papers presented at the Third International Symposium on Laboratory Robotics-1985 held at the Boston Park Plaza Hotel, Boston, MA, USA, October 20-23, 1985. These papers cover a broad interdisciplinary range of topics on the application of robotic technology to laboratory automation. We are indebted to the authors for these contributed papers and their patience in producing this book.

The editors wish to thank Christine O'Neil, Kathleen Coutu and Karen Hughes for their valuable assistance at all stages in the preparation of this volume. Also, we wish to thank Jan Wittmer for the cover design.

This is Volume 2 of a series on Laboratory Robotics and we hope you find it stimulating as well as providing you with a review of current applications.

Janet R. Strimaitis

Janet R. Strimaitis

Gerald L. Hawk

Gerald L. Hawk

CONTENTS

CONTRIBUTORS

James C. Abbott, Paper Products Division, The Proctor & Gamble Company, 6100 Center Hill Road, Cincinatti, OH 45224

H. Abdou, Quality Control Division, E.R. Squibb & Sons, Inc., Georges Road, New Brunswick, NJ 08903

Peter Abley, Beecham Pharmaceuticals, U.K. Division, Clarendon Road, Worthing, West Sussex, England

James H. Addison, Jr., E.I. du Pont de Nemours and Company, Savannah River Laboratory, Aiken, SC 29808

L. G. Alexakos, Mobil Research and Development Corp., Research Services Section, Billingsport Road, Paulsboro, NJ 08066

M. Antloga, The Standard Oil Company, 4440 Warrensville Center Road, Cleveland, OH 44128

L. R. Austin, Los Alamos National Laboratory, P.O. Box 1663 Los Alamos, NM 87545

Kiyoshi Banno, Tanabe Seiyaku Co., Ltd., 16-89 3-chome, Kashima, Yodogawa-ku, Osaka, Japan

R. Barnes, Quality Control Division, E.R. Squibb & Sons, Inc., Georges Road, New Brunswick, NJ 08903

T. J. Beugelsdijk, Los Alamos National Laboratory, P.O. Box 1663 Los Alamos, NM 87545

John E. Brennan, Biomedical Research and Development, American Red Cross, 9312 Old Georgetown Road, Bethesda, MD 20814

James W. Brodack, Mallinckrodt Institute of Radiology, Washington University School of Medicine, St. Louis, MO 63110

B. A. Brown, Pfizer Inc., Central Research, Groton, CT 06390

Eugene L. Brown, Genetics Institute, Inc., 87 Cambridge Park Drive, Cambridge, MA 02140

John E. Brown, Genetics Institute, Inc., 87 Cambridge Park Drive, Cambridge, MA 02140

R. D. Brown, Monsanto Agricultural Products Co., Luling, LA 70070

R. A. Campbell, The Dow Chemical Company, 1803 Building, Midland, MI 48674

William J. Castellani, Department of Biochemsitry, Cleveland Clinic Foundation, Cleveland, OH 44106

B. Cohen , Quality Control Division, E.R. Squibb & Sons, Georges Road, New Brunswick, NJ 08903

Bruce Jon Compton, Pennwalt Pharmaceutical Division, P.O. Box 1710, Rochester, NY 14603

L. B. Coyne, The Dow Chemical Company, 1803 Building, Midland, MI 48674

J. B. Cross, Phillips Petroleum Company, Research and Development, 225 PL PRC, Bartlesville, OK 74004

Ronald Crouch , Burroughs Wellcome Co., 3030 Cornwallis Rd., Research Triangle Park, NC 27709

J. E. Curley, Pfizer, Inc., Central Research, Groton, CT 06390

Haig Donoian , Xerox Corporation, 800 Phillips Road, Webster, NY 14580

Marilyn Dulitzky , Thomas J. Lipton, 800 Sylvan Avenue, Englewood Cliffs, NJ 07632

G. M. Dyches, E.I. DuPont de Nemours and Company, Savannah River Laboratory, Aiken, SC 29808

Anthony J. Dziabo, Allergan Pharmaceuticals, 2525 DuPont Drive, Irvine, CA 92715

David D. Elks, Analytical Development Laboratories, Burroughs Wellcome Co., P.O. Box 1887, Greenville, NC 27834

Helena Franze , Quality Control, Boehringer Ingelheim Pharmaceuticals, Inc., Danbury, CT 06810

Philip L. Fuchs, Department Of Chemistry, Purdue University, West Lafayette, IN 47907

Yoshiki Fujikawa , Tanabe Seiyaku Co., Ltd., 21 3-chome, Doshomachi, Higashi-ku, Osaka, Japan

Frank E. Gainer, Eli Lilly and Company, Lilly Corporate Center, Indianapolis, IN 46285

R. S. Galen, Department of Biochemistry, Cleveland Clinic Foundation, Cleveland, OH 44106

P. Å. Gateff, Paper Products Division, The Proctor & Gamble Company, 6100 Center Hill Road, Cincinnati, OH 45224

Sidney S. Goldberg, E.I. DuPont de Nemours and Co., Building 402, Room 3009, Experimental Station, Wilmington, DE 19898

Michael D. Grady, Campbell Institute for Research & Technology, Campbell Soup Company, Campbell Place, Camden, NJ 08101

M. P. Granchi, Mobil Research and Development Corp. Research Services Section, Billingsport Road, Paulsboro, NJ 08066

Joseph Grano, Jr. , Monsanto Polymer Products Company, 730 Worcester Street, Springfield, MA 01151

Allan Greenberg , Ortho Pharmaceutical Corp., Route 202, Raritan, NJ 08869

P. Grey, Mobil Research and Development Corp., Research Services Section, Billingsport Road, Paulsboro, NJ 08066

J. Habarta, Ortho Pharmaceutical Corporation, Route 202, Raritan, NJ 08869

D. M. Haile, Monsanto Agricultural Products Company, Luling, LA 70070

Marlene A. Hall, Allergan Pharmaceuticals, 2525 DuPont Drive, Irvine, CA 92715

Kevin J. Halloran, Quality Control, Boehringer Ingelheim Pharmaceuticals, Inc., Danbury, CT 06810

E. Halloran, Ortho Pharmaceutical Corporation, Route 202, Raritan, NJ 08869

Harvey J. Hastings, Campbell Institute for Research & Technology, Campbell Soup Company, Campbell Place, Camden, NJ 08101

C. Hatfield, Zymark Corporation, Zymark Center, Hopkinton, MA 01748

Brian P. Hatton, Beecham Pharmaceuticals, U.K. Division, Clarendon Road, Worthing, West Sussex, England

John F. Heaps, Monsanto Polymer Products Company, 730 Worcester Street, Springfield, MA 01151

Darla J. Higgs, Beltsville Human Nutrition Center, Nutrient Composition Laboratory, Agricultural Research Service, USDA, Beltsville, MD 20705

L. J. Hilliard, Mobil Research and Development Corp., Research Services Section, Billingsport Road, Paulsboro, NJ 08066

O. Hinsvark, Pennwalt Pharmaceutical Division, P.O. Box 1710, Rochester, NY 14603

Mei-Hsia A. Huang, Beltsville Human Nutrition Center, Nutrient Composition Laboratory, Agricultural Research Service, USDA, Beltsville, MD 20705

William M. Hurni, Virus and Cell Biology Research, Merck Sharp & Dohme Research Laboratories, West Point, PA 19486

Guy W. Inman, Analytical Development Laboratories, Burroughs Wellcome Co., P.O. Box 1887, Greenville, NC 27834

John H. Johnson, American Critical Care, 1600 Waukegan Road, McGaw Park, IL 60085

R. D. Jones, Phillips Petroleum Company, Research and Development, 225 PL PRC, Bartlesville, OK 74004

Simon S. Jones, Genetics Institute, Inc., 87 Cambridge Park Drive, Cambridge, MA 02140

P. E. Kastl, The Dow Chemical Company, 1803 Building, Midland, MI 48674

John A. Katzenellenbogen, Mallinckrodt Institute of Radiology, Washington University School of Medicine, St. Louis, MO 63110

Michael R. Kilbourn, Mallinckrodt Institute of Radiology, Washington University School of Medicine, St. Louis, MO 63110

Thomas J. Kinzelman, Zymark Corporation, Zymark Center, Hopkinton, MA 01748

Richard M. Kiral, Allergan Pharmaceuticals, 2525 DuPont Drive, Irvine, CA 92715

Linda M. Kline, Biomedical Research and Development, American Red Cross, 9312 Old Georgetown Road, Bethesda, MD 20814

Kenneth A. Klinger, Borg-Warner Chemicals, Inc., Technical Centre, Washington, WV 26181

Charles R. Knipe, Hewlett Packard Company, Route 41 and Starr Road, Avondale, PA 19311

D. W. Knobeloch, Los Alamos National Laboratory, P.O. Box 1663, Los Alamos, NM 87545

R. J. Kobrin, Mobil Research and Development Corp., Research Services Section, Billingsport Road, Paulsboro, NJ 08066

Edward C. Koeninger, Monsanto Polymer Products Company, 730 Worcester Street, Springfield, MA 01151

L. J. Kostek, Pfizer Inc., Central Research, Groton, CT 06390

Gary W. Kramer, Department of Chemistry, Purdue University, West Lafayette, IN 47907

Steven F. Kramer, Biodecision Laboratories, 5900 Penn Avenue, Pittsburgh, PA 15206

B. E. Kropscott, The Dow Chemical Company, 1803 Building, Midland, MI 48674

T. W. Latimer, Los Alamos National Laboratory, P.O. Box 1663, Los Alamos, NM 87545

Harry G. Lento, Campbell Institute for Research & Technology, Campbell Soup Company, Campbell Place, Camden, NJ 08101

Linda Lester, Xerox Corporation, 800 Phillips Road, Webster, NY 14580

Monte J. Levitt, Biodecision Laboratories, 5900 Penn Avenue, Pittsburgh, PA 15206

Tim Lincoln, Xerox Corporation, 800 Phillips Road, Webster, NY 14580

Timothy J. Lux, Beecham Pharmaceuticals, U.K. Division, Clarendon Road, Worthing, West Sussex, England

Hideyuki Mano, Tanabe Seiyaku, Co., Ltd. 16-89 3-chome, Kashima, Yodogawa-ku, Osaka, Japan

E. J. Marak, Phillips Petroleum Company, Research and Development, 225 PL PRC, Bartlesville, OK 74004

M. Markelov, The Standard Oil Company, 4440 Warrensville Center Road, Cleveland, OH 44128

W. Mason, Ortho Pharmaceutical Corporation, Route 202, Raritan, NJ 08869

Masaaki Matsuo, Tanabe Seiyaku Co., Ltd., 16089 3-chome, Kashima, Yodagawa-ku, Osaka, JAPAN

William J. McAleer, Virus and Cell Biology Research, Merck Sharp & Dohme Research Laboratories, West Point, PA 19486

Robert G. Miller, Ortho Pharmaceutical (Canada) Ltd., 19 Green Belt Drive, Don Mills, Ontario, Canada, M3C 1L9

Jean L. Miller, Burroughs Wellcome Co., 3030 Cornwallis Rd. Research Triangle Park, NC 27709

William J. Miller, Virus and Cell Biology Research, Merck Sharp & Dohme Research Laboratories, West Point, PA 19486

Takaaki Miyamoto, Tanabe Seiyaku Co., Ltd., 16-89 3-chome, Kashima, Yodogawa-ku, Osaka, Japan

Mary M. Passarello, Biodecision Laboratories, 5900 Penn Avenue, Pittsburgh, PA 15206

Jean Petin, Laborlux S.A., B.P. 349 L-4004, Esch-Sur-Alzette, Luxemborg

C. E. Pippenger, Department of Biochemistry, Cleveland Clinic Foundation, Cleveland, OH 44106

Frank M. Prozonic, Air Products and Chemicals, Inc., Allentown, PA 18105

Aris Ragouzeos, Burroughs Wellcome Co., 3030 Cornwallis Rd., Research Triangle Park, NC 27709

J. J. Rollheiser, Midwest Research Institute, 425 Volker Boulevard, Kansas City, MI 64110

S. Romano, Ortho Pharmaceutical Corporation, Route 202, Raritan, NJ 08869

S. A. Schmidt, The Standard Oil Company, Sohio Research Center, 4440 Warrensville Center Road, Cleveland, OH 44128

W. A. Schmidt, Midwest Research Institute, 425 Volker Boulevard, Kansas City, MI 64110

D. N. Schneider, Los Alamos National Laboratory, P.O. Box 1663, Los Alamos, NM 87545

Matthew L. Severns, Biomedical Research and Development, American Red Cross, 9312 Old Georgetown Road, Bethesda, MD 20814

Yukio Shimaoka, Tanabe Seiyaku Co., Ltd., 16-89 3-chome, Kashima, Yodogawa-ku, Osaka, Japan

Reiji Shimizu, Tanabe Seiyaku Co., Ltd., 16-89 3-chome, Kashima, Yodagawa-ku, Osaka, Japan

Larry A. Simonson, Department of Chemistry, Framingham State College, Framingham, MA 01701

J. L. Smith, Microbiological Research, Chesebrough-Pond's Inc., Trumbull Industrial Park, Trumbell, CT 06611

Ragu Srinivas, American Critical Care, 1600 Wankegan Road, McGraw Park, IL 60085

W. F. Sowle, The Dow Chemical Company, 1803 Building, Midland, MI 48674

N. D. Stalnaker, Los Alamos National Laboratory, P.O. Box 1663, Los Alamos, NM 87545

K. M. Stelting, Midwest Research Institute, 425 Volker Boulevard, Kansas City, MI 64110

David Stone, Genetics Institute, Inc., 87 Cambridge Park Drive, Cambridge, MA 02140

A. A. Thurston, Los Alamos National Laboratory, P.O. Box 1663, Los Alamos, NM 87545

John Van Antwerp, Quality Control, Hoffmann-LaRoche Inc., 340 Kingsland St., Nutley, NJ 07110

Joseph T. Vanderslice, Beltsville Human Nutrition Center, Nutrient Composition Laboratory, Agricultural Research Service, USDA Beltsville, MD 20705

Darlene A. Vanstone, Genetics Institute, Inc., 87 Cambridge Park Drive, Cambridge, MA 02140

Robert F. Venteicher, Quality Cotnrol, Hoffmann-LaRoche Inc., 340 Kingsland St., Nutley, NJ 07110

Armand Wagner, Laborlux S.A., B.P. 349, Esch-Sur-Alzette, Luxemborg

P. Walsh, E.R. Squibb & Sons, Georges Road, New Brunswick, NJ 08903

Edward H. Wasmuth, Virus and Cell Biology Research, Merck Sharp & Dohme Research Laboratories, West Point, PA 19486

Michael J. Welch, Mallinckrodt Institute of Radiology, Washington University School of Medicine, St. Louis, MO 63110

L. V. Wilson, Phillips Petroleum Company, Research and Development, 225 PL PRC, Bartlesville, OK 74004

Richard Young, Ortho Pharmaceutical Corp., Route 202, Raritan, NJ

W. Zazulak, Pennwalt Pharmaceutical Division, P.O. Box 1710, Rochester, NY 14603

Francis H. Zenie, Zymark Corporation, Zymark Center, Hopkinton, MA 01748

ROBOTICS IN THE ANALYTICAL LABORATORY - A MANAGEMENT PERSPECTIVE

Frank E. Gainer
Biochemical Division
Eli Lilly and Company
Lilly Corporate Center
Indianapolis, Indiana 46285

ABSTRACT

An analytical laboratory manager discusses his justification for use of robotics within his laboratory. The presentation will include rationale for purchase of the first and subsequent robots along with a review of monetary considerations. The effect of robots on the attitudes of laboratory personnel and the expectations of sample submitters can have a bearing on success or failure of a robotic operation. The manager will share his thoughts on this point as well as his approach to meeting the need for a) special training b) new technician skill levels and c) nontraditional laboratory designs required for robotics. The presentation will end with an overview of current and planned laboratory robotic applications.

INTRODUCTION

The Biochemical Division assists the corporation in developing potential new drug substances (NDS) into quality marketable products and in reducing the cost of manufacturing of existing products. Once the potential NDS is targeted to be a product, the responsibility for its development passes from Development, to Technical Service, to Production and to Quality Control components enroute to the Pharmaceutical Division. The potential NDS's are varied and consist of fermented products, semi-synthetic antibiotics and recombinant DNA products. The principal investigators include biochemists, microbiologists, chemists, geneticists, chemical engineers and bioengineers.

The department for which I am responsible provides analytical support to the four previously mentioned areas. The department not only provides support for the different potential NDS types but also responds to the various requests coming from the principal investigators. To say the least, it is a challenge to meet all the possible combinations of analytical needs and requests that come from such a diverse group of products and people.

Figure 1 shows the trend of activity within the department since 1972. As I graphed the raw data, it was clear that the number of products tested and number of assays per analysts showed a slight increase and the number of assays a steady increase, but the number of analysts had plateaued. If these trends continued and we did not respond properly, then we would likely have severe problems within the departemnt, such as personnel fatigue, productivity decrease, high overtime, lack of credibility in assay results, etc.. We had already been heavily committed to automation and were becoming more automated, but we were not increasing the number of assays per analysts. We were making gains,

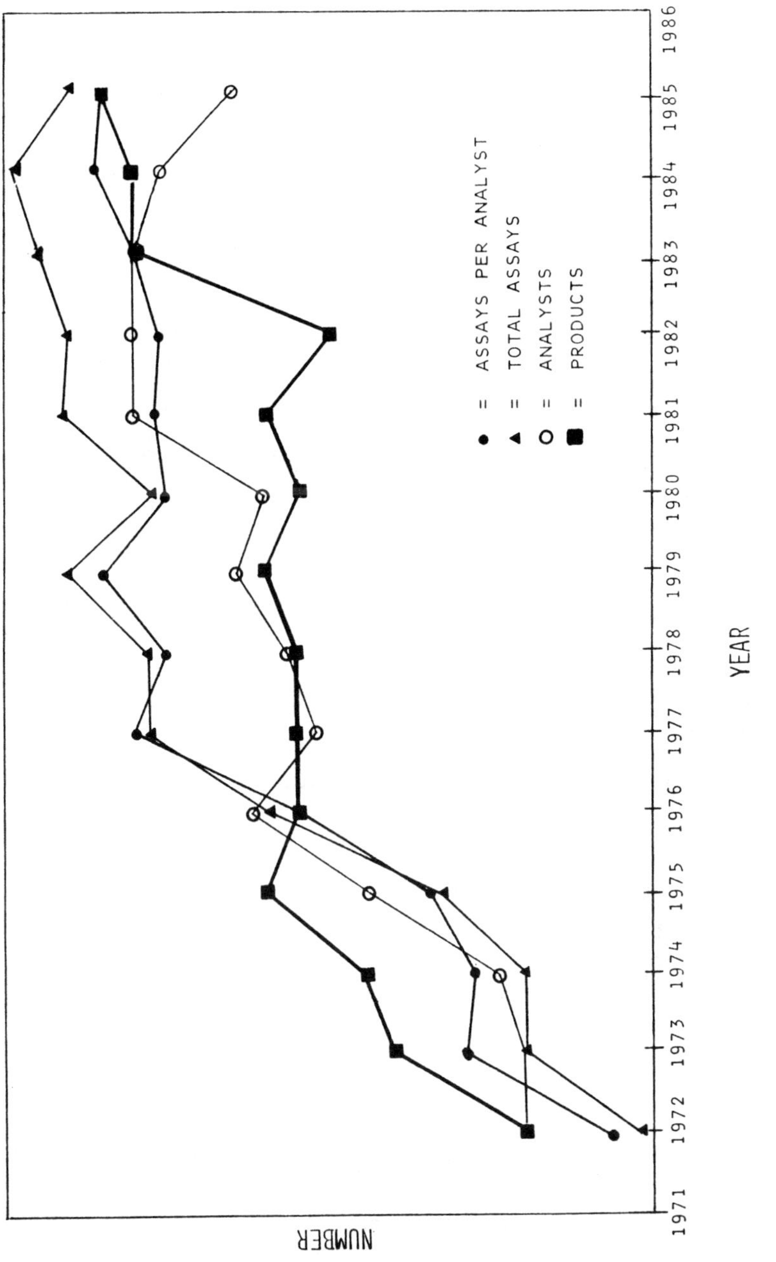

FIGURE 1. Laboratory Trends Since 1972.

but efficiency gains did not offset assay requirements. In fact, some of the perceived problems had already begun to surface. Many lab operations (sampling, injecting, measuring, data reduction) were being performed automatically, but we had no reliable way to automate sample preparation, often the most labor intensive aspect of our assays. We had an ideal set of conditions for justification of robotics.

After considerable discussion during 1982 with my analytical development people, we included funds in the department's Business Plan for a robot for the first time in 1983. In September 1983, we employed a Senior Analytical Chemist to help develop further our strategy for automation, including robotics, in our lab. In February of 1984, funds were approved to purchase our first robot.

JUSTIFICATION FOR ROBOTICS

We made a comprehensive survey of all labor intensive analytical operations to judge their suitability for robotics. By the fall of 1983, we were heavily into the development and production of recombinant DNA products. Sample preparation, prior to assay for most of these products, was both time-consuming and costly. My first thoughts were to choose the most time-consuming and labor intensive sample preparation scheme within the department. Biosynthetic Human Insulin (BHI) fermentation sample preparation was clearly the winner by these criteria. Significant technician time and dollars could be saved by automating the BHI fermentation sample preparation. I told the new chemist that obviously all investigators involved in the BHI project were enthusiastic and hopeful of our success, and he took me seriously.

The chemist performed a thorough assessment of the situation and came back with the recommendation that we not proceed with a project as ambitious as BHI sample preparation with 28 discrete steps. He said that neither he nor Zymark were ready to tackle that complex assignment. I listened to more of his reasoning and agreed to a less complex, but very important project, particularly after he assured me that we were almost sure to fail with the BHI project as initially planned. Table 1 shows the particulars for justification of the first robot for a 9 step recombinant DNA sample preparation scheme.

TABLE 1. Justification for Robotic System to Perform Recombinant DNA Product Sample Preparation.

Procedure - LUO's or sub-classes

1. Receive sample
2. Centrifuge 10 minutes
3. Aspirate supernatant
4. Add reagent
5. Sonicate 2 minutes
6. Stir 16 hours
7. Centrifuge 10 minutes
8. Filter
9. HPLC assay

	Manual	Robot
Sample Load	45	48
Analyst Time	5 1/2 hrs	1 hr

Advantages for Use of Robot

1. Reproducible sample preparation
2. Unattended operation
3. Tow-fold increase in sample throughput

ROI Calculation

Capital Budget:	$41,555.00
Analyst Time Saved:	4 1/2 hrs/day
Annual Dollar Savings:	$20,508.00
Assumed Inflation	6%/year
Nominal ROI	27.38% over 5 years

Everyone in the division was excited about the arrival of our first robot and eager to see it. Many had preconceived ideas about what the robot would look like and their ideas were not too far removed from my 12 year old son's version of a robot as seen in Figure 2. I believe it accurate to say that some people were somewhat disappointed to find that the robot was not standing up and walking around.

This first robot (named Rudy II) was operating and preparing samples routinely within about three months after the equipment was delivered to our lab. During this early period of laboratory robotic developemnt and application, we took special care to observe the mood of technicians, sample submitters, personnel relations people and upper management. Our observations paid off during that three month period, and we felt confident that we had been successful in our first attempt at laboratory automation via robotics.

FIGURE 2. Twelve year old boy's concept of a robot.

As the robot began to prepare our samples routinely around October 1, 1984, I began to promote the use of robotics in the analytical laboratory to others within the corporation and pave the way for additional robot applications. The second robot (named Eugene) was justified on basis of information shown in Table 2, and approval to purchase was granted in November, 1984.

The first two robotic systems were developed to handle sample preparations which had been performed manually so the manipulations had to be altered or adjusted to fit the robot's capabilities. The third robot (named Lizzy), Table 3, was not scheduled for 1985 but resulted from a high priority being assigned to the development of a new product at the plant site.

TABLE 2. Justification for Robotics System to Perform Lipid Assays.

Current Sample Load - 1000 (+)/Month

Description: Solvent extraction followed by drying of the solvent, and weighing of the lipid residue.

Manpower: Manual: 5-6 hours per day
Robot: 1/2-1 hour per day

Cost Advantage: Labor savings: \$22,787 per year.
Equipment cost - \$40,000-\$45,000.

Work Stations Required

Dispensers
Vortexer
Capping Station
Stir Blade Station
Drying Manifold
Balance

TABLE 3. Justification to Purchase ELISA - Robotic Equipment

Description	
ELISA = Enzyme-Linked Immunosorbent Assay	
One of a family of immunological type assays making use of enzyme/antibody/substrate interactions on a 96 well micro-titer plate.	
Estimated Technician Time	
Manual	24 Hours/Day (3 Technicians)
Robot	4 Hours/Day (0.5 Technicians)
Time Saved by Robotics	20 hours/day (2.5 Technicians)
Annual Dollar Savings	$91,150.00
Equipment Cost	$60,200.00
Return on Investment (over 5 years)	80%

There were several other ELISA users within the company but all systems were either manual or semi-automated, at best. We had the opportunity to develop the chemistry and procedural aspects of the assay concurrently with setting-up and fine-tuning the third robotic system. By so doing, we were able to get the system up and routinely running in a relatively short time.

Some points to consider during robotic justifications are presented in Table 4. This list, however, is not intended to be exhaustive. The lower half of the table shows some common misconceptions that we have encountered.

TABLE 4. Important Points to Consider for Robotic Justifications

1. Identify a definite application or need
2. Determine Development Time, Cost and Manpower Needs
3. Investigate Potential for ROI
   a. Technicain Time Savings
   b. Overtime Reduction
   c. Equipment Costs
   d. Annual Dollar Savings
4. Determine Benefit of Capitalization vs. Expensing of Dollars (Equipment) vs. (Salary)
5. Personnel Re-Allocation
6. Multiple or Alternative Use of Equipment
7. Preparation and Submission of Budget
8. Develop Robotic System
9. Make the System Work

Misconceptions

1. One Robot = 1 Technician
2. Robot Startup = Head Count Reduction
3. No Need for Dedicated Development Person; Projects can be Sandwiched Between Other Duties
4. Robot Requires Continuous Chemist Support After SOP
5. Robot Requires Exceptional Technicians for Routine Use

EFFECT OF ROBOTS ON PEOPLE - INTERACTIONS AND ATTITUDES

The impact of laboratory robotics on personnel will be as varied as the number of labs and personnel involved. Management is concerned about an investment return in real dollars. My response is that dollar savings can be realized if use of the robot to perform a procedure requires less analyst time. Use of the robot is justified if the need for the analyst is totally eliminated or if the analyst can be used to perform additional tasks.

Management and personnel are both rightly concerned about manning and are often misled to think that the acquistion of robots leads to an automatic reduction in numbers of departmental personnel. I have been careful not to mislead people nor to allow them to foster wrong information. My response to the man power issue has been that it is unreasonable to ask that a lab guarantee a reduction in personnel upon introduction of its first robotic system. In fact, I say that such a reduction should not be expected during the first 6-12 months after introduction, depending on activities of the lab. What I do guarantee is the fact that over a period of time, the use of laboratory robotics will result in significant technician time savings and increase lab efficiency. This time savings will lead to a reduction in personnel involved in routine procedures or allow the lab to take on additional projects.

Sample submitters are always skeptical of changes to analytical systems which they use. In order to assure them of quality data, we have given formal presentations and live demonstrations of our robotic systems. We have made extensive comparisons between the traditional and robotic sample preparation schemes prior to implementation of routine robotic preparation. As robotic sample preparation is initiated, we have not always begun with all submitter's samples concurrently but have staggered them so as to be certain one submitter is pleased with his/her sample data before accepting samples from another submitter. We believe that our credibility is enhanced through satisfied investigators.

I have found that most technicians will accept major changes in the way work is performed in the laboratory provided they are informed and are given reasonable alternatives for continuing to do challenging work. We have given our technicians opportunities to continue to be challenged through diversification of their daily assignments and in-house training on laboratory tecnhiques. We informed our technicians that the analytical laboratory was changing and probably would continue to change with future innovations in instrumentation and automation. To my knowledge, we did not experience negative reactions from our technicians at the introduction of robotics.

SPECIAL TRAINING AND NEW TECHNICIAN SKILL LEVEL

I have chosen to foster the concept of re-allocation rather than termination of technicians as more robots enter the laboratory. We are operating on the premise that technicians, whose routine work function is now being done by a robot, will either move to a different routine assignment or get involved with trouble-shooting and new project development efforts, possibly even utilizing robotics. To function maximally as a technician in a modern analytical lab requires some awareness and understanding of instrumentation, electronic and computer/microprocessor programming in addition to the traditional chemistry and/or microbiology background. Consistent with this thought, we have elected to employ technicians who have training in electronics and computer/microprocessor programming (Electrical Engineering Technology) but little or no training in chemistry as described in Table 5.

These people have worked out extremely well for us and we are now offering a Heathkit self-taught electronics course to our other technicians on a voluntary basis. The electronically trained technicians are being used as tutors and several people have signed up for the course. The make-up of the course is given in Table 6.

TABLE 5. Statement of Special Technician Need for the Antibiotic Assays Department

The antibiotic assay department is seeking a nonexempt person for a position as technician who has special skills and interests.

The ideal candidate will be:

A. Interested in working as a technician in a chemical and/or microbiological testing laboratory,
B. Knowledgeable of or willing to learn basic chemistry, microbiology and associated good laboratory techniques.
C. Interested in working with analytical instrumentation from several aspects (operation and trouble-shooting),
D. Knowledgeable of modern electronics and enjoy working with electronic, electromechanical and mechanical devices.

Task Statement
The technician shall function in a manner consistent with the corporate technician guidelines and help provide reliable analytical data on samples resulting from development, technical service, production and quality control activities within the antibiotic division at Kentucky Avenue.

Salary Class
A family of salary classes ranging from 15-30 is available to the candidate and he/she can expect to progress in a manner consistent with his/her performance and other guidelines imposed by the TAJC committee.

TABLE 6. Information on the In-House Basic Electronics Course for Technicians and Exempts

Four (4) part Heathkit - Self taught course

DC Circuitry
AC Circuitry
Semiconductors and Solid State
Digital

Supplies - Manuals and trainers provided by department

Length - Est. 15 weeks @ 3 hours per week
4 weeks on company time
11 weeks on employee time

Tutors - Technicians trained in electronics.
(1) hour per week on company time.

Re-Evaluation - End of part 1.

Technician interest - 13 sign-ups plus 2 exempts

Class Size - 4 technicians

It is anticipated that the training received from the electronics course will benefit our traditionally trained technicians through greater versatility and adaptability to a highly automated analytical laboratory. Technician feedback from the course has been positive to date.

ROBOTIC LABORATORY DESIGN

Well before the arrival of our first robot we realized that our traditional fixed laboratory layout was not compatible to robotic use. Robots need approximately a 4' x 7' or 5' x 8' flat work space, ideally with 360 access around the work area. The best design would be to keep permanent fixtures to a minimum, using movable benches and surfaces. This allows flexibility of instrument design, operation, and maintenance as well as easy adaptability to the inevitable changes to this technology in the future. See Table 7 for design features of our planned robotic lab.

TABLE 7. Functional Robotic Laboratory Space Requirements and Design Considerations

Flexible/Modular Lab Features:

1. Elevated Floor - 2' Square Aluminum Panels (solvent resistant covering) on Aluminum Frame and Supports. All services under floor.
2. 9' x 10' Office with Moveable Walls
3. Solvent Proofing on Top of Concrete Sub Floor
4. 6' (2 @ 3') Access Doorway
5. Aluminum Ramp
6. Dual Lighting System
   Fluorescent ceiling and indirect floor mounted
7. Dedicated HVAC System
8. Ventilation Under Raised Floor
9. Dedicated Exhaust (with 4 drop options)
10. Natural Gas detector under elevated floor

The flexible lab not only serves the useful purpose of housing our robotic operation but also is an experimental approach for evaluation of modular lab concepts and designs with the company. The new lab indeed will gives us unrestricted freedom to rotate, relocate and use other analytical equipment (HPLC, GLC, UV/VIS, etc.) more efficiently.

OVERVIEW OF CURRENT AND PLANNED ROBOTIC APPLICATION

We have been quite successful in the integration and utilization of robotics within our analytical laboratory. All of our robots to date are Zymate Systems. Several other laboratories within Eli Lilly and Company are utilizing robotics with varying degrees of success. The secret to success in my lab is two-fold. First, there is a commitment to the concept and potential benefits of robotics. Secondly, I have assigned a competent person to carry out that commitment.

Three robotic systems are now operating routinely. A fourth one (named Olivia) is now being used for experimental or development purposes and others are planned for continued improvements in lab efficiency.

ACKNOWLEDGEMENTS

Dr. Steven D. Hamilton developed the robotic systems with assistance of Mr. Michael L. Alward. Dr. Avinash L. Lagu, Dr. Hamilton F. Niss, Ms. Gwendolyn B. Hurt, and Mr. Raymond L. Hussey were responsible for the chemistry and procedural modificaitons to manual systems for adaptation to robotics.

Zymate is a registered trademark of Zymark Corporation.

MAKING ROBOTS FLY

J. E. Curley
Pfizer Inc.
Groton, CT

ABSTRACT

Robotic systems can be considered advantageous from six points of view: cost or labor savings, improved assay precision, increased rate of data output, improved morale, increased worker safety and better assay technique. The experience gathered from the installation of five robotic systems in support of pharmaceutical product development has led to the proposal of four stages of robotic implementation: defining the goals of the system, developing a plan to meet the goals, constructing the system, verifying that the results from the system are suitable.

The effort required to develop, validate, and implement systems for tablet dissolution, content uniformity assays, and Karl Fischer titrations for water was marked by both successes and setbacks. Establishing a good working relationship with the robot vendor is a key element in assuring a profitable robot application. Cost savings can be dramatic with investment payback in less than a year in favorable cases.

INTRODUCTION

The Analytical Research Departments of Pfizer Central Research now have five robotic systems for use in developing pharmaceutical products.

Three of these are used for tablet dissolution testing. The fourth is used for assaying drug content of tablets. The last determines the moisture content of powders by Karl Fischer titrimetry. From our experience with these systems, I'd like to propose some general guidelines to getting a laboratory robot to function properly. I'll be using examples from the development of our five systems and those reported by other researchers to illustrate various points. It is not my intent to fully describe the technical aspects of these robots. Details on these systems have been or will be published elsewhere.
There are four key stages in getting a robot to "fly". They are:

1. define the mission
2. develop a "flight" plan
3. construct the system
4. validate the results

DEFINE THE MISSION

The first stage is the most important and perhaps the hardest. Why do you want a robotic system? What are the benefits to you? How do you convince your management to invest in a system?

There seems to be six general categories of reasons to go into robotics:

1. cost or labor savings
2. improved assay precision
3. increased data output
4. improved morale
5. improved worker safety
6. improved "product"

Cost/Labor Savings

Cost savings is probably the initial approach for justifying laboratory automation. Upper management likes to hear about less expensive ways of achieving equivalent results. Cost savings can be impressive when there is a high demand for a labor intensive test. The tablet dissolution test described in the United States Pharmacopeia[1] is such a case. The goal of this test is to measure the rate of release of drug from a tablet or capsule. This is a chemistry laboratory procedure used as a quality control test during manufacture and as an indication of bioavailability. The dissolution rate test is perhaps the fastest growing demand on any pharmaceutical testing laboratory.[2]

Pfizer's Analytical Research Department has developed a fully automated dissolution test.[3,4,5] Our procedure can save over 30 hours of laboratory technician time for each hour of setup time a person spends preparing the robot (Table 1). Over the last six months, about 1100 dissolution tests were run on this robotic system. This is a rate of about 45 tests per week. At this pace, about 6,500 technician hours will be saved in a year.

Of course, this labor savings also has its price. The cost of developing this application is estimated in Table 2 as $65,000. The dollar savings for the system in the first year can be estimated as:

6,500 hours saved x $15/hour =	$97,000
less development costs	- 65,000
1st year savings	$32,500

TABLE 1. Labor Savings Through Automated Dissolution Testing.

Manual

45 tests/week x 3 hours/test x 50 weeks = 6750 hours

Robotic

45 tests/week x 1 hour/12 tests x 50 weeks = 187.5 hours

approximately 6560 hours saved

TABLE 2. Estimated Development Costs for Automated Dissolution Testing.

Capital Investment in Equipment:	
Zymate System	$30,000
Dissolution Test Equipment, spectrophotometer, personal computer	20,000
People:	
8 weeks x 40 hours/week x $22/hour =	7,040
8 weeks x 40 hours/week x $15/hour =	4,800
Cost of Money	
$50,000 x 8.5% x 8 months/12 months =	$ 2,800
approximate total =	$65,000

The development costs are recovered in about 8 months of operation. After that, the savings mount up for as long as the need for the test exists. Inasmuch as dissolution testing is now a requirement in nearly 400 established drug monographs,[6] and is frequently a requirement in New Drug Applications and other product registration documents, continued need for the test seems assured.

Improved Assay Precision

The second reason cited for the development of a robotic procedure is improved precision. A robot can treat each sample it handles in the exact manner as the one before. In some assays, the improvement is marked. Lewis, Santarelli, and Malbica[7] reported improved precision when measuring drugs in physicological samples at concentrations in the ng/mL level. Their studies demonstrated that "robotic automation of multistage sample preparation such as liquid-liquid extraction can improve precision by approximately two-fold or better".[7]

Increased Data Output

Increased data output is another good reason to develop a robotic system. Robotic arms tend to be slower than humans at many tasks. However, this slowness is compensated for by a machine's advantage to run day and night, seven days a week. We have seen a great reduction in the turnaround time for gathering tablet dissolution data. Eighteen dissolution rate profiles can be determined robotically in 24 hours. A laboratory technician would require six working days to do the same work. We often use dissolution rate as a criterion for choosing among formulas under development. Being able to gather data at six times the manual rate leads to faster decision making.

Improved Morale

The fourth reason to set up a robotic system is to improve morale. The assays chosen for automation should be the ones that are highly repetitious, not too involved, and frankly boring. Dissolution rate testing can be like this. This is also true of the determination of water in powders by Karl Fischer titrations. Our microanalytical laboratory is requested to run upwards of 200 samples each week for water content. This represents over 50 hours of testing time for a single person. The purpose of automating this procedure was to free up this analyst to learn other techniques and broaden his job. Our laboratory assistants seem happy to give repetitive tasks to a robot and do other, more creative work.

Improved Worker Safety

Improved worker safety can also be a strong selling point for a robotic system. Evelyn Siebert[8] described a Zymate System used for preparing oral contraceptive samples for chromatographic analysis. One justification for the procedure was to minimize health hazards. "Automation of these assays would further reduce the analysts contact with steroids", she reported.

Improved "Product"

The sixth reason to "make a robot fly" is to improve the technique. Hupe and Peters[9] developed the use of robotics in sterile culture techniques. A distinct advantage of the robotic system is that it could handle samples with minimal risk of the contamination associated with manipulation by humans.

Some combination of these six factors should help you choose a system to develop robotically and justify the attempt.

DEVELOP A "FLIGHT" PLAN

After receiving approval for a robotic system, the next step is to make a plan for its development. The first part of this plan is to thoroughly understand the manual method. Convince yourself that the method you wish to automate performs with the accuracy, precision and reliability you require. If this assay has been developed by someone else, make sure that you can perform it yourself. You won't be able to "teach" a robot a method that you cannot do yourself.

Once past that hurdle, it is helpful to develop a flow diagram of the technique you are developing. Start with big steps and break them into smaller units. This is the laboratory unit operations approach advocated by Zymark. These flow diagrams can become the basis for writing the computer code for each block of work. Look for opportunities to develop modules of code that can be used in more than one application. The weighing of a sample is such an example.

This planning stage also includes bench layout. Economy of motion for a particular assay may need to be balanced against a more flexible bench layout that will allow the running of more than one technique.

CONSTRUCT THE SYSTEM

Once the bench is prepared and code written for various laboratory unit operations, stringing the assay steps together begins. This stage of construction can be the most fun or frustrating. Rather than give the details of constructing one of our robots, I'd like to point out some of the mistakes that have been dear lessons to us.

Pick Robust Parts

Once a robotic system is used routinely, it will "work to death" its component parts. Our dissolution robot contains a sequence that involves spraying wash water to clean up test vessels during runs. On a typical day, the power and event controller uses a solenoid valve to turn hot water on and off 600 times. The first solenoid valve we picked was not rated to run at that pace day in and day out. Frequent failures led to using a more rugged solenoid. Similar examples can be cited for small pumps, valves, and stirring motors. We have found that greater reliability can be built into a system by using a pump for each fluid required rather than using a single pump with valves and manifolds. Each pump does less work and there are no valves to fail.

Buying more expensive components is real economy.

Beware of Prototypes

We were interested in building a system that would perform drug content assays on tablets. The end analysis was to be ultraviolet absorption spectrometry. The procedure involves weighing a tablet, extracting the drug from the tablet into solvent on a "vortex" mixer, filtering the extract, diluting, and measuring the UV absorbance.

For convenience, the whole procedure was set up using water as a prototype solvent rather than the actual organic solvent required. When water was used, the automation of the assay was a great success. However, when the procedure was run using the actual organic solvent, methanol, it was a dismal failure. Our pipetting procedure could not reproducibly dilute filtered aliquots of methanolic solutions.

A whole new approach was needed to get that system working.

Look Out for Software Updates

We needed to measure the pH of a solution, so we purchased a plug-in module for our system. This module would not work, because we had an older version of software in our controller. We simply updated our system with new software and the pH module now worked like a charm. But now, our power and event controller, master lab station and centrifuge did not. We had to update the hardware in those modules to correspond to the new software.

This hardware updating is not a particularly difficult task. A little prior-planning would have saved time and aggravation.

Build Confirming Tests into Your System

Our lab robots are blind. This fact becomes troublesome when a particular move is not completed. We have added checks to our programs to assure that filters are on sampling fingers and sample tubes have been picked up by the robot's hand. Failure to check for these types of things can lead to lost assay results.

More serious problems are also possible. One of our robots runs Karl Fischer titrations to determine water content of powders. You may recall that Karl Fischer reagent is a mixture of pyridine, iodine and sulfur dioxide.[10] One evening, the drain tube from our titration vessel cracked. As a result, fresh reagent kept being added to a full titration vessel that could not be drained. The noxious reagent eventually spilled out of the vessel onto the table and floor. This problem has been corrected by adding a level sensor to the titration vessels. If the liquid level rises above a pre-set level, the robotic system shuts down.

The automated dissolution robot uses three liters of hot water to clean test vessels between assay runs. The program contains a check to make sure each vessel is clean before the next assay is run. Two programming oversights on our part led to a very expensive laboratory flood. First, we neglected to include a verification that the wash hand was successfully picked up by the robot arm. Second, the program logic did not limit the number of times that washing of vessels should be attempted.

Late one evening, the robot dropped the vessel wash hand on the robot table pointed in such a way that water arched up from it across the room. Each wash cycle calls for 64-two second bursts of hot water. Following each unsuccessful wash cycle, the robot would sample a vessel, find it dirty and start sparying water again. These sprays of water soaked our brand new robot that was parked adjacent to the one that was malfuncitoning. Considerable damage to the new robot's electronics was done. A night watchman noticed the flood about 2:00 a.m., and the laboratory supervisor was called at home.

We now check for the presence of all hands and allow only one rewash. If the vessels are not clean after two attempts, the system is shut off.

VALIDATE THE RESULTS

The final stage of developing a ro[botic procedure is verify that the results obtained are accurate and sufficiently precise.

An example of this is the evaluation of the robot used for Karl Fischer titrations.[11] Fifty determination of the water content of sodium tartrate dihydrate were performed. The average percent water of these 50 samples was found to be 15.7%. The theoretical water content is 15.66%. We had no complaints about the accuracy. The precision of these determinations was 1% relative standard deviation. This is a wider range than we anticipated. A histogram of results is shown in Figure 1.

The robot was closely watched to determine the cause of the unusually high and low results. The problem was found to be associated with the sample pouring routine. Sodium tartrate dihydrate crystals were sticking to the sample inlet tube. The titration results were then occasionally lower than expected because not all the sample was charged

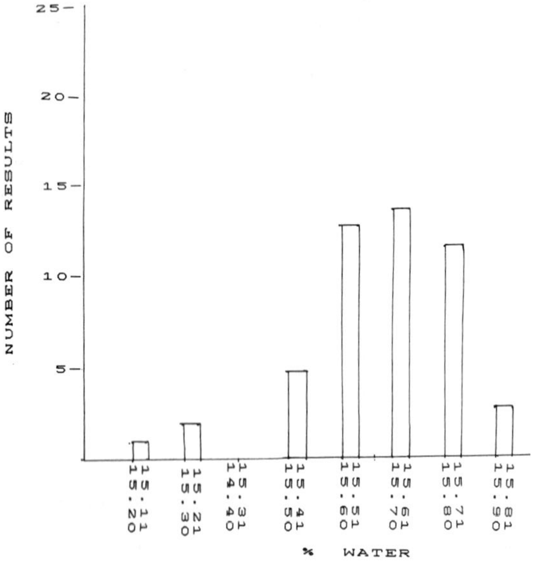

FIGURE 1. Sodium tartrate, dihydrate - Karl Fischer titration for water content. Theoretical 15.66 percent water.

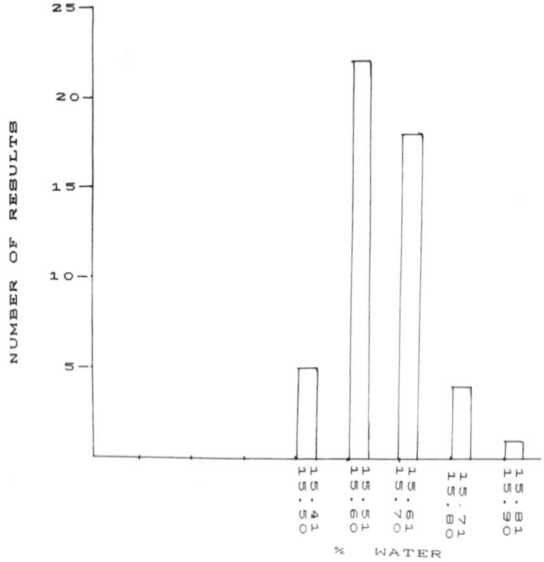

FIGURE 2. Sodium tartrate, dihydrate - Karl Fischer titration for water content. Theoretical 15.66 percent water.

to the titration vessels. After several determinations, the amount of crystal adhered to the inside of the tube would increase to a mass that would be knocked into the vessel from the vibration of the sample port closing. That particular titration result would be higher than theory. To fix the problem, the sample inlet tube was made wider in diameter and shorter in length. Crystals no longer adhered to the tube. The precision was improved. Another fifty samples of sodium tartrate dihydra[te were titrated. The average moisture content was again 15.7%, but the precision was improved to 0.6% relative standard deviation (Figure 2).

Once the robotic system performed with acceptable accuracy and precision, samples of product were titrated. Robotic results were compared to specifications for the product and results obtained by the standard manaul method. Table 3 summarizes results of one such experiment. The agreement between manual and robotic methods was within 3%, which is aceptable.

TABLE 3. Water Content of Pharmaceutical Powder.

	Robotic Values		
1.93	1.96	1.94	1.97
1.96	1.99	1.95	1.98
1.96	1.94	1.92	1.95
1.95	1.93	1.94	1.96
1.96	1.97	1.93	

Average = 1.95 ±0.02
±(0.9% RSD)

Theoretical Water Content = 1.75%

Specification Limits 1.4 - 2.75%

Found by Manual Titration = 1.90%

CONCLUSION

The successful implementation of a robotic system entails four factors:

1. defining its purpose or mission
2. developing a plan for implementing the assay to be automated
3. constructing the robotic system
4. assuring that the system produces results that are sufficiently accurate and precise

Developing a constructive relationship with vendors can facilitate the whole process of getting a robotic system working. The relationship should be candid in both directions. You should be able to offer both positive and negative critiques to your vendor. In like fashion, you should accept both types of criticisms. Robot and instrument vendors can be particularly helpful in achieving the first three stages discussed. The degree to which they can help is diminished in the validation stage. The decision that a robot functions with the precision, accuracy and reliability that you require, is essentially yours to make.

REFERENCES

1. The United States Pharmacopeia XXI, United States Pharmacopeial Convention, Inc., Rockville, MD, 1985, p. 1244.

2. L. M. Sattler, T. J. Saboe, and T. F. Dolan, "Automated Dissolution Testing", Zymark Application Note AP305, Zymark Corporation, Hopkinton, MA 01748, November 1984.

3. J. Curley, in Advances in Laboratory Automation - Robotics 1984, G. L. Hawk and J. R. Strimaitis, eds., Zymark Corporation, Hopkinton, MA 01748, 1984, p. 299.

4. L. J. Kostek, B. A. Brown, and J. E. Curley "Fully Automated Tablet Dissolution Testing", Zyamrk Application Note AP307, Zymark Corporation, Hopkinton, MA, April 1985.

5. L. J. Kostek, B. A. Brown, and J. E. Curley in Advances in Laboratory Automation - Robotics 1985, J. R. Strimaitis and G. L. Hawk, Zymark Corporation, Hopkinton, MA, 1985.

6. The United States Pharmacopeia XXI, in loc. cit. p. xlv.

7. E. C. Lewis, D. R. Santarelli, and J. O. Malbica, in Advances in Laboratory Automation - Robotics 1984, in loc. cit., p. 237.

8. E. Siebert, in Advances in Laboratory Automation - Robotics 1984, in loc. cit., p. 257.

9. D. Hupe and K. Peters, in Advances in Laboratory Automation - Robotics 1984, in loc. cit., p. 91.

10. K. Fischer, Agnew. Chem., 48, 394 (1935).

11. J. Curley and J. Strimaitis, "Fully Automated KARL Fischer Titrations Through Robotics", presented at Parenteral Drug Association, Inc. Spring Meeting, Montreal, June 1985.

THE ROLE OF PEOPLE IN HIGH TECHNOLOGY AUTOMATION

Robert G. Miller
Ortho Pharmaceutical (Canada) Ltd.
19 Green Belt Drive
Don Mills, Ontario, Canada, M3C 1L9

ABSTRACT

In order to compete more effectively in local as well as international markets, corporations are emphasizing company productivity programs. In the laboratory, many of these productivity programs manifest themselves through the automation of manual or labor intensive procedures. Robotics, laboratory data systems, automatic samplers, and sophisticated chromatographic systems are but a few of the high technology alternates available to management to increase laboratory productivity. Part of the driving force to automate is based on the need to increase the productivity of the individual chemist, the need to operate the laboaratory outside of the constraints of the single shift concept with existing personnel, as well as a need to maintain the short term capabilities of certain laboaratory functions where the personnel critical mass is small.

As the automation within the laboratory becomes extensive the role of people within the high technology laboratory must change. Management must have a clear plan for the new role of the laboratory chemist. The proper balance between human and technological resources, consistent with the long term needs of the company, must be assessed.

Management must not be too hasty to the clip the short-term "productivity coupon" by exchanging people for technology on a one to one basis but rather they must develop a rational strategic plan that builds on the strength of their people, as well as their technological base.

Such a strategic plan is being assembled at Ortho Pharmaceutical (Canada) Ltd. The basic concepts of that plan forms the basis of this presentation.

DISCUSSION

As part of a corporate quality and productivity program, I had the opportunity to view several presentations by Tom Peters recently. His video presentations based on In Search of Excellence[1] and A Passion for Excellence[2] were supplemented by a special presentation he made to the Presidents and Managing Directors of all Johnson & Johnson Companies in June 1985.

One is left with the singular idea that people make it all happen. Many of these companies who emphasize people are also major users of high technology and therefore, must develop a strategic plan that clearly places the role of people and high technology in perspective.

As part of our corporate strategic planning process, we at Ortho have found it necessary to address the question "What will the role of people be in our high technology environment?".

It is our contention that productivity is achieved through a balance between high technology and innovative, motivated people who care enough to do it right the first time, on time. These people must feel secure and know that the reward for successfully implementing cost effective high technology is not unemployment. The factory of the future will have automation, but it will also have people. Even in countries where labor is cheaper than in North America, automation is an important economic tool. Japan leads the way with at least three times the number of robotic systems operating as in the U.S.A.

We have had automation at Ortho Canada for many years. Automatic samplers and computer systems did not appear as people-threatening as our robotic system.

The Zymate Laboratory Automation System is an example of programmable flexible automation and is capable of replacing a chemist quite effectively in certain analytical environments. If the chemist who implements this system renders himself or a fellow worker redundant, then the incentive to succeed is negligible. How then do companies implement high technology and how do they clip the productivity coupons while still maintaining a highly motivated people work-force?

In our company it is recognized that corporate knowledge and people are extremely important assets that match the worth of our physical resources. It is people that make a business work and not an assembly of equipment. It took intellectual input to plan, organize and direct all of the processes requiring automation at Ortho Canada, and it will take even more people input to refine and optimize those processes. Therefore, it is obvious that a company must maintain a core of highly motivated and capable individuals who can get things done. Tom Peters spent some time talking aobut the project champion and the need for such individuals in his book <u>A Passion for Excellence</u>.

In our examination of the issues relating people and high technology to productivity we identified three focus areas. The corporate issues (Table 1) dealt with business environment, technological impact, and the corporate position on matters of importance. The middle management issues dealt with the impact of the high technology on essentially our first line of supervision. The last area of focus dealt with the impact of high technology on the work-force.

TABLE 1. Corporate Issues.

1. Technological Awareness
2. Corporate Identity
3. Productivity = Effeciency?
4. Senior Management Support
5. Product Life Cycles are Becoming Shorter
6. People Need More Information Faster
7. Engineering Technology Doubles in Less Than 5 Years
8. Computer Technology Doubles in Less Than 2 Years
9. Greater Demand for New Products in Less Time
10. Effect of Technology on People

A company must have an individual or group of individuals who understand the latest technology in their area of focus and know how that technology will impact on the success of the business. They must be those project champions who can catch management's attention so that beneficial technology is actually implemented.

The corporate identity scenario reminds me of the spirit of the champions creed that carries the message "if you think you can't then you can't" and that "it is not always the fastest or strongest who wins, but the person who thinks that he can". If your company feels that high technology is for everyone else, then it probably is. The successful implementation comes from companies and people who make it work. The productivity/efficiency issue will be dealt with in detail a little later, but it represents management's philosophy on the degree of re-investment of the benefits achieved through automation back into people resources.

I have never met a successful business person who did not attempt to understand and analyse all of the factors related to the success of his business proposal. However, when it comes to high technology proposals management can be impatient and demonstrate a reluctance to understand all of the ramifications, both good and bad, associated with the high technology proposals.

Albert Einstein once said that "all things should be made as simple as possible, but not simpler". To me this means that top mangement must commit themselves to the goal of understanding the impact of high technology on their business and then support the champions.

The points 5 through 9 in Table 1 are today's business realities. Technology is having a major impact on how we do business now and how we will do it in the future.

The last item is how technology will affect the output of your work-force. If an adverserial relationship between potential technological improvement arises then these technological improvements will not be successfully implemented if they depend on those people to implement.

The Canadian Postal System is a prime example of the introduction of high technology into a less than sympathtic environment. After many tens' of millions of dollars the productivity per worker is less than it was before the so-called high technology improvements.

TABLE 2. Middle Management Issues.

1. Project *vs*. Product Management
2. What is the Meaning of Information?
3. Worker/Manger Ratio Will Increase
4. Boundaries Between Manager & Worker Duties Will be Less Clear?
5. Increase in Technological Skills
6. Increased Demand for Administrative Skills
7. Increased Planning Needs

The effects of high technology really begin to manifest themselves as we examine the effect of the middle and junior management groups (Table 2).

The first point suggests a return to broader issues as high technology office and plant automation will be capable of doing the detail work. In fact, the latter will be product to project to process management.

The second point is really the meat of the issue. High technology will gather the information through sophisticated data acquisition systems, correlate these data via expert computer systems and present these data in a user friendly form. High technology via systems such as robotics will do many of the tasks now carried out by people.

The laboratory chemist and supervisor will take on new roles as the boundaries defining their jobs begin to overlap. Each will be concerned with the meaning of information rather than the collection of data.

In our Quality Assurance laboaratory robotic systems prepare samples for HPLC and an on-line computer systems reduce the data to a user friendly format. Our Quality Assurance staff now cannot only economically satisfy the regulatory requirements of pharmaceutical manufacturing but can effectively concern themsleves with the meaning of analytical information. When I assumed the additional responsibility for our Pharmaceutical Development function at Ortho Canada, it became quite obvious that I and most Q.A. chemists had a lot to learn about pharmaceutical processing. If we wish to increase process reliability and eliminate unnecessary analytical work arising out of a business uncertainity, then additional focus will be required to optimize the manufacturing and testing processes. In our experience to date we have found this focus on process reliability to be of major economic benefit to us.

The last three points on Table 2 indicate that new skills will be required by middle management to handle the impact of high technology.

TABLE 3. Direct Labor Issues.

1. Job Security
2. Retraining/Qualifications
3. Shift in Role - Maintenance vs. Production
4. Higher Level Skills Exercised

Finally we have to address the operational impact of high technology on the direct labor work-force (Table 3). The number one issue is always job security. Consequently, how does a company implement labor saving high technology into an environment that will be receptive. At Ortho Canada we have a policy that "clips the productivity coupon" only through natural employee attrition. This means that we are managing this issue for the long term rather than the short term. No one will lose their job as a result of the implementation of high technology proposals, and this is clearly understood by all. However, if someone leaves a relevant area through job promotion, retirement, or by changing companies, then the company may elect not to replace that position, may replace or maintian headcount by creating a job elsewhere or whatever.

The result is that people then make the high technology work because they know the future of their company in which they have a long-term stake may depend on the successful implementation of the high technology.

The preparation of the human resource base is critical to the successful implementation of any high technology venture. The corporation must have some direction as to the new role of the laboratory chemist'after automaton.

What then will the role of the high technology chemist be in the modern technology? In addition to gaining new technical skills, the emphasis in our organization will be to operate the high tech systems and to take the output of these systems and relate the output to our business needs. In other words, what is the meaning of analytical information as it pertains to our regulatory and business needs.

The chemist in our laboratories will be exposed to more aspects of the manufacturing processes. Again, all of this is made possible by automation and a shift in emphasis from "a gatherer of data" to an individual who knows the meaning of information.

If we contrast the non-technological based laboratory with the high tech laboratory, we see that the former is one which emphasizes the collection of information manually. The review of paper-based laboratory data is extremely labor intensive. Therefore, it is rarely done. On the other hand, the high tech laboratory facilitates the capability to determine the meaning of information by automating all aspects of the analytical process. The automation carries itself through the paperless laboratory where integrated software packages can aid in manipulation of data.

The rate of introduction of high technology is extremely important as we have seen from our case study. At Ortho Canada the physical and human resource base had 13 years to develop before robotics was introduced. In our case the rate was determined to a great extent by the availability of the technology rather than our ability to absorb it. Nevertheless, a reasonable time period should be allowed to introduce major technological advances into a laboratory.

The profile of the "typical" robotics user was of interest to us and consequently, with Zymark's help we conducted a survey in the spring of 1985. We surveyed 25 users that represented between 5 and 6% of the users. Industry, Government, Hospitals, and pure research Institutes were surveyed. Research and quality assurance applications represented the bulk of the users. It was quite evident that for the most part these laboratories were already sophisticated. Most of them have advanced data reduction capabilities. Many had a considerable investment in automation, particularly in chromatographic and spectroscopic instrumentation. This would confirm the Ortho Canada exeprience that a reasonable automation base usually is in existence before a robotics system is introduced. As we have seen, in order to implement high technology successfully, the technology should be introduced over a reasonable time period. The skill level of your staff must be raised appropriately. A clear policy on displacement resulting from efficiency gains using automation should be visibly in place. Finally, the corporation should have a broad corporate strategy to provide some direction productivity vs. efficiency question.

If corporations consistently replace labor with its equivalent output in automation, then I feel that they are losing one of the major benefits of automation. The capability to be able to give people higher level tasks centering around "the meaning of information" is one of the major benefits of automation.

The company must also understand the return on investment criteria are more difficult to quantify. Most high tech companies discount the requirements by 5 to 10 percentage points in order to capture the hidden and difficult to quantify benefits. The ability to re-invest your creative work-force to build the future is difficult to quantify, but would in reality be one of the major benefits of high technology.

Finally, the company must have faith in its champions of high technology and be willing to risk funds for potentially beneficial technology that is either unproven or does not exist.

When the system does not exist, (for example, a specific configuration of robotics and special tooling for a production environment), then the company should set some funds aside to finance conceptual and design studies along with selected prototyping of the key design features. The company must also realize that all ventures will not be successful and that a "fear of failure environment" will crush any innovation within the corporation. It should reward, by support, those innovators with a proven track record.

In summary, the management at Ortho Canada has developed a number of individuals who have focused their expertise in various high tech areas and these project champions are being supported by senior management.

A strategic plan exists as a subset of our formal total strategic plan that outlines where we want to be, who will do what, and when to get us there.

We have visibly stressed the displacement policy to encourage innovation and enterprenurship in our organization.

While there is no published guarantee process to succeed in successfully implementing high technology, we at Ortho feel that the approach that we are taking is effective, and all management measurement tools to date confirm it.

REFERENCES

1. T. J. Peters and R. H. Waterman, Jr., In Search of Excellence, Harper and Row, New York (1982).

2. R. Peters and N. Austin, A Passion for Excellence, Random House, New York (1985).

STRATEGIC TRENDS IN LABORATORY AUTOMATION - 1985

Francis H. Zenie
Zymark Corporation
Hopkinton, MA 01748 USA

ABSTRACT

Leading chemical, pharmaceutical, food, energy and biotechnology companies face intense, world-wide competition. To meet this challenge, they must develop innovative new products and efficiently manufacture these products to high quality standards.

This strategy requires increased laboratory support which demands far better utilization of their most limited resource - qualified scientists and technicians. Laboratory automation is necessary to achieve this strategy.

The emerging laboratory robotics technology complements earlier advances in instrumental techniques and data handling - thereby creating an opportunity to automate entire laboratory methods as integrated systems. This concept of a Methods Management System (MMS) bridges the gap between individual automated instruments and LIMS systems. The goal - MAKE TIMELY, QUALITY DECISIONS BASED ON VALID DATA!

Successful laboratory automation projects have four key requirements - motivated people, proven chemistries, disciplined planning and creative implementation.

When these are met, laboratory automation will be effectively implemented and will substantially contribute to the organization's strategic foundation on goals!

INTRODUCTION

Most industrial laboratories have greater demands for analytical support than they can meet. Early developments in laboratory automation such as autosamplers and simple data handling devices gained rapid acceptance, because they replaced manual tasks where people were poorly suited to perform these tasks - and, these immediate needs allowed many laboratories to justify the moderate investment.

Today, improving laboratory productivity has reached a strategic urgency in many organizations. Our leading chemical, pharmaceutical, food, energy and biotechnology companies face intense, world-wide competition. To meet this challenge, their strategies demand the following:

1. Develop innovative new products often tailored or optimized for a defined (specialty) use.
2. Efficiently manufacture these products to the highest possible quality standards.

The feature article in the July 22, 1985, issue of Business Week magazine is titled "The Race for Miracle Drugs - A New Generation of Custom Remedies is Revolutionizing the Industry". The article states:[1]

> "For the first time, drugmakers are attacking not just the diseases but the fundamental causes of the disease. And they are quickly turning their discoveries into a new class of customized drugs that, by being more precise than anything now available, will be far more effective."

More recently, the Monsanto Company agreed to acquire G. D. Searle & Company for $2.7 billion. This is part of a major restructuring by Monsanto to substantially reduce its dependency on bulk petrochemical and commodities in order to implement a strategy built on specialty chemicals, pharmaceuticals and agricultural products.[2]

These emerging strategies place many demands on their respective organizations, important among them:

1. Provide increased laboratory support for Research, Product Development and Quality Control.

2. Improve productivity of the skilled scientific and technical staffs.

Improved productivity is not primarily motivated to save money but to better utilize a very limited resource - qualified scientists and technicians. Fortunately, there are many tasks currently performed manually which can be transferred to automated instruments, thereby, freeing people to make more valuable and challenging contributions. Remember, however, the easy advances in laboratory automation are behind us. The next step requires higher level systems capabilities - but the potential benefits are well worth the investment and are essential to the strategy.

To achieve any strategic goal, commitment and action must begin now! Powerful technologies, such as those used in laboratory automation, require training and experience to gain their full potential.

We hear the phrase "If it ain't broke - don't fix it" more and more in our organizations. Although this may be valid in some situations, it can become an excuse to blindly maintain the status quo while slipping into mediocrity. A far more positive theme is "You don't have to be sick - to get better". This challenges us to improve before problems become crises.

LABORATORY AUTOMATION - SYSTEM INTEGRATION

In this early phase of laboratory automation, we're accustomed to fragmented laboratory operations where days, or sometimes weeks, elapse between sample submission and the final result. This must change!

The wheel is a component or tool with limited intrinsic value. Several wheels, however, when utilized in bicycles, wagons, automobiles or trucks become part of a system with unlimited value.

Today, laboratory instrument manufacturers, as well as scientists working in industrial and academic laboratories, are developing and enhancing tools to improve laboratory productivity. We can now begin creatively linking these new tools together into integrated systems.

Traditionally, we've separated analytical methods into three distinct functions (Figure 1). In this representation, we've lost sight of the primary purpose for the analysis, that is, MAKE TIMELY, QUALITY DECISIONS BASED ON VALID DATA!

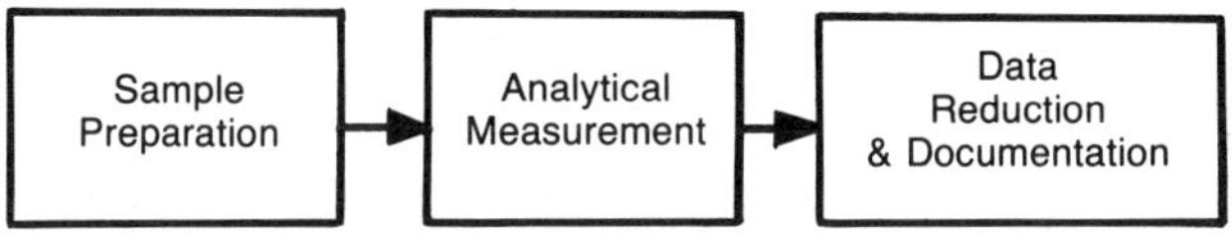

FIGURE 1. Analytical Methods.

Figure 2 illustrates real-time systems integration with validation and decision making as the primary focus. This level of integration is the foundation for Methods Management Systems (MMS).

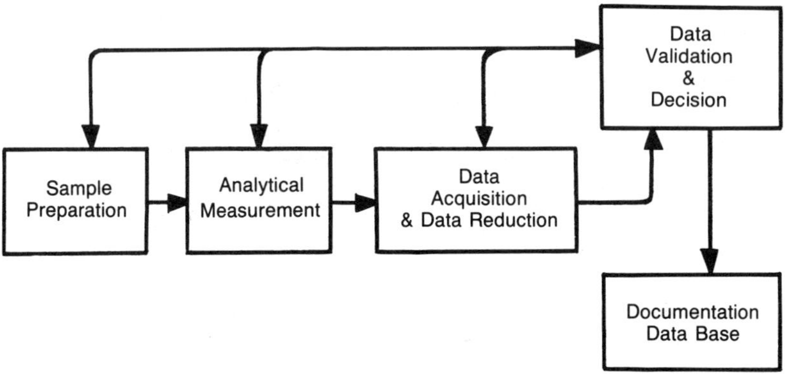

FIGURE 2. Methods Management System (MMS)

Regardless of the degree of automation, people must control the process and make any judgements required by extraordinary conditions. Methods Management Systems bridge the gap between automated instruments and higher level information management and data base (LIMS) systems. MMS systems make possible real-time systems integration.

MMS Example - Titrations: Titrations, for example, are one of the most powerful and widely used analytical procedures. Manufacturers of titration equipment recognized the repetitive nature of these procedures and developed automated titrators which automatically dispense reagents and calculate results. The new titrators significantly improve precision but still require a technician to prepare the sample (if necessary), introduce the sample into the titrator, specify the method (when required), push the start button and finally clean and initialize the titrator for the next sample.

Today, laboratory robotics and automated titration technologies can be combined into integrated systems.[3] The robotics system controller stores a library of titration methods for unattended operation. The robotics system then prepares a sample, sets the titration parameters in the titrator, introduces the sample and initiates the titration. Typically, the robotics system prepares the next sample while the current sample is being titrated. If the methods require burette changing, the robotics system can automatically reposition burettes. When the titration is complete, the robotics system acquires the calculated results from the titrator and validates the data with respect to acceptance criteria programmed in the robotics system controller. Finally, the robotics system will clean and initialize the titrator and, as required; rerun the sample, run a standard, proceed to the next sample or stop and request Help.

Integrated MMS systems, as shown in Figure 2, requires compatible interfaces between each instrument or subsystem. Differing application needs determine the specification for these interfaces and where systems integration takes place. For example, applications may be either Data Intensive or Control Intensive with the following implications:

Data Intensive Applications - Large numbers of data points are acquired and processed using sophisticated software to determine analytical results. In these applications, systems integration may best be performed by the data processing workstation.

Control Intensive Applications - Limited data points are acquired and processed using straight forward mathematical equations. In these applications, data reduction and systems integration may best be performed directly by the control workstation.

LABORATORY ROBOTICS - THE NEXT STEP

Laboratory robotics emerged in the early 1980s as a new approach to improve sample handling and sample preparation technology - to the level attained by laboratory instruments and computers. The first such system, The Zymate Laboratory Automation System, was introduced in the Spring of 1982. Following hundreds of successful installations, laboratory robotics is now recognized as the next step in laboratory automation.

Prior to laboratory robotics, justifying an investment in automating laboratory procedures required large numbers of identical, repetitive operations. Today, rapidly improving computer technology, particularly powerful microprocessors, makes available easy to use and low cost programmable computers. **Robotics, then, is the extension of programmable computers which allows computers to do physical work as well as process data.**

In a practical sense, laboratory robotics is far more than a programmable mechanical arm. It is a programmable system for performing laboratory procedures - automatically and unattended. Combined with other instrumental technologies, it makes possible true Methods Management Systems.

Laboratory robotics goes beyond information management - it interfaces with physical operations in the laboratory. Initially, the goal was to emulate people, but experience shows that creative approaches buildingon the strengths of laboratory robotics lead to more effective systems. Warren P. Seering, Associate Professor of Mechanical Engineering at MIT, describes the importance of creativity in his article, "Who Said Robots Should Work Like People", which discusses the lagging progress in industrial robotics.[4]

"... robots, most people thought, would have manipulators that swung like human arms, grippers that grasped objects like human fingers, and sensing abilities comparable with human senses. Made in the image of humans, these robots would soon outperform humans in cost and efficiency.

This strategy hasn't worked. It was fatally flawed because it assumed that humans are optimally designed to perform manufacturing tasks and therefore deserve to be emulated. This is not true. Humans are designed to throw stones, pick berries and climb trees.

Suppose, for instance, the task of sewing shirts was still being performed by hand, and that you were given the job of automating the process. You would probably not propose that the task be performed by a robot using a needle exactly the way a human does."

In an evolutionary way, future laboratories, laboratory equipment and laboratory procedures will be designed to work effectively as part of laboratory robotics systems.

CREATIVITY - KEY TO SYSTEM IMPLEMENTATION

Rarely can we achieve something of value without an investment. Laboratory automation and laboratory robotics offer great potential value but requires an investment in people as well as money. Most purchasing procedures require that funds be available before a purchase order is issued. Why not also require adequate staffing?

Successful laboratory automation requires commitment of qualified people with sufficient time to design and implement the system. If these people lack experience, additional time must be available for training and familiarization prior to implementation - and, this must be quality time. Laboratory supervisors responsible for daily operating results should not be asked to implement a major laboratory automation project in their spare time.

Programming laboratory automation is analagous to training people to perform similar work. Automation generally requires more disciplined planning but, when complete, has permanent value. People, on the other hand, require training or retraining with each assignment change.
Once adequate funds and people are available, the following four requirements are key to all successful automation projects:

1. Motivated People
2. Proven Chemistries
3. Disciplined Planning
4. Creative Implementation

Motivated People

Major laboratory automation projects are strategic investments similar to product or process development. While some benefits come quickly, the strategic benefits will accumulate over time and grow to be substantial.

Questions regarding the impact on people's jobs and careers should be discussed openly since the goal isn't to obsolete people but to free them from repetitive tasks so they can make more valuable contributions. There should be a sense of shared risk between management and people assigned to these projects. The project's importance should be clear to everyone and the implementation team should be protected from short term problems and interruptions. Above all, recognize and reward success!

Proven Chemistries

Variability in analytical results is often automatically blamed on "human error". Variations in chemistry due to reagents, standards, adsorbants and filters may also be the cause. More reliable chemistry will lead to more precise results using automation.

Disciplined Planning

When provided incomplete instructions, people improvise to obtain acceptable results. Automated systems require complete, detailed instructions but, with proper planning and programming, will deliver consistant results.

Manual procedures are typically a series of tasks sequentially linked together where the contribution of each step to the final result may be lost or obscure. The best approach to automation system planning is to invert the orientation and start with the desired result and systematically break it down into functional procedures with individual operations derived from the functional requirements.

Creative Implementation

While using Proven Chemistries, we must stretch our thinking beyond direct emulation of the manual procedure. Creatively building on the strengths of automated equipment, permits greater precision and productivity than possible through direct emulation. This is best illustrated by three brief examples.

Pharmaceutical Tablet and Capsule Handling - One of the most graphic demonstrations of laboratory robotics is automated assay of pharmaceutical tablets and capsules. Immediately, the problem of tablet handling becomes apparent - they may be large, small, round, cylindrical or other shapes and may be oriented randomly with respect to the robot. A complex robotics problem was solved by placing the tablets or capsules in small test tubes, located in racks, where the robot picks up the uniform tube and pours the tablet into a tared tube on the electronic balance.

Environmental Analysis of Trace Organics - As a leading manufacturer of synthetic chemicals, Dow Chemical has a major program of monitoring the working environment for potentially hazardous organic contaminants. A common approach is to draw a quantitative air sample through a small glass column packed with an appropriate solid sorbent. The organics, which are adsorbed on the sorbent, may be extracted using a suitable solvent and then analyzed by gas chromatography.

In the manual procedure, the glass tube is cut in half and the charcoal emptied into a test tube. The necessary eye/hand coordination would be difficult to emulate with a laboratory robot. Mark Dittenhafer, Linda Coyne and Judy Warren of Dow's Analytical Laboratories decided to run anexperiment to determine the organic recovery with the sorbent bed left intact in the original glass column[5]. The results were excellent and, today, hundreds of samples are run routinely using laboratory robotics. While preserving valid chemistry, they creatively developed a reliable, efficient automated system.

Testing of Specialty Nonwoven Fabrics - The Procter & Gamble Company manufactures high quality diapers using cellulose fibers to draw moisture away from the infant. Following absorption by the cellulose fibers, a specialty nonwoven fabric isolates the infant from the moisture. A "strike-through and rewet" test was developed and is used routinely to test these properties.

The first step in automating this procedure is to pick up individual sample sheets of the nonwoven fabric and place them on the test fixture. Again, critical eye/hand coordination is required to reliably pick up individual sheets from the sample stack. The automated solution was proposed by Jim Abbott, Lesa Jenkins and Carol McLaughlin of P&G's Paper Product Development Laboratories.[6] They offered to insert standard staples into two diagonal corners of each square sample and alternate the diagonals down through the stack. A special robotic hand equipped with programmable electromagnets can then reliably pick up sequential samples.

Two important ideas are illustrated by this example:

1. The approach of inserting staples and picking up samples using electromagnets probably could only be created by end users because it might appear foolish or even stupid if proposed by the system vendor.

2. P&G actually introduced a new manual step in order to run the rest of the procedure unattended. The new manual step is performed quickly during the daily setup.

STRATEGIC TRENDS SUMMARY

Laboratory robotics is part of a strategic evolution in laboratory operations. Last year, some of these trends were reviewed at the International Symposium on Laboratory Robotics.[7] Since then, progress has been made in several areas and some new requirements have emerged. The following is a 1985 summary of these strategic trends.

Laboratory Layout and Work Organization

Laboratory walls are being removed. The laboratory of the future will be open with islands devoted to integrated systems. Laboratories will be more decentralized with clearer responsibility for final results including precision, cost and turnaround time.

With automated sample preparation, procedures are being serialized rather than performed in batches as had been done manually. Serialization leads to improved staff and equipment utilization, uniform sample history and faster availability of results.

Laboratory Staffing

Automation specialists play an essential role in the laboratory. System integration is a complex function and laboratory personnel require specialized technical support. As this technology is more widely used, the specialized knowledge will be dispersed within the organization.

System Integration

Intelligent System Modules: We often forget the essential tasks performed by people in our laboratories. Automated systems require instruments and modules capable of unattended operation which means the human tasks must transferred to the instrumentation.

For example, most laboratory centrifuges require the operator to load and unload samples, identify and remember sample locations, sense destructive conditions such as out balance and set time and speed for each run. A truly automated centrifuge must communicate with its system controller and perform all these functions.

Methods Management Systems: Sample preparation, analytical measurement and data acquisition will be integrated into Methods Management Systems. Data will be acquired and validated followed by automatic method correction as required.

LIMS Networks: MMS systems will be networked into higher level Laboratory Information Management Systems for overall laboratory administration and data base management.

Reliability: System reliability will be increased through use of automatic verification (CONFIRM) techniques. Positive sample identification will be confirmed throughout the procedure. Automatic data acquisition and validation insure reliable operation just as vision systems improve reliability of industrial robotics.

System/Application Software: Software integration is equally important as hardware integration. Application software modules provided for Laboratory Unit Operations and other instrumentation will have to be compatible.

Laboratory Disposables

New laboratory disposables with automatic dispensing will be developed for more efficient and reliable operation within an automated system. Disposable will truly be part of the system and "special" techniques used by skilled technicians may designed into the disposable.

Automated Methods Development

With the growing use of integrated analytical systems, new methods will be developed around this technology. Improved methods validation will be possible because of the ease of performing sensitivity experiments.

Once developed, these methods can be easily delegated or transferred to other laboratories for routine operation.

Automated Research and Product Development

Research and Product Development often require multiple experiments under varying parameters. In many ways, this extends the role of laboratory automation building on its power to perform repetitive tasks in an experimental protocol. The addition of physical property sensors and process control capability will further extend the system boundaries.

Customer/Vendor Partnerships

Laboratory instrumentation vendors recognize their growing role as system architects - compared to their traditional role as makers of laboratory tools. Most laboratory personnel, while skilled in chemistry, have limited experience in mechanical and electrical engineering and advanced computer techniques.

Effective laboratory automation requires technical support. Many customers will create strong internal support organizations while others will look to the automation vendor for support. Ultimate system responsibility must remain with the end user.

Cooperative Relationships Between Instrumentation and Computer Manufacturers

The absolute need for system integration requires new behavior from instrumentation, laboratory equipment and computer manufacturers. While continuing to compete for customer business, we must cooperate to achieve hardware and software compatibility.

CONCLUSION

These are exciting times to be a chemist and to participate in the laboratory revolution. We're still in the pioneering phase but our experience will be the foundation for even more rapid strategic progress.

REFERENCES

1. "The Race for Miracle Drugs," Business Week, McGraw Hill, Inc., New York, NY, (July 22, 1985), pp 92-97.

2. "Monsanto's New Regimen: Heavy Injection of Drugs and Biotechnology", Business Week, McGraw Hill, Inc, New York, NY (December 3, 1984) pp 64-69.

3. Fully Automated Titrations, Zymark Corporation Publication LA 350. Zymark Corporation, Hopkinton, MA.

4. Seering, W. P., "Who Said Robots Should Work Like People?", Technology Review, MIT, Cambridge, MA (April 1985) pp 59-67.

5. Coyne, L. B., Dittenhafer, M. L., and Warren, J. S., "Robotic Desorption of Air Sampling Tubes", Presented at American Industrial Hygiene Association Meeting, Los Vegas, NV May 1985.

6. Abbott, J. C., Jenkins, L. A., and McLaughlin, C. A., "Laboratory Robotics Applied to the Testing of Nonwoven Fabrics: The Strike-through and Rewet Test", Presented at LabCon New England/84 Woburn, MA October 1984.

7. Zenie, F. H., Advances in Laboratory Automation-Robotics 1984, G.L. Hawk and J.R. Strimaitis, eds., Zymark Corp., Hopkinton, MA, pp 1-16.

THE AUTOMATED BENCH: HOW DOES YOUR ROBOT COMMUNICATE WITH ANALYTICAL INSTRUMENTS?

Charles R. Knipe
Hewlett-Packard Company
Route 41 and Starr Road
Avondale, Pennsylvania 19311

ABSTRACT

As laboratories become more automated, coordination between the sample preparation and the analysis tasks grows increasingly important in order to optimize the usefulness and efficiency of the instrumental tools which are available in each of these areas. Robotics provides a flexible system for automatically preparing samples for various analyses - either by a sequential mode or by an interwoven mode that combines several preparation methods simultaneously. On-line coupling of such a versatile preparation system to analytical instruments provides the user with the capabilities to alter preparation methods or sequences immediately with no user interaction.

However, to avoid costly and potentially disappointing results, the user must be cognizant of what is necessary and feasible to configure a system that will allow sample preparation and analysis to be integrated into a useful and viable package. Several examples will be presented that illustrate the various levels of complexity and sophistication required for this communication.

INTRODUCTION

The tasks in the analytical chemical laboratory can be divided into three general areas: 1) sample preparation, 2) sample analysis and 3) data handling. Laboratory automation over the past 10-15 years has primarily been directed at the sample analysis and data handling areas. Only recently (the last 2-3 years), has the problem of automating the sample preparation and handling been addressed through the use of robotics and dedicated sample handling systems - i.e. autodilutors, pipetters etc. These latest developments have provided the last pieces necessary for a totally automated laboratory bench.

In the near future, samples will be automatically prepared and analyzed on a routine basis. Analytical results and sample information will be stored in a data base, and a final report will be generated - all with little or no human intervention at any point during the life of the sample. Currently, many laboratories already have the capabilities to control analytical instruments automatically and feed the data into a laboratory information management system (LIMS). The final piece to the puzzle is tying the automatic sample preparation system to the chemical instrumentation and data handling networks so that they all work in unison and can be configured to each user's particular needs.

The purpose of this paper is to present an overview of the considerations necessary for integrating a robotics system into an automated laboratory. General interfacing concepts are presented followed by actual examples representing two levels of communication between a robotic sample preparation system and a gas chromatographic analysis system.

INTERFACING OVERVIEW

Interfacing a robotic sample preparation system directly to an analytical instrument requires three areas of interactions:

1. Physical lay-out
2. Electrical interface
3. Communication protocol

A minimum requirement of any fully automated system is that the robotic manipulator be able to load a sample into the analytical instrument directly. The instrument must be located physically within the working area of the manipulator and have a sample holder that allows the manipulator to place and remove samples in and out of the holder easily. Since these requirements are peculiar to each individual laboratory, it is assumed that the physical layout and sample hand-off is possible or can easily be modified by the user for special cases.

Once the instrument is placed in a position so that it is capable of accepting a sample from the robotic system, an electrical communication path is required to pass control and data information between the two systems. Actual implementation of this electrical communication path may range from a few wires to a more sophisticated bus structure. Electrical compatibility requires specifying voltages, currents, and possibly impedances that are used by both systems. Also, to avoid timing problems, the duration of the signals should be defined and should be compatible with the systems involved to handle even worst case conditions. Generally, the voltage levels will be TTL signals, RS-232 compatible (+15 volts), or relatively low voltage levels ranging from -24 volts to +24 volts. If the electrical signals are compatible, the next potential obstacle is assembling or purchasing the required cable and connectors. The manufacturer of the robotic system and the manufacturer of the analytical instrument should have documentation describing the various interfaces and interconnections which they supply with their equipment. Ideally, the user will be able to purchase interconnecting cables with the proper pin-out and mating connectors. This can be particularly tricky with RS-232 cables, the user needs to be aware of which componenets in his system have Data Communication Equipment (DCE) ports and which have Data Terminal Equipment (DTE) ports.[1] It may be necessary to assemble "custom" cables or modify existing cables to provide the correct signal path for the particular configuration being used.

This leads to the final problem of establishing the communication protocol between the two systems. Even if "standard" (e.g. RS-232 or IEEE-488) interfaces are used, this only assures that when the systems are plugged together, nothing goes up in smoke; and does not guarantee that the systems will communicate with one another. Communication protocols consist of synchronization signals, timing standards, control and status information and transmission packets with header information and data encoded in a known format.

The software protocol used in processor-controlled robots and analytical instruments varies widely; there are no industry standards. However, what is standard is the format and representation of the information as it exists in the digital domain. Data is generally stored in one of three formats: ASCII (American Standard Code for Information Interchange), BCD (Binary Coded Decimal), or Binary.[2] If the components that are being intefaced together do not use the same format, one of the units will have to have enough software flexibility to translate all of the communication conversations into a common format that both can understand. Besides transmitting the data, other ancillary information is sometimes added for error checking and dialog instructions, as well as headers to indicate the type of information contained in the transmission packet. The data may also be followed by a number representing a checksum to validatae the integrity of the information. Finally, a termination character or sequence may be present to indicate the end of that packet of information.

This discussion presents only a few of the high points and most frequently encountered situations. In order to illustrate how some of these concepts can be utilized, two examples are given which couple a gas chromatography (GC) system with an automatic sample preparation system. The first example depicts a simple control and status configuration of limited sophistication but allows the two systems to work together and coordinate sample hand-off while operating independently. The second example is more extensive and shares data about the sample between the two systems. A personal computer has been added which acts as a translator for the robot and GC but also provides additional data and file manipulation capabilities and closes the loop between sample preparation, sample analysis and data handling.

STATUS AND CONTROL EXAMPLE

One example of using an automatic sample preparation system in the analytical laboratory is the preparation of samples for gas chromatographic analysis. The preparation methods generally include dilutions, extraction steps, addition of internal standards, and proper capping and storage of prepared samples. The tasks of sample preparation and sample analysis can easily be coupled using only simple control and status signals. In this example a Zymate Laboratory Automation System passes prepared samples to a Hewlett-Packard gas Chromatography (GC) system and then starts a GC analysis using one control and one status line.[3,4]

The Zymate System includes a Z120 Zymate controller, Z110 robot module, Z830 power and event controller and additional modules for the sample preparation (Figure 1). The Hewlett-Packard gas chromatography system includes a HP5890 Gas Chromatograph, HP7673A Automatic Injector, and a HP3392A Integrator which communicate with one another over an INET (Instrument Network) communications loop (Figure 2). After a sample has been prepared, the Zymate System must do two things: 1) sense the status of the GC system to determine if it is ready to accept a sample and 2) signal the GC that a sample has been placed in the sample turret and a GC run should begin. The HP3392A integrator has a remote port which provides the status information and the "start" control of the GC and can be connected to the Z830 power and event controller using a HP Remote Start Cable (Figure 3). The "Ready Output" line from the integrator remote port (providing a TTL signal representing the status of the GC system) is connected to one of the input lines of the Z830

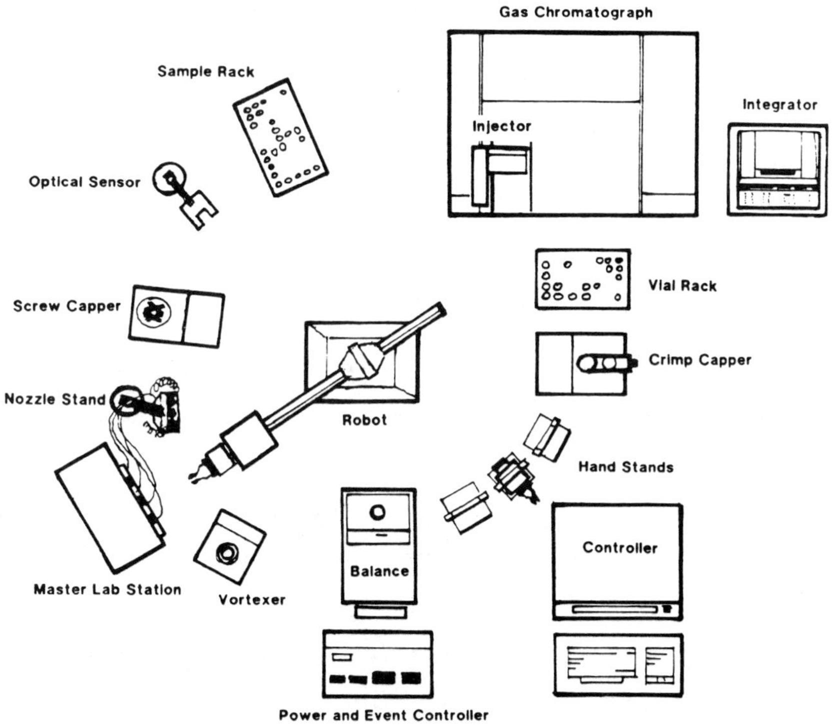

FIGURE 1. Zymate System layout for GC analysis.

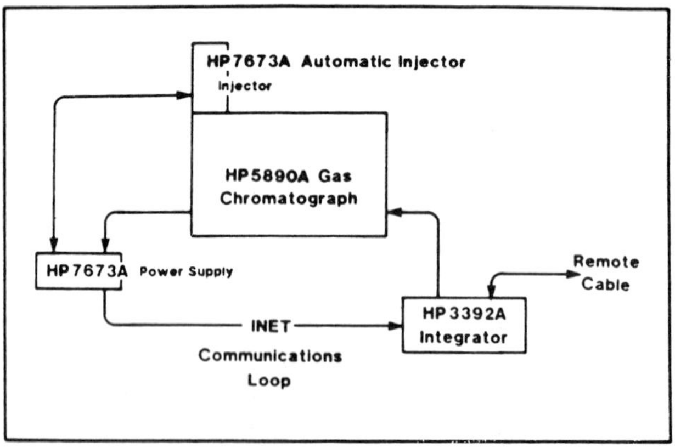

FIGURE 2. Schematic of the gas chromatographic system configuration.

power and event controller, thereby enabling the Zymate System to monitor the status of the GC system. When the GC system is "ready", the Zymate System can place a sample in the GC autoinjector turret and start an analysis by closing a switch on the power and event controller which is connected to the "Start 3392" line going to the integrator via the remote port. This generates a TTL signal that begins a GC analysis, during which the integrator controls the automatic injector and GC and is unavailable to the Zymate System until the analysis is complete. While the GC system is doing the analysis the Zymate System can continue preparing more samples, passing them to/from the GC, and activating the GC system as required.

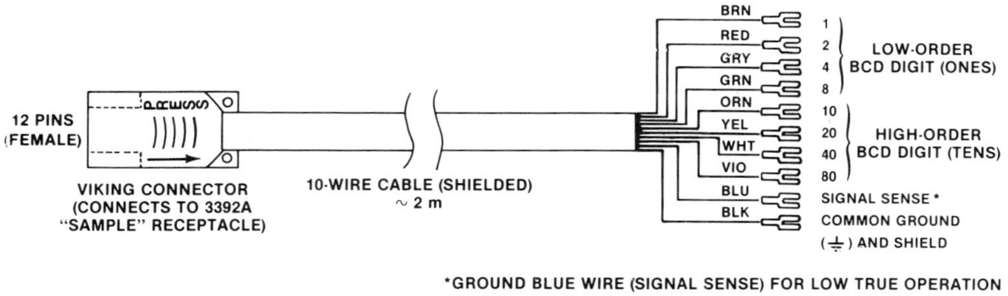

FIGURE 3. General Purpose Remote Start Cable, HP Part No. 03392-60830.

PARAMETER PASSING EXAMPLE

The previous example provides a means of coordinating the tasks of sample preparation and sample analysis but does not allow for the sharing of information about the sample between the two systems. The analytical instrument often uses information generated during the preparation step either for internal calculations or to set setpoints appropriate to that particular sample. Typically, this information is input into the instrument by manual switch settings, keyboard entry, or a "remote data port". The configuration and type of port will determine what information can be transferred and how that will take place.

The GC system described above communicates internally over a two-wire INET networking loop. The HP3392A integrator acts as a controller on the loop and stores pertinent setpoint and control information about each of the devices on the loop in a workfile. The information in this workfile can be accessed, and, if necessary, edited via a RS-232 "computer" port on the HP3392 integrator. The Zymate controller also uses an RS-232 module card for communication with other RS-232 devices or computers. Unfortunately, the software protocols used by the two systems are not compatible and cannot be altered to let the integrator and Zyamte controller communicate directly. However, a computer can be used as an intermediate translator not only to handle the protocol differences but also to add additional capabilities concerning file management and data handling. The following example describes just such a system where a HP-150 personal computer system with dual serial ports acts as the "translator" between the GC and sample preparation systems described earlier (Figure 4).

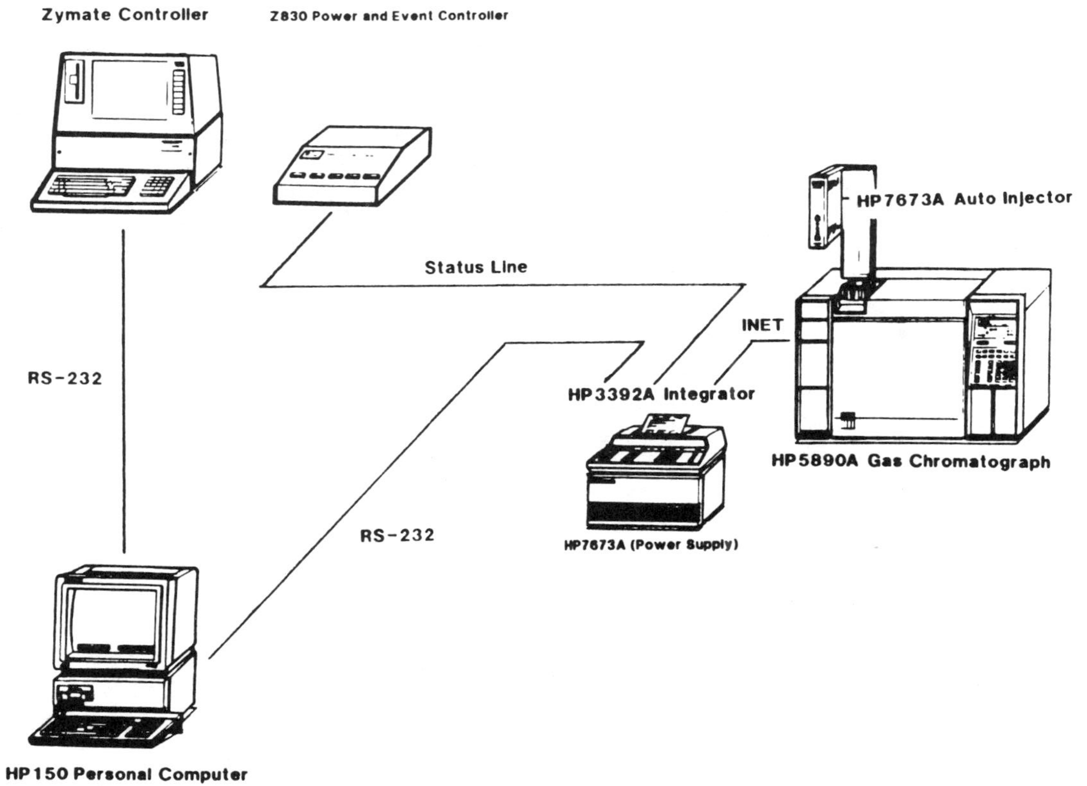

FIGURE 4. Communication Links between the Zymate Laboratory Automation System, the HP Gas Chromatography System and the HP-150 Personal Computer.

As outlined before, one of the first requirements of actually configuring the overall system is determining the necessary cables with the appropriate connectors for the RS-232 communication. In this particular example, a 12-pin Viking to 25-pin "D" RS-232 cable was necessary for the HP3392A integrator to HP-150 computer interface and a null modem was used between the Zymate controller and the HP-150. The baud rate, the ASCII format, number of start and stop bits, and any active hardwire handshake must be established. Table 1 shows the settings used for this particular system.

TABLE 1. Settings for the GC/HP-150/Zymate System.

	HP-150 to HP-3392A integrator	HP-150 to Zyamte Controller
Baud Rate	9600	1200
ASCII code	7 bits	7 bits
Stop bits	1 bit	1 bit
Hardwire Handshake	none	none

Next, the problem of software protocol needs to be addressed. The HP3392A integrator and the Zymate controller each has its' own particular dialect to handle the software handshake and data packeting. The Zymate controller transmission packet begins with a message type identifier (transmit or receive), an identification code, the length of text being transmitted, the message text, the checksum and a termination sequence (carriage return, linefeed). The computer must be able to interpret this transmission packet and respond accordingly with an "acknowledge" or "not acknowledge" packet, or transmit a data packet of its' own. The HP3392A integrator uses a slightly different format; the message begins with a four letter code, followed by the message text, and terminated with a carriage return, linefeed. To identify the sender of each message, the host computer prefaces each message with an escape character followed by an upper-case C, while messages originated by the integrator have no preface prior to the four letter code. Short diagnostic programs should be written to confirm that the interfaces are indeed operating properly. Putting in extra time and effort on these diagnostic routines along with good documentation will pay dividends in the long run, particularly for possible future system reconfiguration or easier system troubleshooting.

Finally, an executive routine has to be written on the computer that will coordinate the data transfer between the sample preparation system and the analytical system. The Zymate System still monitors the status of the GC system as in the previous example (using the power and event controller and the "Ready Output" line) but relinquishes responsibility for starting the analysis to the HP-150 computer. The program on the computer waits until the Zymate System sends any required weights, volumes, or sample identification information over the RS-232 interface. The information is not sent until a sample has been placed in the automatic injector to insure that a sample is present prior to a start injection command. The program then incorporates this information into a workfile that the integrator will use to calculate the amounts of the components of interest in that sample. The computer then sends down a remote start command and waits until the integrator indicates that the run is complete. The results are printed out at the integrator and sent to the HP-150 computer for archival storage and integration into other database routines. The program then waits for information on the next sample or a message that all samples have been analyzed from the Zymate controller.

SUMMARY

In actuality, there is no one correct answer to interfacing robotic systems and analytical instruments. The most important step is to first outline, in as much detail as possible, what is required of the combined systems, in terms of answering the laboratory needs. The answer to these needs will include some combination of control, status, error and data lines. The resulting configuration will be dependent upon the equipment involved, the time and funds available to implement it, and the expertise of the people defining and designing the system. Interfacing options will become clearer as robotic systems establish themselves as an integral component of the automated bench.

REFERENCES

1. Leibson, S., Byte, 7(5), 202 (1982).
2. Leibson, S., Byte, 7(6), 242 (1982).
3. Randal, L. G. and Poole, J. S., "The Electrical and Mechanical Handshake between a Hewlett-Packard 5890A Gas Chromatograph and a Zymark Automated Sample Preparation System", Hewlett-Packard Technical Paper No. 105 (1985).
4. Zymark Corporation, "Interfacing the HP5890A Gas Chromatograph with the Zymate System", Zymark Corporation P/N 38884 (1985).

IMPROVING THE FLEXIBILITY OF AN ANALYTICAL ROBOTIC SYSTEM BY USE OF PROGRAMMING COLUMN SWITCHING, SOLVENT SELECTIONS, AND ROBOTIC COMPUTER CONTROL OF PROGRAMMABLE HPLC EQUIPMENT

John Van Antwerp and Robert F. Venteicher,
Hoffmann-LaRoche Inc.
340 Kingsland Street,
Nutley, NJ 07110

ABSTRACT

Our goal of optimizing the flexibility of an analytical robotic system, particularly for multiple HPLC applications, was enhanced by interfacing with the Zymate controller an Autochrom column switching device, a programmable HPLC pump, a UV-visible HPLC detector, and a solvent selector valve. As currently configured, the Zymate controller can select among 5 HPLC solumns and a purge valve position, and through the use of solenoids, can change mobile phases for new methods or wash cycles for orderly shut-downs and new start-ups as programmable options with minimal operator involvement. The Zymate controller can select between UV-visible or fluorescence detection and for UV-visible detection, can select the monitoring wavelength. The controller also determines which detector signal is monitored for data acquisition, can turn on/off the HPLC pump and select among several flow rates including the ability to start-up and shut-down flow at a controlled rate of change to protect columns such as those used for GPC. Programmable down-stream control by the robotic system consitutes a significant achievement towards implementing highly flexible, routine, multiple HPLC methods.

INTRODUCTION

High volume, dedicated analytical robotic testing is an obvious, direct application already realized over the past several years in many laboratories.[1-3] Applications of dissolution testing,[4] content uniformity testing,[5] analysis of pourable powders,[6] and numerous other uses are already widely known. However, an important aspect of routine Quality Control testing is the need to be flexible and readily convertible between several analytical applications within a single workday, which is an important goal for many robotics laboratories. Ideally, several different products involving a small number of samples could be analyzed during the workday and one type of product involving a large number of samples analyzed during the evening or weekend when analysts are not in the laboratory. If a laboratory receives 50 samples a day, usually these samples would consist of 5-10 different types rather than all of the same kind. In terms of inventory control, cost containments, and meeting production and sales requirements, quick completion of testing and product disposition are extremely important. This report describes a robotic system that has been improved by interfacing an automatic column switching device, a solvent selector, and a programmable HPLC system which represents a significant improvement in flexibility particularly for multiple HPLC applications. Emphasis has also been placed on implementing applications where analytes of limited solution stability are being applied to robotic analyses and which are designed in serial mode to enhance the reliability and accuracy of the anlaytical procedures.

EXPERIMENTAL

The assembled robotic system consists of the Zymate controller, the robotic arm, 2 power and event controllers (PEC), a master lab station (MLS), a modified New Brunswick orbital shaker, a Zymark vortexing station, a Mettler AK-160 analytical balance, a Zymark HPLC injector station, an Autochrom automatic column switching device, a Waters 4-way solenoid switching valve, a Waters 590 programmable pump, a Perkin Elmer LC-10 fluorescence HPLC detector, a Waters 490 programmable multiwavelength UV/visible detector, a Sola constant voltage transformer, and a Topaz uninterruptable power supply. The detectors of the system are interfaced to the department data acquisition computer by an A/D converter.

OPERATION DESIGNS

The system is configured as shown in Figure 1. The unit operations for the analytical routines can be performed in serial, batched, or combined serial and batch modes depending on which scenario would be most reliable and/or efficient. For example, the sample weighings or subdilutions could be performed in a batch mode for a solution-stable analyte while column selection, mobile phase conditioning, and standard injections for system suitability testing are being done. For solution-labile analytes, sample weighings could be performed during start-up, however, extractions and subdilutions would be performed following HPLC system equilibration. For highly unstable analytes, total serialization might be the only analytically reliable and efficient mode of operation.

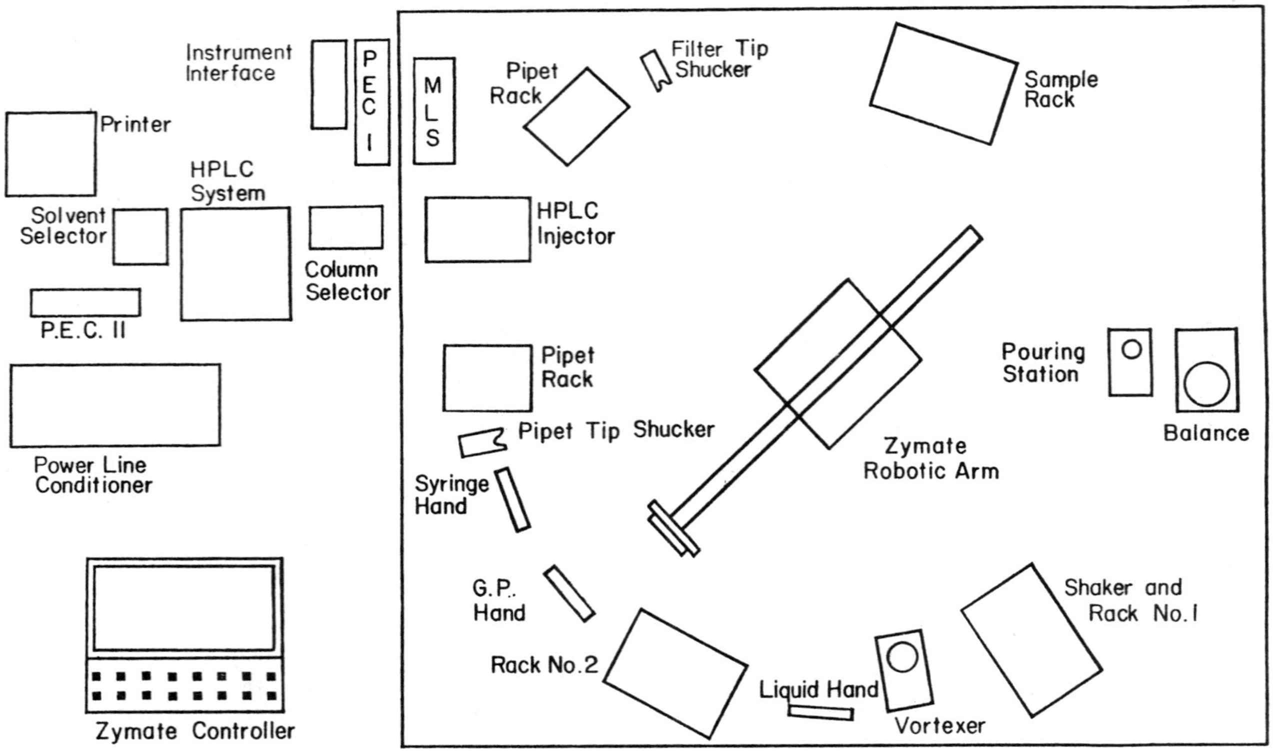

FIGURE 1. Diagram showing the module and instrument arrangements of the robotic system for HPLC.

PROGRAMMABLE COLUMN SWITCHING

The column switching routines are executed upon programmable commands of the EasyLab computer program by the Zymate System controller. The controller can select the exact position desired by random access through a series of "if...then" statements. The system can select from among five different HPLC column positions or the purge valve position. A diagram representing the valve arrangement is shown in Figure 2. Interfacing the column switching valve was done by connecting the parallel output cable supplied with the valve to six of the inputs on the PEC. The valve is pneumatically operated and is logic controlled. A copy of the output logic table is shown in Table 1.

Table 1. Computer Output Logic Table of the Autochrom Column Switching Device

Column	Red	White	Green	Orange	Blue	Brown	Black
1	1	0	0	0	0	0	G
2	0	1	0	0	0	0	R
3	0	0	1	0	0	0	O
4	0	0	0	1	0	0	U
5	0	0	0	0	1	0	N
6	0	0	0	0	0	1	D

In the EasyLab program, the controller checks to see if the given input corresponding to the correct column position is high. If not, it instructs the valve to switch by means of a contact closure on the PEC. This arrangement insured that the valve must be in the correct position

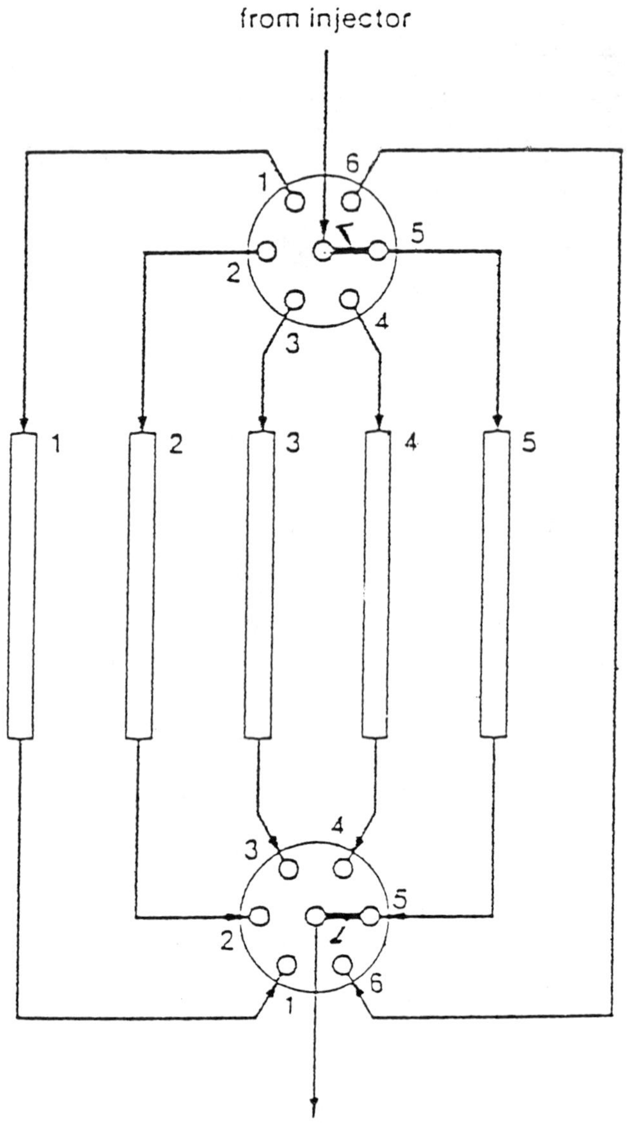

Figure 2. Representation of the valve selector positions of the Autochrom automatic column switching device for HPLC.

before the EasyLab program can proceed thereby eliminating damaging results if the high pressure air line or the output switches fail.

SOLVENT SELECTION

Solvent selection was achieved using a Waters solvent selection valve interfaced to the PEC by means of two output switches. This valve was chosen because it was logic controlled and could select from four different solvent reservoirs using only two switches instead of the four that would normally be required. A diode was placed across the solenoid load to prevent damage to the switches caused by inductance.

DETECTOR SELECTION

The two detectors were set up in series with the fluorescence detector downstream of the UV detector in order to protect its flow cell from excessive backpressure. The use of low dead volume tubing to connect the two detectors to be on line without the use of a valve to direct the effluent of the column to the correct detector, thereby simplifying the system design and programming. The power to each detector was controlled by switched AC receptacles on the PEC allowing the detectors to be turned off when not being used. In addition, the 490 detector has a provision allowing it to be put in a standby mode which keeps the electronics on but the xenon lamp off, thereby maximizing lamp life. Since analog signals from both detectors were multiplexed through switches on the PEC allowing the Zymate controller to select the correct signal to be sent to the data acquisition computer.

DETECTOR WAVELENGTH

In order to achieve maximum flexibility, a Waters Model 490 programmable wavelength detector capable of changing wavelengths and auto-zeroing by external control was interfaced to the robotic system. Wavelength changes were made by contact closure of output switches on the PEC which initiated a previously programmed series of commands within the detector.

HPLC PUMP FLOW RATE SELECTION

A Waters Model 590 programmable solvent delivery system was interfaced to the robotic system in the same manner as the 490 detector. This allowed for the selection of three different flow rates and also allowed the rate of change of the flow rate to be controlled which was necessary when using the GPC column. In addition a wide range of flow rates were permitted enabling rapid purges and clean-out for system start-ups and shut-downs.

SYSTEM INITIALIZATION

When the system is initialized, the controller first sends a command to the column switching valve to insure that the valve is in the "purge" position. When this is verified the solvent selector valve is switched to allow the pumping of an intermediate solvent such as THF (tetrohydrofuran). The pump is started and the HPLC system is purged with the THF with no column in the liquid stream. The injector is cycled between the "load" and "inject" positions to insure proper rinsing of the sample loop. The pump is stopped and the appropriate solvent is selected and pumped through the system with the column selector in the "purge" position. Then, the injector is again cycled between the "load" and "inject" positions to flush the loop. The pump is then stopped, the proper column is selected, and the pump is started at the desired flow rate. The proper detector is turned on and the correct analog signal is selected to be sent to the A/D converter. When the analyses are completed, the pump is started again. Once the system is washed out, the reverse of the start-up procedure is done bringing the system to an orderly shut-down with an intermediate solvent in the system and the column selector in the "purge" position. This procedure allows normal, reverse phase, and gel permeation chromatography to be carried out on the robotic system.

DISCUSSION

The combination of solenoid valve control and automatic column switching greatly expanded the flexibility of the robotic system to improve the usefulness of the system for a variety of analytical tasks within the same working day. This was made possible for the most part by the large number of inputs and outputs available on the PEC and the flexibility of the operating system. Among the applications being added to the original system are the routine analysis of a bulk water soluble vitamin by HPLC with UV detection at 254 nm, analysis of a drug in bulk and in premixes by small molecule gel permeation chromatography with fluorescence detection, and analysis of a solution-labile vitamin complex by HPLC analysis and UV detection. Whereas, the vitamin complex application will be carried out totally in a serial mode because of limited lifetime of one constituent in solution, the other applications are a combination of serial and batch modes to maximize the effeciency during column selections, purges, and equilibrations. For example, while samples can be weighed and extracted during system start-up, final dilutions and filtrations will be performed during the chromatographic analysis of samples in between HPLC injections.

Applications expected to be added in the near feature are the analyses by HPLC for some additional vitamins in bulk froms and in premixes using low wavelength UV detection at 200 nm which were formerly done using microbiological assays.

SUMMARY

By combining the general robotic unit operations of weighing, extraction and/or dilution, subdilution, filtration, and injection into the Zymark HPLC injector station with programmed column switching, solvent selection, and use of programmable HPLC equipment, an improvement in flexibility and productivity was achieved. The additional aspect of achieving orderly start-ups, purges, and shut-downs for cleaning out buffer systems for preserving pumps, columns, and injectors during unattended operation is also significant. For routine Quality Control testing, it is now more feasible to rapidly change between several HPLC analyses with minimal operator involvement. The goal of performing 3-4 different types of analyses within the same day with additional overnight operation for "dedicated" applications is also possible. Such advancements in flexibility might well increase by 3-4 times the productivity improvements already achieved by implementing analytical robotic applications. Such a versatile system should also be applicable in the future for unattended methods development, solvent optimization, or extraction optimization routines.

REFERENCES

1. R. Venteicher and J. Van Antwerp, in Advances in Laboratory Automation-Robotics 1984, G.L. Hawk and J.R. Strimaitis, eds, Zymark Corporation, Hopkinton, MA, p 275 (1984).

2. W. Jeffrey Hurst, ibid, p. 117.

3. J.E. Taylor, ibid., p. 83.

4. J. Curley, ibid., p. 299.

5. E. Siebert, ibid., p. 257.

6. J.B. Cross and E.J. Marak, ibid., p. 181.

A SHARED ROBOTIC SYSTEM: AUTOMATED PIPETTE CALIBRATION AND PIPETTE TIP FILTER ASSEMBLY

James H. Addison, Jr. and Gregory M. Dyches
E. I. duPont de Nemours and Company
Savannah River Laboratory
Aiken, South Carolina 29808

ABSTRACT

At the Savannah River Laboratory a Zymate Laboratory Automation System has been developed to perform two completely independent tasks within one workcell. One opeartion is the precise calibration of pipettes; the other is the assembly of a filter in a pipette tip. Since neither task requires full robot time, the shared system is an economical means of automating both processes. These are tedious, repetitive, time-consuming tasks. Human operators fail to yield constant results. Automation insures a repeatable process which increases product quality.

In the pipette calibration application, Pipetman-F fixed volume (in the range of 2 to 1000) microliter pipettes, manufactured by Instrument Co. Inc. (Emeryville, CA), are automatically calibrated. Each pipette is calibrated gravimetrically by an iterative process of pipetting, weighing and adjusting pipette volume. The entire procedure is automated in a manner that mimics the human operator's pipette handling. Integrated pneumatics enable the aliquoting of samples. The Mettler balance determines the weight of the dispensed sample. Pipette adjustment is accomplished via a stepper motor interfaced to an extra drive circuit in the master laboratory station. Software controls the direction and number of steps that the motor moves and iteratively finds the point of precision calibration.

For the task of installing sediment filters in pipette tips, the tips and filters are oriented by vibratory bowl feeders, after which all parts handling operations are performed by the robot. The robot moves the tip from the bowl feeder to a heating block, then to an assembly station. Upon completion of assembly, a vacuum test is performed to verify correct operation. Finally, the completed assembly is placed in an accept or reject bin.

This paper discusses the development of both applications and their implementation in the same workcell.

INTRODUCTION

Very often a robotic system is not required to operate 24 hours a day, seven days a week in order to complete its required tasks. In fact, a robot may be idle as much time as it is operating. While the robotic system may be justified for its given task, it is not being used to its full potential. Where possible, a shared robotic system can improve overall efficiency. A shared robotic system refers to the use of a single robotic system to perform two or more complete applications. The shared system significantly reduces the original automation cost since one robot and controller are used for two or more procedures. The cost of peripheral hardware is also reduced when some of the same equipment is shared for both applications.

At the Savannah River Laboratory (SRL) a Zymate Laboartory Automation System has been demonstrated in a shared robotic system to complete two unique applications. The first application is the calibration of Rainin Instrument Co. Pipetman pipettes. This application uses the Zymate robot, several standard peripheral modules, and some original equipment interfaced to the control system. The second application is the assembly of sediment filters in the ends of disposal pipette tips. This application uses the Zymate robot for material handling and the power and event controller to operate peripheral equipment used in the assembly task. The complete system diagram is given in Figure 1.

In this shared robotic system, cost savings are achieved. Neither of these tasks requires dedicated use of the robot, and some of the same hardware is used for both applications. Both are tedious, repetitive, time-consuming tasks, and human operators fail to yield constant results. Thus, an individual robot system would be justified for each task. However, combining the jobs in a shared workcell provides an effective, efficient solution.

DISCUSSION

Automated Pipette Calibration

Rainin Instrument Co. Pipetman precision microliter pipettes are used routinely throughout the Savannah River Plant and Laboratory (SRP/SRL) in many wet chemistry methods. These single volume pipettes (Figure 2) dispense volumes ranging from 2-1000 uL. The pipette has a hand-contoured handle, tip ejection mechanism and thumb operated two-stop plunger. A notched rotating calibration sleeve requires a special keyed tool to adjust delivered volume. Although factory calibrated when purchased, periodic calibration is required to maintain needed precision and accuracy (e.g., bias and precision limits for 50 pipettes are $\pm$0.75% and 1.00% at 95% confidence level, respectively). Calibration is an iterative gravimetric procedrue requiring the technician to repetitively pipette and weigh distilled H_2O and adjust the pipette calibration sleeve until the dispensed volume is within limits. The procedure is tedious, time-consuming, and results vary with

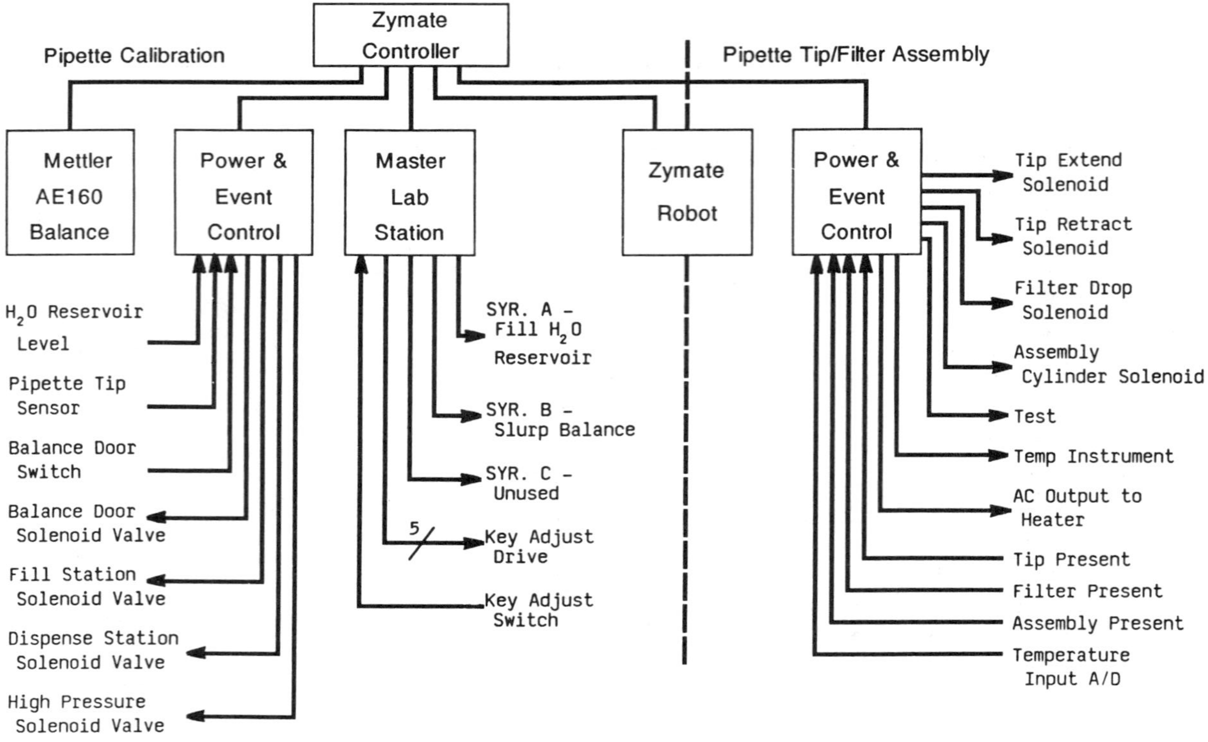

FIGURE 1. Shared application system diagram for pipette calibration and pipette tip/filter assembly.

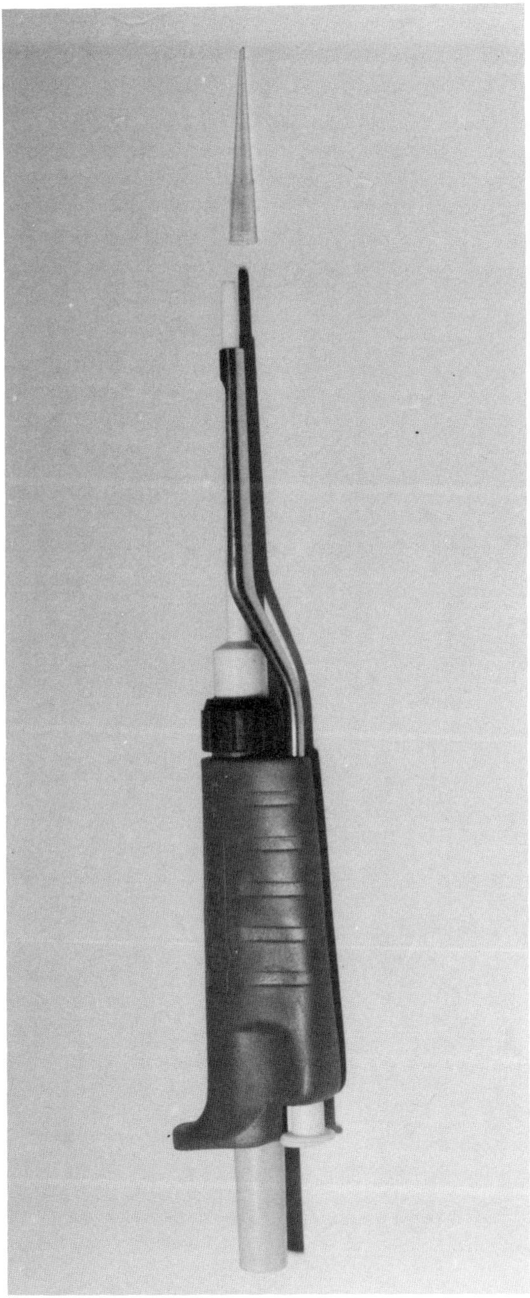

FIGURE 2. Rainin single volume pipette.

each technicians's technique. The benefits sought from automating this procedure are:

- equal or better precision and bias than that of the best technician,
- uniform technique (plunger speed, immersion depth, etc.),
- better use of limited technician time,
- elimination of special training and "qualification certification" for individual technicians.

The system (Figure 1) is based on the Zymate robot and controller, master lab station (MLS), and power and event controller (PEC). A Mettler AE160 balance with pneumatic door opener is used to weigh pipetted volumes to within 0.1 mg. A special pneumatic control circuit (Figure 3) interfaced to the PEC allows control of the Mettler door opener as well as two small bore air cylinders used for depressing pipette plungers. Two regulators in this circuit provide a "low" pressure (typically 20-25 psi) for depressing the plunger to the first stop, and a "high" pressure (typically 45-50 psi) for depressing the plunger to the second stop. One cylinder (Figure 4) is located above an H_2O reservoir (fill station) and the second cylinder is located above the Mettler balance pan (dispense station). A capacitive proximity sensor monitors the H_2O reservoir level. The level of this reservoir is kept constant by using one syringe of the MLS to transfer H_2O from a larger reservoir. A second MLS syringe is used to periodically pump H_2O from the container on the Mettler balance.

A special key adjust station (Figure 4) is comprised of an Airpax stepping motor interfaced to an unused stepper drive circuit (reserved

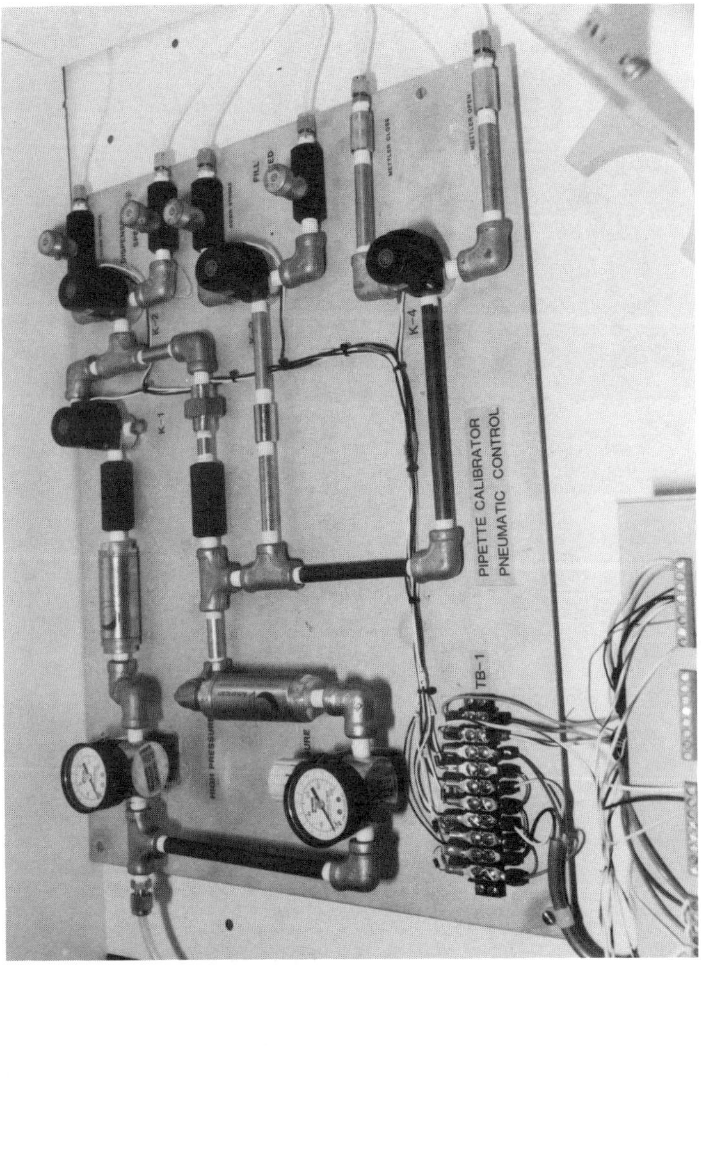

FIGURE 3. Pneumatic control circuit.

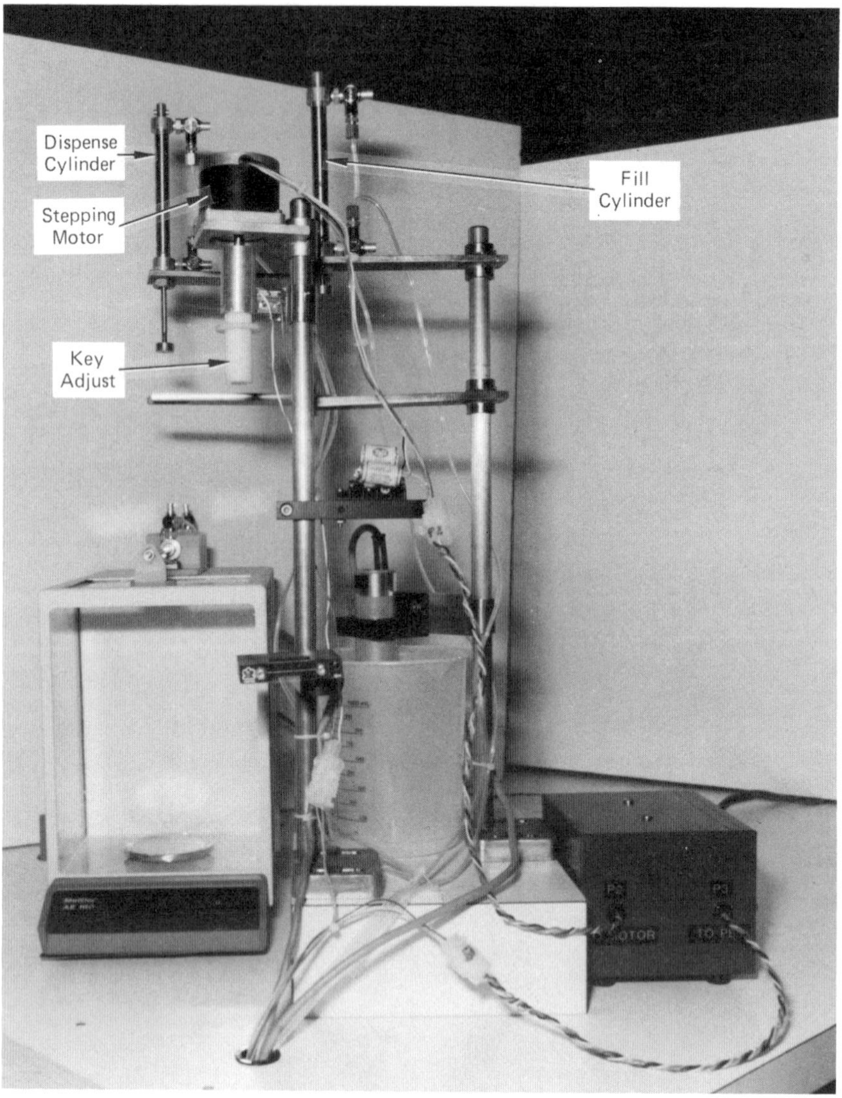

FIGURE 4. Fill/Dispense Cylinders and Key Adjust.

for turntable operation) in the MLS. The stepping motor shaft has a spring-loaded notched tool that mates with the pipette adjustment sleeve. A microswitch with roller actuator senses when the tool is engaged with the sleeve.

A pipette storage rack and pipette tip rack store the pipettes and three sizes of disposable tips. An optical thru-beam sensor is used to confirm proper tip attachment. A special urethane rubber gripper molded to contour to the pipette handle surface is mounted on the Zymate high torque hand. The gripper provides the surface area contact required to prevent slippage during operation.

In a typical operation sequence, the robot picks up a pipette from the rack, attaches a tip, and fills the pipette at the fill station. Then the pipette is dispensed and weighed on the balance. Filling and dispensing is repeated 10 times (software selectable). If the pipette is out of limits, it is adjusted and the procedure is repeated until the dispensed volume limit criteria is met.

In addition to robot point sequencing, controller applications software operates pneumatic sequencing, tests sensor inputs and operates the key adjsut stepping motor. Balance data are kept in array storage for statistical analysis. Spurious balance readings due to electrical or air current interference are detected by calculating $\bar{x}$ and σ for a set of 10 readings of the same sample (Figure 5). If any value exceeds the $\bar{x} \pm 1\ \sigma$ limits, it is discarded and a new $\bar{x}$ calculated. This method is used each time a weighing operation is requested. A similar analysis is

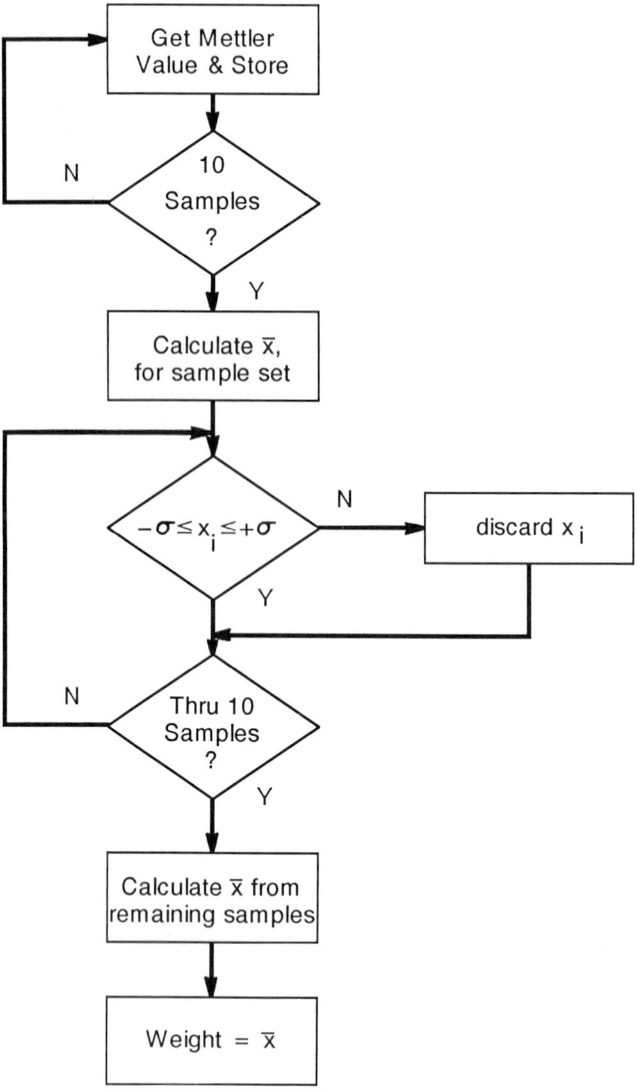

FIGURE 5. Mettler balance read operation for pipette calibration.

done on a series of weight values from 10 pipettings to determine the average delivered volume of that pipette. A density correction factor is necessary due to H_2O density temperature dependence. The difference between actual and desired volume is used to calculate the number of steps and direction for driving the stepping motor. The above process is repeated until the delivered volume is within acceptable limits.

Development of this procedure is underway, and no calibration results are currently available. Preliminary results indicate that the speed may be slower than expected due to the large number of balance readings and statistical calculations performed on each data set. One adjustment iteration currently takes five minutes. Development work is continuing to speed up the procedure via optimizing calculations, minimizing the number of statistical sample set elements, or possibly using the Zymate System high speed math option in conjunction with a shorter weighing integration time.

Once application programming is optimized, experiments will measure the robot system's calibration accuracy and compare it with that of the best technician.

Pipette Tip/Filter Assembly

At SRL/SRP disposable pipette tips are routinely used for various chemical analyses. For some specific jobs, stainless steel sediment filters are installed in these tips (Figure 6). Currently, the filters are placed in the pipette tip by the following manual procedure:

- Drop filter in tip

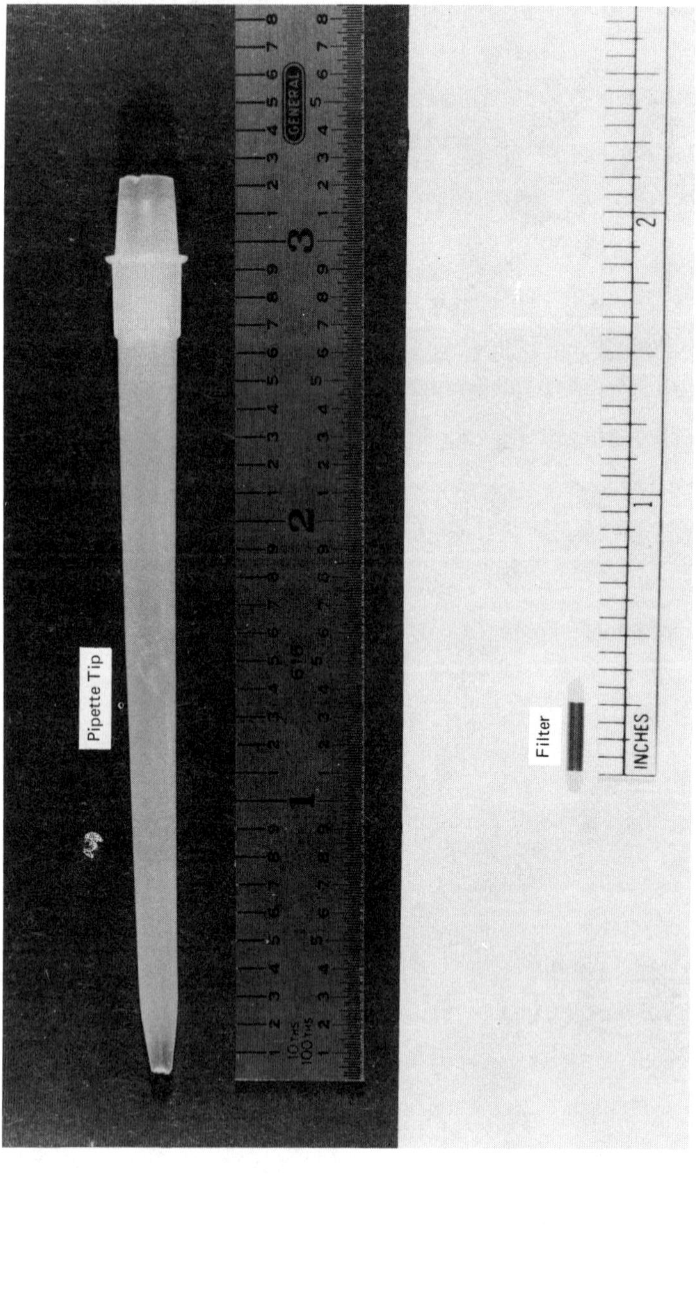

FIGURE 6. Disposable pipette tip and filter.

- Heat assembly in block heater
- Remove assembly from heater
- Invert assembly over inserter rod
- Push filter to end of tip

This is a repetitive task which results in an inconsistent, unsatisfactory final product. Since the position of the filters affects the chemical analyses, it is important to deliver consistent results. The use of a robotic system will provide the required quality results. Often, in order to meet user demand, these tips are assembled on overtime. Therefore, the robot will concurrently respond to the need for temporary manpower, and provide a return on investment. The benefits sought from automating this procedure are summarized as:

- Eliminate the need and cost of overtime
- Increase quality by a repeatable, uniform technique

The automated system is based on the Zymate robot and controller and the power and event controller. A dual vibratory bowl feeder system was specified and custom made to orient and present the pipette tips and filters (Figure 7). Pipette tips are isolated for robot pick-up using a shuttle mechanism. Presence of the tip at the pickup point is confirmed by an optical thru-beam sensor. The filters are aligned end to end (vertical axis) on a gravity track in order to be "dropped" into a tip. The presence of each filter is also confirmed by an optical sensor. A double-acting cylinder releases one filter at the time to a pipette tip.

A Lab-Line block heater is used to heat the pipette tips for easier

FIGURE 7. Dual bowl feeder system.

filter insertion, also resulting in less splitting of tips due to insertion. The temperature of the block is monitored using a temperature transducer that is interfaced to the power and event controller's A/D input.

The assembly station (Figure 8) consists of a holding fixture with a retro-reflective optical sensor to verify the presence of the assembly. An air cylinder coupled with an insertion rod is used to consistently push the filter to the pipette tip end.

A test device is being designed that will measure the air flow through the completed assembly. This will determine if the final assembly is defective, due to a clogged filter, split tip, or a malfunction in any previous part of the procedure.

During a typical operation sequence, the robot grasps a tip from the block heater, verifying proper pick-up with an optical sensor (the same one used to verify pipette tip attached in the pipette calibraton application). The controller then checks for a filter present at the drop point, and having received the proper input, the robot moves the tip in position to recive the filter. The drop mechanism is activated and deposits the filter in the tip. The assembly is now taken to the assembly station. When the assembly present signal is detected, the controller activates the cylinder to insert the filter. The completed assembly is then taken to the test station for air flow testing. Based on the test results, the final assembly is placed in the accept or reject bin. The controller now checks for a tip present at the pick-up

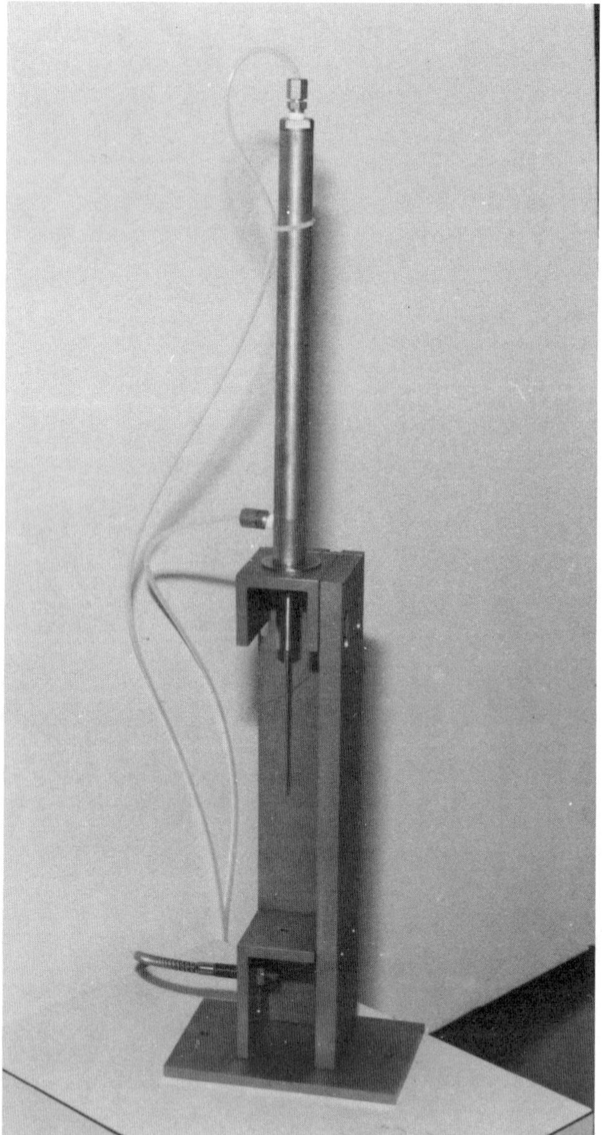

FIGURE 8. Assembly station.

point and grasps the tip once its presence is verified. The tip is placed in the vacated heater position, and the procedure is repeated.

Since each step of this application relies on the previous one, presence of the proper materials is verified before performing the next task. If materials are not sensed to be in their proper position, software determines what operations should be repeated to obtain the desired state. These steps also incorporate timers, so that in the event a required state cannot be achieved in a reasonable amount of time, the operator will be alerted. Also, if the air flow test reveals several consecutive "reject" assemblies, the operator will be notified since this is usually an indication of a hardware problem.

The "fine-tuning" and development of this application is continuing. Currently, the robot system results are not sufficient to compare with technician's results. However, the Zymate System has proven to be a good system for material handling in this assembly task. This is enhanced by its "used friendly" input and output control via the power and event controller.

SUMMARY

The Zymate Laboratory Automation System is an effective and efficient solution to a wide range of applications that should not be limited to wet chemistry. It provides a low cost, user-friendly robotic system to perform the required tasks for pipette calibration and pipette tip/filter assembly. The shared robotic system allows multiple procedures to be executed on the same robot system, thus minimizing the hardware overhead per application. The shared system makes the automation of non-dedicated application more economical, and at the same time, the robot utilization is optimized.

ACKNOWLEDGEMENTS

The information contained in this article was developed during the course of work under Contract No. DE-AC09-76SR00001 with the U.S. Department of Energy.

INTERFACE OF ZYMATE LABORATORY AUTOMATION SYSTEM WITH A VARIAN FT80 NMR SPECTROMETER

Aris Ragouzeos, Ronald C. Crouch, Jean L. Miller
Burroughs Wellcome Co.
3030 Cornwallis Rd.
Research Triangle Park, NC 27709

ABSTRACT

A Zymate Laboratory Automation System has been interfaced to a Varian FT80 NMR spectrometer. The robot prepares and introduces the samples while data from the spectrometer is stored on floppy disk. The Zymate-NMR interface permits unatteded overnight operation, increasing throughput and removing the necessity for handling toxic compounds.

EXPERIMENTAL

Automation of Nuclear Magnetic Resonance (NMR) analyses by an FT80 NMR spectrometer under control of a Zymate System is described. An NMR instrument consists of a magnet and an electronics console. The magnet houses a probe with glass insert into which a sample tube is introduced and spun. The console's electronics lock onto the deuterium signal of the solvent (the dispersion mode of the deuterium signal is integrated and the results used to correrct the magnetic field to maintain a constant ratio to the system frequencies). A radiofrequency pulse is applied to the sample, the free induction decay is acquired and this data is stored on disk. These funcitons have been interfaced with the Zymate controller (computer). All computer I.O. is through the E-bus interface paralled port, device address 075. Two execute lines, Exec 577 and Exec 677, are wired to the same port for robot communication. The robot has been taught to prepare and change samples. Data acquistion, sample preparation and sample changing have been combined into one continuous automated operation with a rate of ca. 7 min/sample.

Analysis is initiated by a signal from the console's disk that data has been filed. The robot's arm with gripper hand drops over the probe and grasps the sample tube. This tube is 5 mm in diameter, 8 in. long and is inserted into a spinner turbine at a specified depth.

When the robot removes the sample from the probe the spectrometer senses a loss of spin, the lock drops out, offsets and waits. Removal of the tube and turbine from the probe is verified by a photoelectric cell and a sense line to the spinner tachometer. Failure to remove the sample from the probe leads to an abort program that protects the probe. When the next sample is inserted into the probe, the spin sense is checked. If it is not spinning, the robot will lift the sample a few cm and drop it back in the probe. If the sample spins, the automated locking routine programmed into the FT80 is signaled to begin. Lock sweep is done by positive or negative voltage applied to the flux stabilizer on the lock receiver card. Sample locking is communicated to the Zymark

System by the FT80 and sample preparation begins. Failure to lock or spin will be recorded on the printer and the spectrometer will wait for the next sample. Data collection begins after lock is established. The FT80 has been programmed to automatically scale the receiver gain to avoid overloading the ADC. A photo of the system is in Figure 1.

Sample preparation begins with removal of the tube from the spinner. The fit is tight so the turbine must be secured to allow the robot to remove the tube. This is done with a standard door bolt operated by the robot. After placing the tube in a tube rack, the robot moves to the compound rack. Samples are in 1 dram screw cap vials. Capping and uncapping are done by the capping station. Uncapping is verified by tripping the verification switch. Capping is verified by the programmed torque test.

The vial is carried to the nozzle of the syringe station and 0.6 mL of a deuterated solvent ($DMSO.d_6$, $CDCl_3$ or D_2O) is added. The solution is heated by a heat gun and mixed on a vortexer. The gripper hand is then replaced by the syringe hand. A tip with a filter is attached and its presence checked by a verification switch. The solution is drawn into the pipette tip, the filter is removed and the removal verified by a verification switch. The solution is then injected into a 5 mm NMR tube, the tip is removed, the removal verified and the gripper hand is retrieved. The tube is then capped and inserted into the spinner. The fingers are brought over and down on the tube to set proper depth. The robot then awaits a signal from the FT80 that a spectrum has been filed. The process is then repeated.

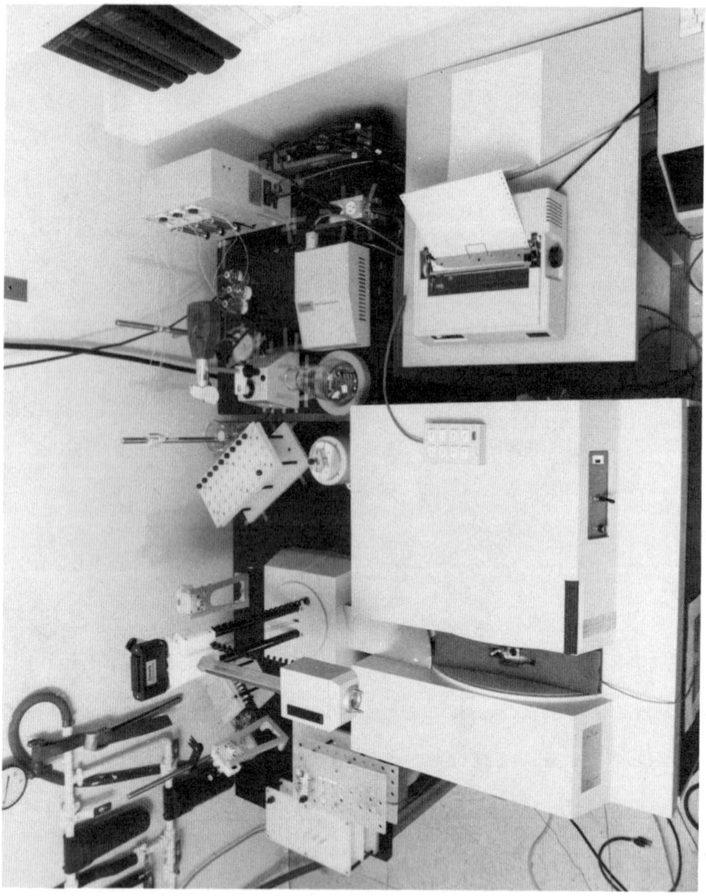

FIGURE 1. Zymate System interfaced to prepare samples for a Varian FT80 NMR Spectrometer.

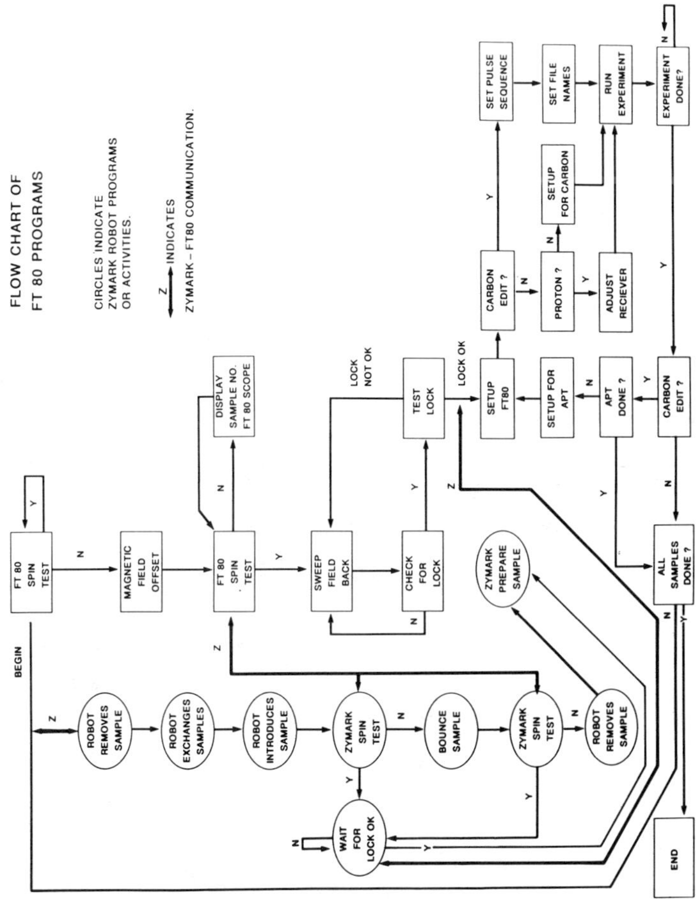

FIGURE 2. Flow chart of FT80 programs. Circles indicate Zymark Robot programs or activities. Z indicates Zymark-FT80 communication.

The FT80 was programmed in assembly language. Figure 1 is a flow chart of the FT80 programs, which required less than 650 words of memory. In addition to standard ^{1}H and ^{13}C spectra, provision has been made for a Carbon Edit experiment. This experiment has two modes of operation. The first involves an APT (Attached Proton Test) where the signals from methynes and methyls are of opposite phase from methylenes and quaternaries. The second mode involves running both the APT and a second experiment which uses the same pulse sequence as the APT (a spin-echo experiment) but without modulation of the protons. This yields a standard ^{13}C spectrum which can be used, along with the APT, in the FT80 add-substract routine to edit the spectra into two subspectra.

Details are available describing the FT80 programming. A paper is being published which will contain all pertinent information.

SUMMARY

This automation has increased throughput fourfold for ^{13}C and ^{1}H tasks. It has also removed the necessity to handle toxic compounds. The system can and often is used as a dample changer only.

PREPARATION OF HERBICIDE SAMPLES FOR HPLC ANALYSIS BY ROBOTICS

Sidney S. Goldberg
E. I. Du Pont de Nemours and Co.
Agricultural Chemicals Department
Experimental Station
Wilmington, Delaware 19898

ABSTRACT

Automation of analytical procedures is the key to increased productivity in the laboratory. While portions of analytical procedures have been automated for years, sample preparation has been performed manually. The advent of laboratory robotics offers an opportunity to automate this portion of analysis.

A Zymate laboratory robot was used to prepare herbicide samples for liquid chromatographic analysis. During the sample preparation, the robot executes the following operations: weighs, uncaps and caps vials, dispenses solutions, dissolves the sample, draws an aliquot, filters, and transfers the prepared sample to a vial for HPLC analysis.

Setup and programming of this application will be discussed. Several problems encountered will be highlighted.

INTRODUCTION

Today's chromatography laboratory is equipped with microprocessor controlled instruments, automated sample injectors and laboratory computers. Typically, once a technician loads the sample into the injector, the rest of the operation is automated. Output from the instruments is sent directly to a computer which calculates and reports the results.

The innovations of the past 10-15 years have made the analytical laboratory much more productive. However, it has been said of analytical services: "The more information you give them, the more they want." Even though the analytical laboratory is providing greater number of results than ever before, the demand for these services is continually increasing.

To meet this demand, we purchased a Zymate Laboratory Automation System to automate the sample preparation portion of analysis. The robot was installed in a high-volume laboratory which assays herbicide samples. Our intention was for the robot to perform routine operations, freeing the technician for other more productive tasks.

Development of the robotic procedure was divided into three phases. In the setup and teaching phase, the robot modules and racks were laid out on the bench (Figure 1). The table was designed to minimize movement of the robot arm from module to module. The robot was "taught" the positions needed for the procedure. The second phase involved programming the robot. A program consists of a series of positions. Robotic procedures consist of a series of these subprograms. Descriptive position and program names make programming simpler.

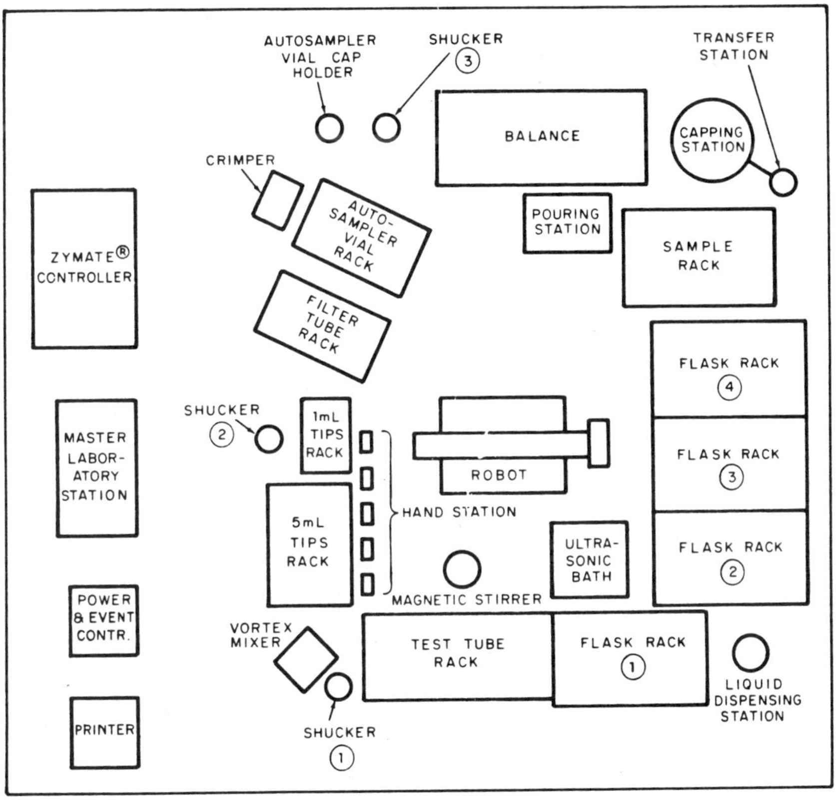

FIGURE 1. Layout of robotic system for herbicides by HPLC.

Finally, once programming was completed, the precision and accuracy of the procedure was confirmed versus the manual procedure.

EXPERIMENTAL

The Zymate robotic system used for this procedure was composed of the following modules: the robot arm and controller, master laboratory station, power and event controller, capping station and vortex mixer. Five robot hands are needed to perform the operation: a large general purpose hand, small general purpose hand, syringe hand, and two blank hands with nozzles attached. A Mettler AE160 electronic balance, interfaced to the Zymate controller, was used for weighing. A pneumatic crimper (Wheaton Industries, Millville, New Jersey) was used to seal aluminum caps on autosampler vials. An ultrasonic bath (Branson Cleaning Equipment, Shelton, Connecticut) was used to dissolve samples.

The capping station (Figure 2) was mounted in an inverted position. This allows the sample vial cap to be held in the capper jaws, rather than the vial itself. The possibility of contamination from material on the vial cap is eliminated because the cap is never left at any location. Disposable Acro LC13 filter discs (Gelman Sciences Inc., Ann Arbor, Michigan) were used to filter solutions.

PROCEDURE

The side door of the balance is opened by an air-actuated piston controlled by the power and event controller. Using the large gripper hand, a 125 mL erlenmeyer flask with a magnetic stir bar is transferred to the balance. After the weight of the flask is tared, the flask is

FIGURE 2. Inverted capping station.

placed on the pouring station. The large gripper picks up a sample vial and places it on the transfer station. The hand is inverted 180 degrees and regrips the vial. This repositions the vial to obtain greater clearance for the vial inside the capper's jaws (Figure 2).

The vial is uncapped, leaving the cap in the capper's jaws. The vial is returned to the transfer station and the gripper hand returned to the 0-degree position. The uncapped sample vial is taken above the erlenmeyer flask at the pouring station and the sample is poured into the flask. The sample is weighed and compared versus a preset minimum sample weight desired. If the minimum weight is not obtained, additional sample is added. Otherwise, the flask is taken to the dispensing station and 50 mL aliquot of diluent is added.

The flask is placed in an ultrasonic bath to facilitate dissolution of the sample. Periodically, the flask is placed on a magnetic stirrer. The combination of these devices insures dissolution of the sample within a reasonable time period. Once dissolved, a 5 mL aliquot is removed from the flask and transferred to a test tube. The test tube is taken to the dispensing station, where internal standard solution and additional diluent are added. The test tube is vortexed to mix the solutions.

Using the syringe hand, 1.5 mL of solution is transferred to an empty 3 mL tube with a 0.45 micron filter disc attached to the luer-tip of the tube (Figure 3). A blank hand with a fitted nozzle removes the tube from the rack and moves above an autosampler vial. The filtered

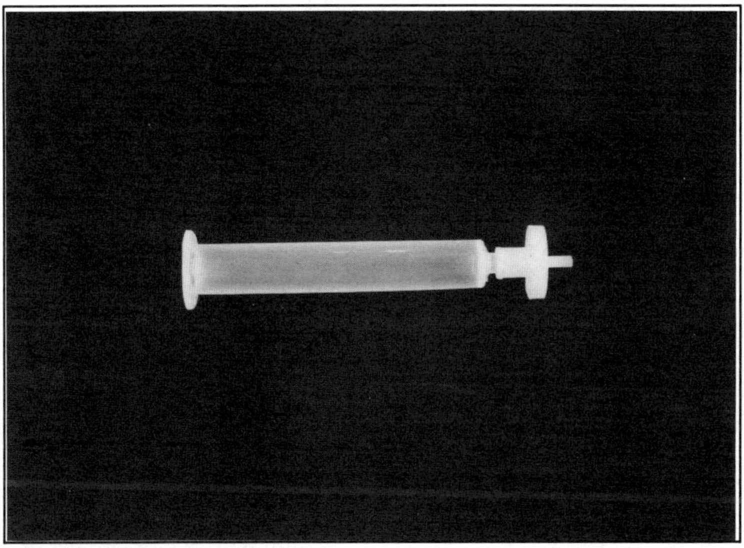

FIGURE 3. Filtering tube.

solution is dispensed into the vial by nitrogen through the nozzle. The filled autosampler vial is capped and sealed by a pneumatic crimper, described previously.[1]

After the samples are prepared by the robot, they are loaded onto the autosampler of a Hewlett-Packard 1084B liquid chromatograph for analysis.

RESULTS AND DISCUSSION

After the programming phase of the robotic procedure was completed, we attempted to test the system with an authentic sample. Six replicates of an herbicide sample were prepared twice by the robot. The results after HPLC analysis were poor. Precision for the robot prepared samples was ±5% RSD compared to ±1% for manually prepared samples.

The source of poor precision was isolated to a pipetting operation where a 5 mL aliquot of sample solution is taken by the robot. During this operation, a blank hand with a nozzle picks up a 5 mL tip and verifies its presence. The nozzle on this hand is connected to a 10 mL gas-tight syringe on the master laboratory station by plastic tubing. The pipette tip is immersed into the solution and 5 mL drawn into the tip. The hand is raised and the solution is dispensed into a test tube.

In order to isolate the problem, a short test procedure was written. [The robot pipets a 5 mL aliquot of acetonitrile-water solution into a tared erlenmeyer flask. The flasks are reweighed and the weight of pipetted solution determined. The precision of the pipetting operation using the scheme described above was ±1.4% RSD. The imprecision in this step would be magnified when dealing with authentic samples.

To improve the precision of this operation, the pipetting procedure was modified. In the current procedure, the pipette tip is attached tightly to the nozzle. The hand is placed above the flask. Before the tip is immersed in the solution, 0.5 mL of air is drawn into the syringe, eliminating backlash in the syringe. The tip is lowered.into the solution, and 5 mL is drawn into the tip. The tip is raised above the solution and the aliquot is dispensed back into the flask, prewetting the disposable plastic tip with solution. This step is crucial to eliminate pipetting errors.

The syringe volume is reset to zero, and 0.5 mL of air is redrawn into the syringe. The tip is lowered into the solution a second time, and a 5 mL aliquot is drawn. The tip is raised out of solution. A 0.1 mL of air is drawn into the bottom of the tip. This air space prevents any solution from dripping during transfer to the test tube. Larger volumes of air tend to rise to the top of the tip, losing the air space. A thin layer of stopcock grease was applied daily to the sides of the nozzle to minimize any air leakage.

The test procedure described above was repeated using the modified pipetting operation. The precision improved to $\pm$0.2% RSD.

In order to insure the accuracy of the pipetting step, various volumes pipetted with the master laboratory station were compared with manually pipetted volumes. Aliquots of acetonitrile-water solutions were weighed in triplicate and the average weight calculated (Table 1). Weights of dispensed volumes using the master laboratory station were consistently less than those obtained manually (Figure 4). In light of these results, the master laboratory station was calibrated.

TABLE 1. Aliquot Weights of Acetonitrile-Water Solutions.

Volume (mL)	Weight (gm) MLS	Weight (gm) Manual	Difference (Manual-MLS)
3.00	2.67	2.76	-0.09
4.00	3.57	3.68	-0.11
5.00	4.49	4.61	-0.12

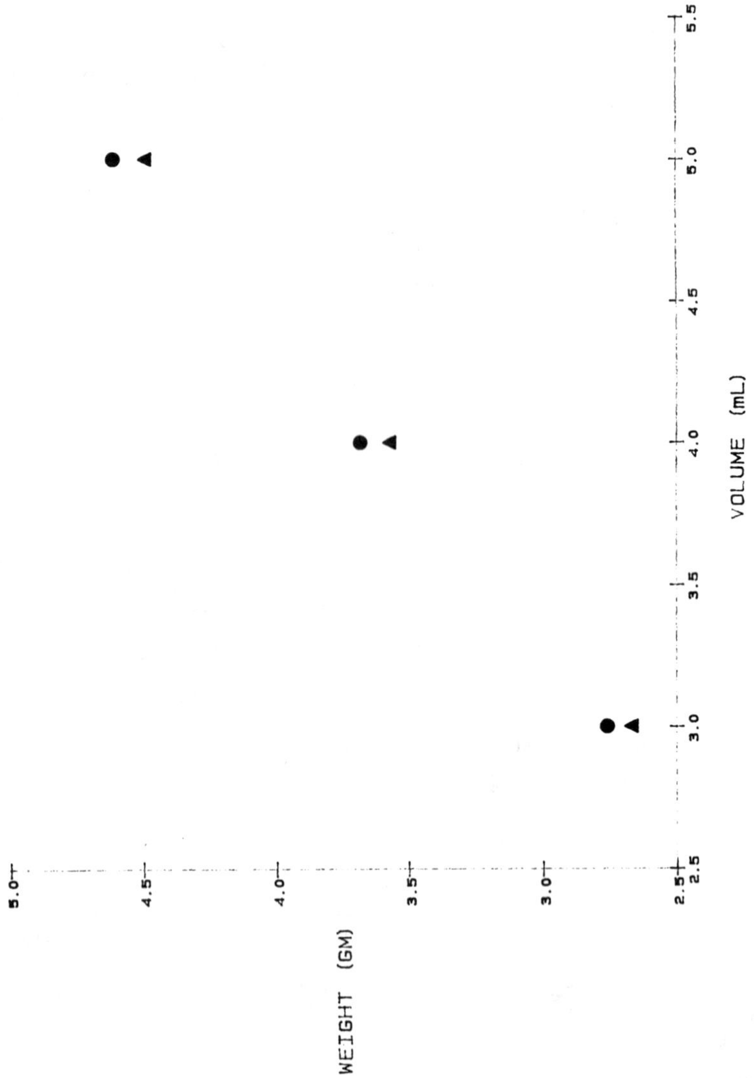

FIGURE 4. Comparison of dispensed volumes (● MANUAL ▲ ROBOT).

After these improvements were made, the robotic system was retested. Replicate samples of two herbicides were prepared by the robot and compared to results obtained manually (Table 2). Robotic results for Herbicide 1 were equivalent to those obtained manually. Although the robotic results for Herbicide 2 are higher than the manual procedure, they are still within the acceptable range.

TABLE 2. Comparison of Robotic and Manual Procedures.

Herbicide	Robot			Manual		
	Average	S.D.	%RSD	Average	S.D.	%RSD
1	77.6	±1.2	±1.6	77.7	±0.7	±1.0
2	76.1	±0.4	±0.6	75.2	±0.5	±0.7

A comparison of the day-to-day reproducibility of the robot indicates no significant differences (Table 3).

TABLE 3. Day-to-Day Reproducibility of Robotic Procedure for Herbicide Sample.

	Average	%RSD
Day 1	77.6	±1.6
Day 2	77.8	±0.8

A sample can be prepared by the robot in approximately 30 minutes. Fifteen minutes is alloted for ultrasonication and stirring of the sample to insure that the sample dissolves. During an eight-hour period, the robot can prepare 15 samples. A technician could prepare a similar number during that period. The robot, however, can work overnight preparing samples for analysis the following day, doubling the number of samples ready for HPLC analysis.

SUMMARY

A robotic procedure was described for the preparation of herbicide samples for HPLC analysis. Precision problems encountered were isolated to the pipetting procedure. A modified pipetting procedure was described. Results were equivalent to those obtained manually.

ACKNOWLEDGMENTS

I would like to acknowledge the assistance of Steve Miller and Mike Feroah of Zymark Corporation. Their insight was extremely helpful in developing this application. In addition, I would like to thank Mark Watson for reviewing this paper.

REFERENCES

1. Lewis, E. C., Santarelli, D. R. Malbica, J. O., in Advances in Laboratory Automation-Robotics 1984, Hawk, G. L. and Strimaitis, J. R., eds., Zymark Corporation, Hopkinton, MA, p. 237.

ROBOTIC PREPARATION OF WATER SAMPLES FOR TRACE HERBICIDE ANALYSIS

D. M. Haile and R. D. Brown
Monsanto Agricultural Products Company
Luling, Louisiana 70070

ABSTRACT

An automated method for the preparation of water samples for Triallate herbicide analyses in the parts per billion (ug/L) range is described. This method utilizes the Zymark Laboratory Automation System to transfer an aliquot of the wastewater sample to a specially designed culture tube. After the addition of an isooctane internal standard solution, the culture tube is capped and vortexed to affect the extraction of the Triallate into the isooctane. After sonication to break any emulsion formed, the isooctane layer is transferred to a crimp cap autosampler vial to await automated analysis by capillary gas chromatography with electron capture detection.

INTRODUCTION

Analytical methods can be viewed as consisting of a series of manipulations whereby the sample is first processed into a usable form, followed by a series of steps whereby the analysis is performed. Until recently, laboratory automation was directed toward automating the analysis portion of a given method. These approaches usually involved the addition of automatic sampling devices to analytical instrumentation or the development of a specific piece of instrumentation to automate a specific analysis. These approaches typically do not include the sample preparation part of the methodology. The introduction of robotics into the laboratory makes it possible to automate the sample preparation, as well as the actual analyses. Since most of the instrumentation at Monsanto's Luling facility is already automated, the ability to automate sample preparation procedures holds the potential to achieve significant cost savings in the laboratory.

In order to evaluate the applicability of robotics to the Luling laboratory, a Zymate System was obtained in early 1984. A sophisticated sample preparation scheme was chosen to fully test the capability of the system.

EXPERIMENTAL

The solvent-extraction sample procedure for the parts-per-billion determination of Triallate (2,3,3-trichoroallyl diisopropyl thiocarbomate) in water was chosen for the first application.

Briefly, this procedure involves:

1. Transfer an aliquot of sample to a 50 mL screw cap culture tube
2. Add internal standard - isooctane extraction solution
3. Shake well to perform extraction
4. Sonicate to break any emulsion formed
5. Transfer isooctane extract to a GC autosampler vial

This method was automated using the Zymate System. The layout of the system is shown in Figure 1.

The Zymate System performs this application as follows:

The crimper hand picks up a GC autosampler vial from the rack, removes the uncrimped cap from the vial using the cap holder on the crimp capper. The uncapped vial is placed on a holder located on the screw cap holder next to the capper station. The crimper hand is exchanged for the capper hand. A culture tube is removed from the rack and uncapped at the capper station.

The capper hand is exchanged for the remote dispensing hand. After obtaining a 10 mL pipet tip from the tip rack, two 10 mL aliquots of sample are obtained from an aluminum foil sealed sample bottle and transferred to the uncapped culture tube previously placed in the capper station. After removing the pipet tip, the wrist is rotated 180 degrees and 2 mL of the internal standard isooctane solution is added to the culture tube.

The remote dispenser hand is exchanged for the capper hand. The culture tube is capped and placed in the vortex station. After vortexing for 20 seconds, the culture tube is held in a sonic bath for one minute to break any emulsion formed. The culture tube is then placed in the capper station and uncapped.

The capper hand is exchanged for the dispenser hand. The wrist is

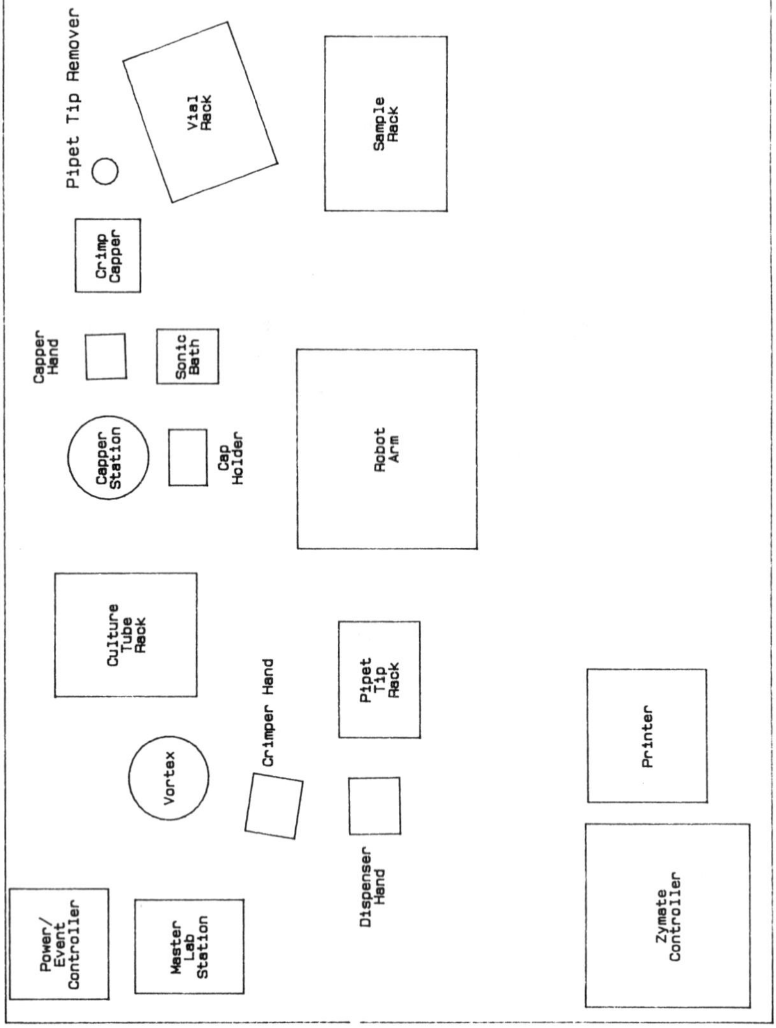

FIGURE 1. Bench layout of Zymate System for triallate herbicide analysis.

rotated 180 degrees so the cannula can be used to remove 1 mL of the isooctane extract layer. The collected extract is dispensed to the autosampler vial previously placed in the holder near the capper station.

The dispenser hand is parked. The capper hand is attached and the used culture tube is capped and returned to its rack.

Finally, the crimper hand is attached and the vial containing the isooctane extract is capped using the crimp capping station and replaced in the vial rack.

The program is repeated until all samples have been extracted.

Two problems were encountered in automating this procedure. The biggest problem was related to the chemistry of the extraction process and involved the physical mixing of the isooctane and water layers during the extraction step. Since Zymark does not have a wrist action shaking station, the extraction had to be performed using a vortex mixer. The initial results indicated that a low extraction efficiency ocurred with vortex mixed extractions due to inefficient mixing. This problem was overcome by installing a glass or Teflon baffle in the screw top test tube.

With the addition of the baffles, the mixing efficiency was greatly improved and the extraction efficiency was found to be identical to the manual procedure.

The other problem encountered during the installation of this method was the capping of the small screw cap GC autosampler vials using the capper station. The vial cap's diameter was smaller than the minimum size required for the Zymark capping station. This problem was partially overcome by enlarging the effective cap diameter with a piece of rubber tubing. This approach solved aabout 95% of the the capping malfunctions. However, the use of the crimp capping station for autosampler vials eliminated the only remaining problem with automating this sample preparation scheme.

RESULTS

To evaluate the robotic method, a comparison test between the manual and robot prepared extraction was conducted. Over a 30 day period, extracts of a 44.24 ug/L standard solutions were prepared by both techniques. The prepared extracts were analyzed via Capillary - ECD Gas Chromatography. The instrumentation was standardized with manually prepared extracts. The results are summarized in Table 1.

TABLE 1. Summary of Comparison Test on Replicate Analyses of a 44.24 ug/L Triallate Standard.

	Manual	Robot
Average Conc. of Triallate Found - ug/L	44.4	41.4
Standard Deviation	1.70	1.67
Relative Standard Deviation	3.8%	4.0%

The data indicates that both techniques exhibit comparable precision (3.8% manual vs. 4.0% robot). The difference in the average amount found is largely due to minute differences in the relative amounts of isooctane internal standard and sample delivered in the manual vs. robot method. The ratio of the isooctane to sample in the manual method is exactly 0.100, where as the ratio for the robot method was measured as 0.1031. Correcting the robot average for this difference, raises the robot value to 43.1. The resulting difference is felt to be insignificant. The differences caused by the slight variation in the volume ratio is self-correcting when the instrumentation is calibrated with robot prepared standards. The comparative analysis of actual water samples further indicates the two techniques are equivalent.

SUMMARY

This project was undertaken with the intent of investigating the ability of the Zymate System to perform a sample preparation scheme that was fairly detailed and error prone. If the robot could perform a sample preparation scheme which required a technician with a high skill level, then the vast number of "simpler" schemes common to a manufacturing plant laboratory could also be automated. Having successfully demonstrated that the Zymate System indeed has the capability to perform this task, laboratory efforts are now directed to designing systems that will provide substantial cost savings.

Zymark and Zymate are registered trademarks of the Zymark Corporation.
Teflon is a registered trademark of the E. I. DuPont Company.

ROBOTIC APPLICATIONS WITHIN DOW'S HEALTH AND ENVIRONMENTAL SCIENCES LABORATORY

B. E. Kropscott, L. B. Coyne, R. A. Campbell,
P. E. Kastl, and W. F. Sowle, III
The Dow Chemical Company
Analytical and Environmental Chemistry Research Laboratory
1803 Building
Midland, Michigan 48674

ABSTRACT

Studies conducted within Health and Environmental Sciences (H&ES) at The Dow Chemical Company involve a range of disciplines including industrial hygiene, toxicology and environmental chemistry. Thus, analytical chemists within H&ES are faced with variety of sample preparation problems and large numbers of standard-format analyses.

Examples of automated sample preparation via the Zymate Laboartory Automation System will include: 1) sample preparation and method optimization for subsequent chromatographic analysis, and 2) whole tube desorption, a new technique for analyte desorption from charcoal tubes in industrial hygiene surveys.

INTRODUCTION

The Health and Environmental Sciences (H&ES) Laboratory at the Dow Chemical Company includes a variety of disciplines including toxicology, industrial hygiene, and environmental chemistry. Toxicology is the study of the fate and effect of chemicals in biological systems (from the cellular level up to and including man). Industrial hygiene is a study of the prevention and evaluation of potentially hazardous exposures in the workplace. Environmental chemistry is a study of the fate and effect of chemicals in the environment (water, soil, and air). Each of these disciplines rely on analytical chemistry support.

Yearly, the analytical chemists prepare and analyze thousands of samples which provide necessary information for H&ES disciplines. Two Zymate Laboratory Automation Systems have been installed in H&ES to automate sample preparation, thereby relieving the analytical chemists of some of their workload. Since not all sample preparations can be or should be automated (such as soxhlet extractions or limited one-time extractions, respectively), robotic applications were identified which would be cost effective and have a significant impact on the analytical workload - such as solvent extractions, air sampling tube desorption, and various types of extraction column chromatography.

This paper will present two examples of automated sample preparation via the Zymate System. The first application is sample preparation and method optimization associated with toxicological dietary studies. The second application is whole tube desorption, a new concept for analyte desorption from charcoal tubes used in industrial hygiene surveys.

<u>MATERIALS AND EQUIPMENT</u>

Two Zymate Laboratory Automation Systems were purchased from the Zymark

Corporation (Hopkinton, MA). The Zymate System components include:

2 microprocessor/controllers with application software

2 laboratory robots

7 operational hands

2 master lab stations (solvent handling)

2 power and event control stations

2 capping stations

1 centrifuge

1 linear shaker

3 vortex mixers

2 printers

2 balance interfaces

1 teaching module

10 sample and hardware racks

Schematics showing the distribution of the system components are displayed in Figure 1 (toxicology) and Figure 2 (industrial hygiene).

Samples and solvents were weighed on an electronic balance, model PC440 or PE360 (Mettler Instrument Corporation, Princeton, NJ). Chlorpyrifos was supplied by The Dow Chemical Company (Midland, MI). Certified laboratory rodent chow was obtained from the Ralston Purina Company (St. Louis, MO). Analytical grade solvents (methylene chloride, hexane, toluene and carbon disulfide) were obtained from Fisher Scientific Company (Midland, MI). One gram PCB air sampling tubes were obtained from SKC Inc. (Eighty-four, PA). Kimax glass screw-top test tubes (20 x 150 mm) and Pyrex 50 mL centrifuge tubes were purchased from Fisher Scientific Company (Midland, MI). Disposable 3 and 6 mL chromatography

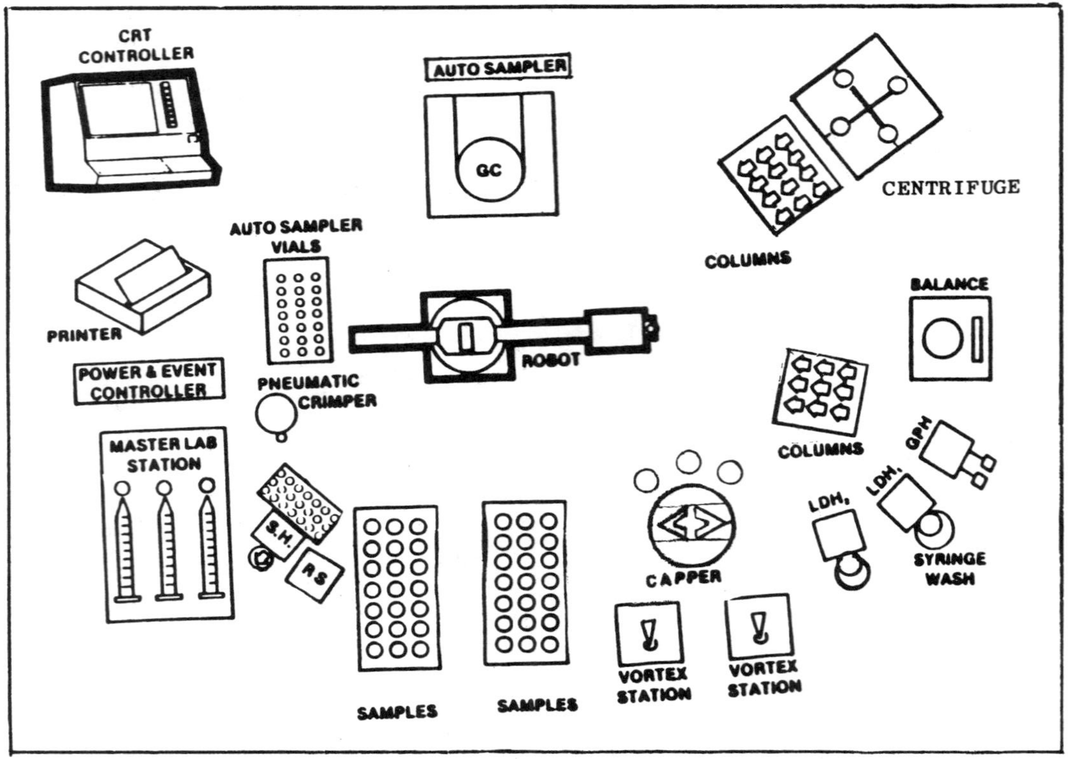

FIGURE 1. Schematic of Bench-top Layout for Toxicology Applications. S.H. = Syringe Hand, R.S. = Rest Station, LDH.1 = Liquid Distribution Hand 1, LDH.2 = Liquid Distribution Hand, GPH = General Purpose Hand.

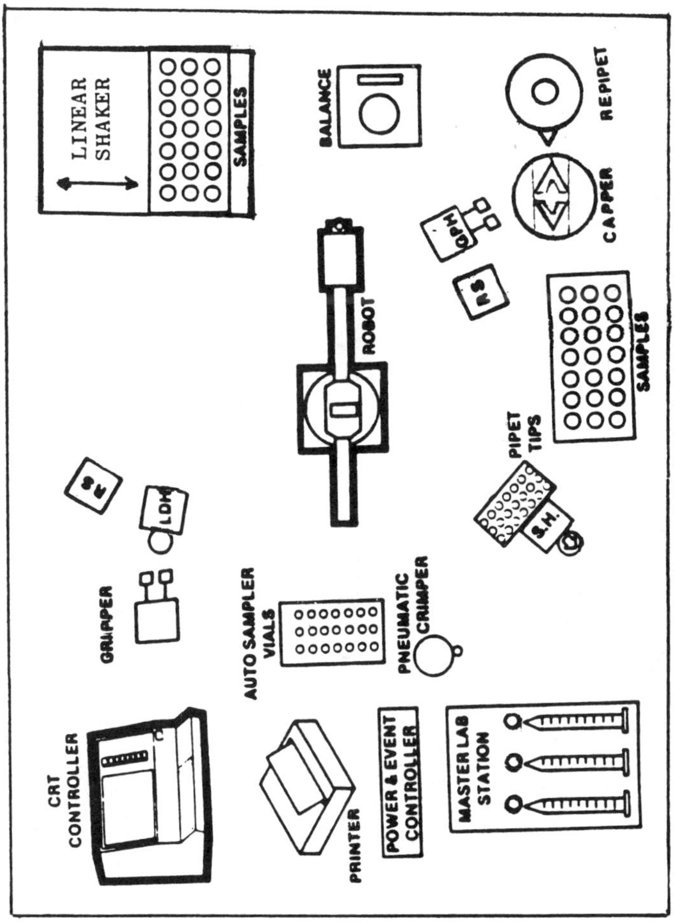

FIGURE 2. Schematic of bench layout for industrial hygiene applications. S.H. = Syringe Hand, R.S. = Rest Station, GPH = General Purpose Hand, LDH = Liquid Distribution Hand.

columns were obtained for J. T. Baker Chemical Company (Phillipsburg, NJ) and Analytichem International (Harbor City, CA). Disposable blue pipet tips (101-1000 uL) were obtained from Alltech Associates (Deerfield, IL). One mL autosampler vials and caps were purchased from Anspec Company (Ann Arbor, MI). Autosampler vial caps were crimped via a Wheaton Air Powered Crimper (Wheaton Scientific, Millville, NJ). Sample extracts were analyzed using a Varian 3700 gas chromatograph equipped with an electron capture detector and a flame ionization detector (Varian Instrument Inc., Sunnyvale, CA).

ROBOTIC PARAMETERS

The laboratory robotics system is well-suited for method developemnt and processes a variety of parameters to optimize recovery. The parameters include:

1. Uniform sample handling (sample history)[1]
2. Extraction timing
3. Number of extractions
4. Composition and amount of solvents
5. Solid/liquid extraction chromatography

Samples can be prepared in a batch (as a group), serially (one-at-a-time) or with a combination of the two modes. Using a batch mode, samples are prepared with the robot executing a single operation (such as adding a solvent or mixing) to the entire sample set before proceeding to the next operation. Batch operations have an advantage if samples degrade in a matrix (such as biological samples). Using a serial mode, the robot prepares samples one-at-a-time, with the same timing, and with uniform physical handling. Uniform sample history can help the analyst distinguish if variations in recovery are caused by the analyst's technique or if the variation is due to a chemical or physical property (such as column overloading or binding).

The effect of extraction time and number of extractions on recovery can be evaluated with the use of 8 system times and programming software (Do Loops). These parameters should be used to determine whether increasing the time of agitation or whether increasing the number of extractions would make any difference in percent recovery.

The percent composition and amount of solvents can be adjusted via the master lab station (MLS). The MLS consists of 3 syringes which are used independently or in conjunction with each other for delivery of a variety of solvents and solvent mixtures.

A variety of normal phase, reverse phase, and ion exchange extraction chromatography columns are commercially available. These columns can be evaluated singly or stacked (Figure 3) to determine which packing is best suited for sample cleanup.

TOXICOLOGY APPLICATION

In toxicology, a variety of studies are performed to evaluate a compounds' potential for inducing chronic toxicity or oncogenicity in biological systems. One example was a dietary study in which chlorpyrifos in rodent chow was administered to mice at a concentration of 15 ug/g. Chlorpyrifos (0,0-diethyl 0-(3,5,6-trichloro-2-pyridinyl)

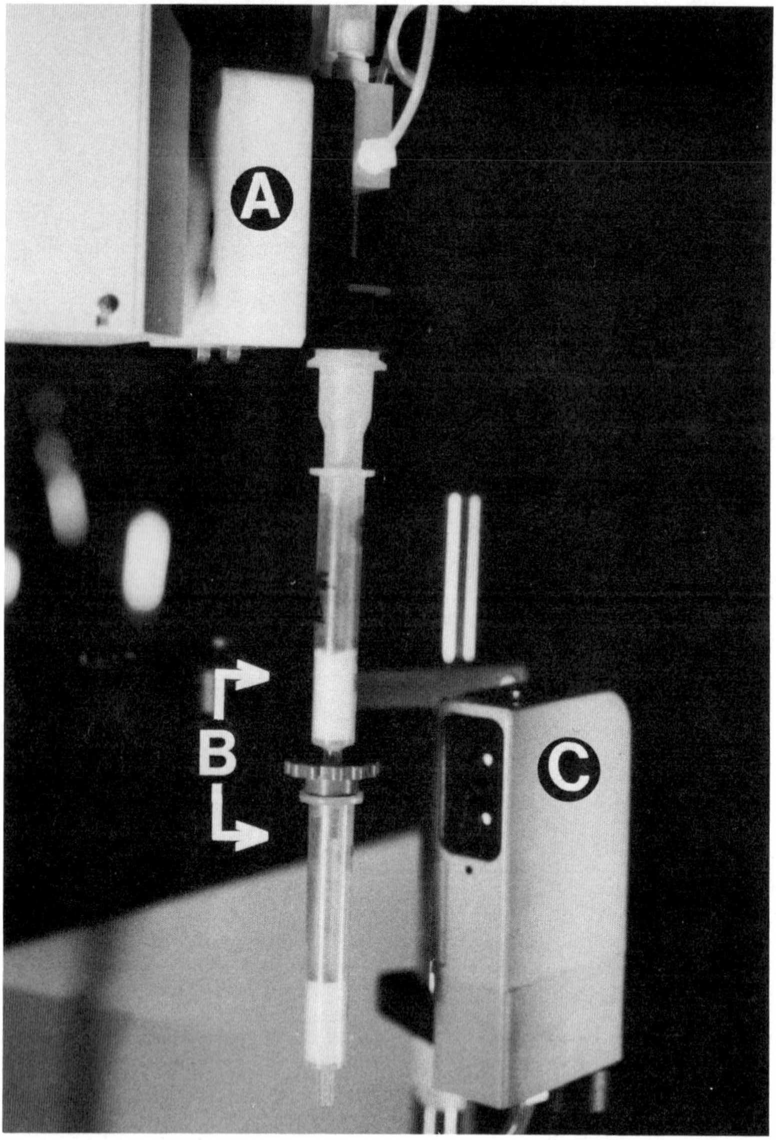

FIGURE 3. Robotic arm verifying the extraction columns have been successfully stacked. A = Robot arm with liquid distribution hand attached; B = extraction columns (stacked); C = optical sensor.

phosphorothioate) is the active ingredient in DURSBAN insecticide which is used to control a broad spectrum of insect pests.

A manual procedure had been developed to extract and quantitate chlorpyrifos to document that the diet was properly mixed. This procedure used many analytical techniques commonly used in support of toxicological studies required for registration of agricultural products.

Some method development and parameter evaluation were necessary to automate the manual procedure via robotics. In the example of extracting chlorpyrifos from feed, a slight improvement in % recovery was observed with an increase in extraction time (Table 1).

TABLE 1. Effect of Extraction Time of % Recovery (15 ug/g Chlorpyrifos in Feed).

Time (min.)	% Recovery
1	90
2	95
5	98

No difference in % recovery was observed for a single or double extraction using a 2 minute extraction time (Table 2).

TABLE 2. Effect of Multiple Extractions on % Recovery (15 ug/g Chlorpyrifos in Feed Using a 2 Minute Extraction).

Number of Extractions	% Recovery
1	95
2	95

No adjustment in the amount or composition of solvents or the type of extraction chromatography columns was needed in this example to obtain >95% recovery.

One modification was made to the Zymate System to improve the physical handling of samples. An air actuator was mounted on the vortex's tube holder to keep the tube from spinning in the vortex station. This allowed vigorous and uniform sample agitation.

Using the information gathered during the evaluation of robotic parameters, the manual extraction procedure was converted, automated and optimized using robotics.

METHOD (Toxicology)

Spiking standards were prepared by accurately weighing chlorpyrifos into a volumetric flask and diluting to volume with hexane. Spiked samples were 5 grams of rodent chow fortified with 1 mL of an appropriate spiking solution (or hexane) to provide target concentrations of 0,3,18, or 90 ug/g. The hexane from the spiked samples was allowed to evaporate before extraction.

Reference standards were prepared by accurately weighing chlorpyrifos into a volumetric flask and serially diluting with toluene.

In the manual method, 3 aliquots of spiked feed from each concentration were extracted with 20 mL of hexane and agitated with a wrist-action shaker for 20 minutes. The feed was then centrifuged at 1800 rpm for 1 minute. A 1 mL aliquot of supernatant was pipetted onto a 6 mL Florisil column. Chlorpyrifos was eluted from the column with toluene by gravity into an appropriate volumetric flask and brought to volume with toluene. The flask was capped, inverted several times, and shaken by hand to insure mixing. An aliquot from the final extract was transferred to a screw-top autosampler vial.

In the robotic method, 4 aliquots of spiked feed from each concentration were extracted with 20 mL of hexane and agitated with a vortex mixer for 5 minutes. The feed was then centrifuced at 2000 rpm for 1 minute. A 1 mL aliquot of supernatant was pipetted onto a 6 mL Florisil column. Chlorpyrifos was eluted from the column with toluene and a nitrogen purge. The final extract was collected in a clean test tube, weighed to calculate the delivered volume of toluene, and agitated to insure mixing. Aliquot from the final extract was pipetted into an autosampler vial and capped with a pneumatic crimper.

Both sets of final extracts were analyzed using a gas chromatograph equipped with an electron-capture detector. A Hewlett Packard 3357 Data System was used to integrate peak areas and quantitate concentrations via external standard calculations.

RESULTS AND DISCUSSION

Using the optimization parameter information, the % recovery for chlorpyrifos in feed (3-90 ug/g) was 100±2%.

Results of feed extraction (optimized) are listed in Table 3. Table 4 lists the differences between the robotics and manual method.

TABLE 3. Percent Recovery for Chlorpyrifos/Feed Extraction.

Concentration ug/g	Manual Method[a]	Robotics Method[b]
3	89±3	101±2
18	94±1	98±2
90	90±1	99±3
$\bar{M}$ ± S.D.	91±2	100±2

[a] $\bar{M}$±S.D. of 2 determinations of 3 aliquots per concentration.

[b] $\bar{M}$±S.D. of 2 determinations of 4 aliquots per concentration.

TABLE 4. Differences in Chlorpyrifos/Feed Analysis: Robotics Versus Manual Method.

Robotics	Manual
Serially handled	Batch handled
Test tubes	Volumetrics and bottles
Vortex mixer	Wrist-action shaker
Florisil extraction column	Florisil extraction column
Disposable pipet tips	Pipetsand syringes
5 minute extraction	20 minute extraction
Centrifugation @ 2000 rpm	Centrifugation @ 1800 rpm
100±2% recovery	91 ±% recovery

Very little modification to methodology or equipment was necessary to convert the manual method to robotics. The difference in mean % recovery (100±2% for robotics versus 91±2% for manual) was probably caused by the difference in the physical agitation of samples. The 5 minutes of sample agitation using a vortex mixer (robotics) was significantly greater that 20 minutes of agitation using a wrist-action shaker.

This example for toxicology demonstrated the use of a laboratory robot to both develop and automate standard-format analyses.

INDUSTRIAL HYGIENE APPLICATION

Industrial hygiene (IH) monitoring surveys are used to evaluate concentrations of airborne compounds in the workplace. One example of a desorption method associated with an IH monitoring survey is the desorption of methylene chlorde form activated charcoal with carbon disulfide. Methylene chloride is a volatile liquid used for various applications (such as paint removal, parts degreaser, and solvent extraction) and is readily available in the laboratory.

The second application of robotics for H&ES involved whole tube desorption, a new technique in analyte desorption from charcoal air sampling tubes.[2]

METHOD (Industrial Hygiene)

Ten uL of liquid methylene chloride (13.4 mg) was injected directly on the front charcoal-bed of 10 air sampling tubes. Air was drawn through the charcoal/air sampling tubes using a vacuum manifold at 100 cc/min for 3 hours. Methylene chloride was desorbed from the charcoal with 20 mL of dry ice-chilled carbon disulfide. Half of the spiked sample set was desorbed using a validated industrial hygiene method,[3] which required the charcoal to be removed from the glass sampling tube. The other half of the spiked sample set was desorbed with the charcoal in the tube (whole tube desorption). Both sample sets were agitated on a linear shaker for 30 minutes (Figure 4). Final extracts were analyzed using a gas chromatograph with a flame ionization detector. A Hewlett Packard 3357 Data Ssytem was used to integrate peak areas and quantitate concentrations via external standard calculations.

RESULTS (Industrial Hygiene)

The results for the whole tube versus sorbent desorption of methylene chloride from charcoal are listed in Table 5.

TABLE 5. Whole Tube Versus Sorbent Desorption of 13.4 mg Methylene Chloride From Charcoal Air Sampling Tubes.

	Whole Tube	Sorbent
% Recovery[a]	101±4	99±4

[a]mean±S.D. of 2 determinations of 5 aliquots each.

DISCUSSION (Industrial Hygiene)

Whole tube desorption is a new technique for preparing air sampling tubes used in industrial hygiene[2] and was designed with robotics specifically in mind. Since the robotic system could not remove the

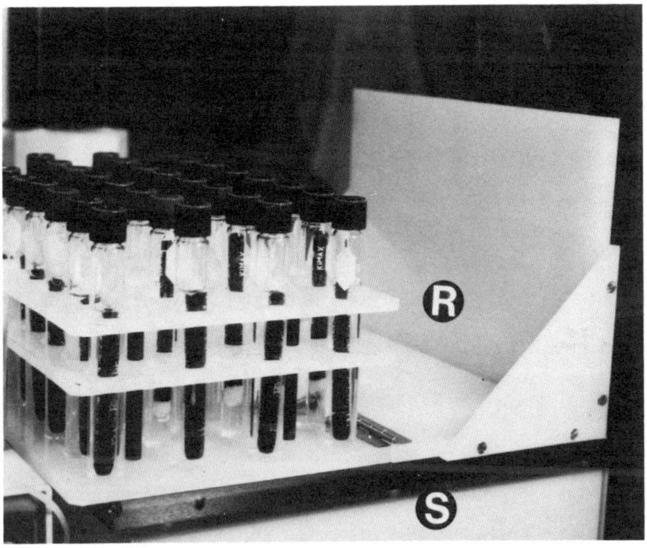

Rack at Loading and Unloading Position

Rack at Shaking Position

FIGURE 4. Hinged Rack (Designed for Industrial Hygiene's Whole Tube Desorption Technique). R = Hinged Rack, S = Linear Shaker.

charcoal sorbent form the glass sampling tube without additional mechanical fabrication, desorbing the entire air sampling tube with chilled solvent inside a test tube was investigated. The recovery of spiked samples desorbed using the whole tube desorption were similar to spiked samples desorbed using a validated industrial hygiene monitoring method, which required the removal of the charcoal sorbent prior to desorption. In addition, whole tube desorption reduced preparation time and increased the safety of the operation.

CONCLUSION

The Zymate Laboratory Automation System provided the accuracy, precision and reliability to automate and optimize the feed extractions. Thinking in terms of robotics, whole tube desorption has been developed as a viable means of desorbing analytes from charcoal tubes. Laboratory robots should not be thought of as a replacement of the analyst, instead the robot should be used as an analytical tool for developing and automating analyses thereby relieving the analyst from repetitive laboratory operations to keep pace with the H&ES analytical needs.

ACKNOWLEDGEMENTS

The authors thank Dave Williams for the photography of the robotic system and Linda Roy for assistance in preparing the manuscript.

REFERENCES

1. Zenie, Francis H. in Advances in Laboratory Automation - Robotics 1984, G. L. Hawk and J. R. Strimaitis, eds., Zymark Corporation, Hopkinton, MA, 1984, p. 8.

2. Coyne, L. B., Warren, J. S., Cerbus, C. An Evaluation of a New

Desorption Technique for Air Sampling Sorbent Tubes, American Industrial Hygiene conference, Las Vegas, NV, May 19-24, 1985.

3. Norwood, S. K., Coyne, L. B., Park, C. M., Poettcker, G. Validation of a Personal Monitoring Method for Airborne Methylene Chloride, Unpublished Dow Report, 1983.

Zymark and Zymate are registered trademarks of Zymark Corporation.
Dursban is a registered trademark of Dow Chemical Co.

LABORATORY ROBOTICS APPLIED TO CHEMISTRY FOR TOXICOLOGY

J.J. Rollheiser, K.M. Stelting, and W.A. Schmidt
Midwest Research Institute
425 Volker Boulevard
Kansas City, Missouri 64110

ABSTRACT

Laboratory robotics generally provides improved productivity and technical quality in analytical chemistry. When applied to dosage formulation studies for toxicology testing, potential advantages of robotics include improved precision and reliability (e.g., reproducibility over time and between laboratories) for analyses used to confirm dose levels and to evaluate the homogeneity and stability of dosed feed blends. Robotics also offers safety advantages when working with feeds dosed with potentially toxic chemicals. The use of a Zymate Laboratory Automation System for analysis of several dosed feed blends will be discussed. Unit operations in these automated analyses include extraction, centrifugation, filtration, dilution and chromatographic analysis. The results of manual vs. automated methods are presented.

INTRODUCTION

Midwest Research Institute has provided chemistry support for toxicology research for more than 12 years. During the past year, laboratory robotics has become part of an ongoing effort to improve efficiency and productivity of our operations. For years we have used "simple" automation such as repipettors, automatic titrators and autosamplers for specific laboratory operations or to do simple analyses. Like many laboratories, we have taken advantage of recent advances in computer technology, particularly in using microcomputers (e.g., for instrument control, data acquisition and data handling). Conventional, dedicated automated analysis systems have not been used because there is not a continuing need for the same analysis in our laboratory. Our need is to develop and to do many different analyses or to adapt our previously developed methods for new matrices, different concentrations, etc.

In deciding which toxicology support operations to automate via robotics, potential applications were considered in terms of workload, turnaround time, safety and technical issues (e.g., complexity of the analysis, sample integrity requirements, etc.). After careful consideration, we chose to focus our laboratory robotics development on dosage formulation analyses.

Dosage formulation studies involve analysis of mixtures of test chemicals with carriers for administration to animals. This work requires the development and validation of an analysis method which is then used to verify homogeneity of the mixtures, determine stability of the mixtures under various storage and dosage conditions, and verify accuracy of the dose concentration level over test periods up to 2 years. A typical toxicology test requires approximately 450 dosage analyses for each chemical over a 2-year period.

Analysis of dosage formulations was chosen as our first robotics application for toxicology support because:

1. The workload is relatively high (i.e., 450 analyses per chemical, minimum of 25 chemicals per year).
2. Analyses for different chemicals require essentially the same laboratory unit operations.
3. Sample history is important for stability studies. The serial processing capability of robotic systems assures that each sample has the same history—a decided technical improvement over batch (manual) processing.
4. Sample preparation is time-consuming due to the complexity of feed matrices. The sample preparation capability of the robotic system considerably reduces staff time required for this operation.
5. Automation through robotics improves the safety of the operation by decreasing personnel exposure to toxic or potentially toxic chemicals.
6. Robotics offers improved reproducibility. The Zymate System's ability to exactly duplicate a procedure (e.g., shaking speed, time, etc.) eliminates analyst technique-dependent variations, thereby improving the quality of the data.

EXPERIMENTAL AND RESULTS

Dosed Feed Analysis - High Performance Liquid Chromatography

The Zymate Laboratory Automation System was first used to chromatographically analyze animal feeds dosed with various chemicals. The system consists of the following peripherals: centrifuge, capping station, master lab station, shaker, HPLC injector, uninterruptable power supply, and a power and event controller. The system is arranged on a 4 x 7 ft grid (Figure 1).

The procedure begins by manually weighing triplicate portions of dosed feed into large test tubes. The robot continues the procedure by performing extractions, clarifications, dilution (including addition of an internal standard), and chromatographic analysis. In the extraction step, the dosed feed is shaken with an extracting solvent. Clarification and dilution involves centrifuging a portion of the extracts to remove particulates, then diluting the sample. Finally, the robot injects the samples and standards it has prepared for chromatographic analysis (HPLC or GC).

This robotic system was used to determine the homogeneity of dosed feed blends, using replicate analyses of grab samples taken at different locations in the blends. Concurrent HPLC analyses were done manually and by the robotic method and results compared for methyleugenol (a flavoring agent), disulfiram (an alcohol deterrent), and di-(2-ethylhexyl) phthalate (a plasticizer), also known as a DEHP.

The results in Table 1 are for triplicate analyses of grab samples taken from right, left, and bottom locations in a twin shell blender. "Automated" refers to the robotic method and "manual" to the standard manual method (which used larger volumes of solvents to compensate for precision/accuracy limitations when manually pipetting small volumes).

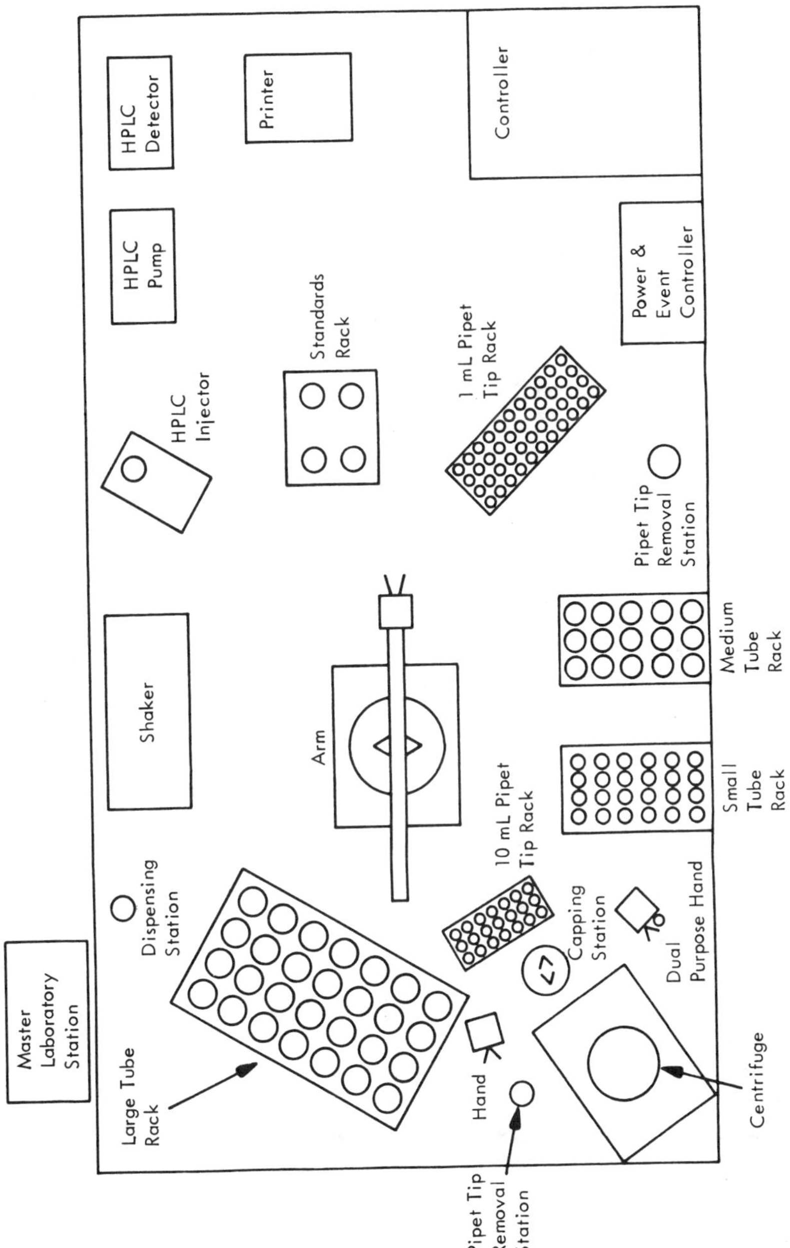

FIGURE 1. Table layout for robotic analysis of dosed feed.

TABLE 1. Automated Homogeneity Evaluation by High Performance Liquid Chromatography.

CHEMICAL	SAMPLE LOCATION		AUTOMATED FOUND (%)	SD[a]	SIMULATED FOUND (%)	SD[a]	MANUAL FOUND (%)	SD[a]
Methyleugenol	Right		98.0	1.0	97.0	1.0	96.7	0.7
	Left		98.5	0.4	96.0	2.0	96.7	0.2
	Bottom		98.5	0.4	95.9	2.4	96.2	1.0
		$\bar{x}$	98.3	0.6	96.3	1.7	96.5	0.7
Disulfiram	Right		100.1	1.4	101.3	2.5	99.8	0.4
	Left		100.0	1.5	97.4	2.0	100.7	1.5
	Bottom		101.1	1.5	97.7	1.4	99.3	0.8
		$\bar{x}$	100.4	1.4	98.8	2.6	99.9	1.1
DEHP	Right		98.5	1.3	98.6	0.3	98.3	0.4
	Left		97.7	1.0	97.9	0.7	97.8	0.3
	Bottom		96.9	1.0	96.2	1.6	96.5	0.3
		$\bar{x}$	97.7	1.2	97.6	1.4	97.5	0.9

a Standard deviation

SUMMARY

Dosed Feed Analysis - High Performance Liquid Chromatography

The HPLC results, evaluated at the 95% confidence level, indicated that for methyleugenol precision of the robotic method is better than that of the equivalent manual method and equivalent to the precision of the standard manual method. For disulfiram and DEHP, the precision of the robotic method is not significantly different from that of either the equivalent manual or standard manual methods. For all three chemicals, the accuracies of the three methods were equivalent.

EXPERIMENTAL AND RESULTS

Dosed Feed Analysis - Gas Chromatography

Dosed feed mixtures were also anlayzed by the robotic system using gas

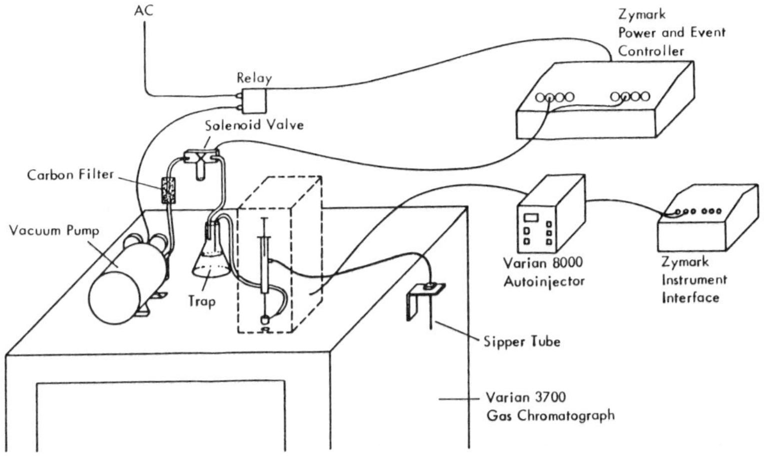

FIGURE 2. Schematic of GC-Zymate System Interface.

chromatography. For these analyses, a Varian Model 8000 autoinjector was interfaced to the Zymate System as shown in (Figure 2).

The injection procedure begins as the robot arm moves a test tube containing the diluted feed extract under the sipper tube. The Zymark instrument interface then triggers the Varian autoinjector to begin its normal injection cycle. At the same time, the power and event controller turns on the vacuum pump through the power relay. The solution in the test tube is then drawn up the sipper tube, through the syringe barrel and needle, out the waste well, and into the trap. After the syringe is flushed for approximately 45 sec, the power and event controller actuates the solenoid valve, which releases the vacuum and then turns off the vacuum pump. The normal Varian injection cycle continues from this point.

Several chemicals, including benzyl acetate (a scenting agent and solvent) were analyzed by gas chromatography using this system. Results for the robotic analysis method are given in Table 2.

TABLE 2. Automated Homegeneity Evaluation by Gas Chromatography.

CHEMICAL	LOCATION		SAMPLE FOUND (%)	SD[a]
Benzyl Acetate	Right		94.9	1.5
	Left		94.7	1.1
	Bottom		96.6	0.6
		$\bar{x}$	95.4	1.4

a Standard deviation

SUMMARY

Dosed Feed Analysis - Gas Chromatography

The results of the GC analyses indicated that the accuracy and precision of the robotic method for benzyl acetate were comparable to those previously obtained with the standard manual method. Only minor modifications to the autoinjector were required, allowing it to be converted back to normal configuration in less than one hour.

EXPERIMENTAL

Analytical Methods Optimization

After verifying the feasibility of using robotics to perform dosed feed analyses with previously developed analytical methods, the next logical step was to begin using the robot to do analytical methods development.

Since methods development is the most time-consuming step in dosage formulation studies, automating this step would have a major impact on the efficiency of our operation. It would significantly improve our ability to meet contractual project deadlines. The resulting improvement in response time would allow client laboratories conducting animal studies to better plan for animal procurement, quarantine, etc. Our initial goal in using robotics for methods development was to automate optimization of extraction parameters for dosed feed analyses. This procedure should produce an optimum system for more than 70% of the wide variety of chemicals tested (e.g., pharmaceuticals, dyes, industrial chemicals) and should quickly screen out the 30% of chemicals requiring chemist intervention in development.

In the development of optimum extracting parameters, a problem frequently encountered is low recovery of test chemicals from dosed feed which has been stored up to 3 weeks during stability studies. To accurately interpret stability test data, we must be able to determine if this low recovery is due to chemical decomposition/reaction or to "binding phenomena" in which the intact chemicals is still present in the feed but not extractable. Our experience indicates that if the composition of the extracting solvent is varied (i.e., by lowering pH, adding modifiers such as water or organic solvents, etc.), binding can usually be reversed and the stability of the chemical on feed can then be reliably assessed.

To optimize extractions, parameters are varied systematically as follows:

1. Solvent - typically methanol or acetonitrile.
2. Solvent pH - addition of HCl in concentrations ranging from 0.5 to 2.0%.
3. Solvent modifiers - addition of water at concentrations from 10 to 50%.
4. Extraction time - varying from 2 to 15 min.

Methods optimization by robotics requires more computing power than the robot controller possesses. Our robotic system was therefore expanded to include an IBM PC/XT, equipped with a hard disk drive and interfaced to the Zymate Controller (Figure 3). A program which allows the IBM PC/XT to run in a multi-tasking mode is used. In the foreground, a BASIC program which is central to the optimization system makes decisions and passes instructions to the robot controller; in the background, chromatographic data reduction software calculates peak areas. The integrator interface acquires the chromatographic data from dosed feed analyses in real time, digitizes it, and passes it to the IBM/PC. Upon receipt of a signal from the Zymate Controller, the BASIC program reads peak area data from the hard disk, does statistical calculations and comparisons, makes decisions on the next experiment to be performed and communicates instructions on the parameters for the next experiment to the robot controller. The robotic system then analyzes another set of dosed feed samples and the decision-making process is repeated.

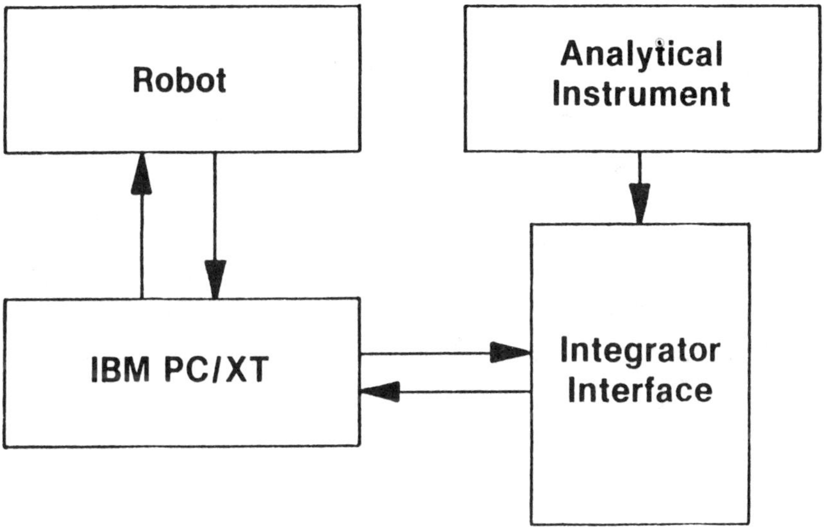

FIGURE 3. Schematic of Methods Optimization System.

The optimization procedure (Figure 4) begins with preparation of standards and extraction of feed samples by the robot, using both acetonitrile and methanol. Recovery for the spiked feed samples is calculated by the BASIC program, and the solvent giving the highest recovery is used for the remainder of the method optimization procedure. If recoveries are equivalent for the two solvents, acetontrile is chosen as the default solvent because it typically gives fewer chromatographic interferences from the feed extract.

Once an extracting solvent is chosen, dosed feed samples which have been stored for 3 weeks at $-20^{O}C$ and at room temperature are analyzed. If the recovery of chemical from samples stored at room temperature is lower than that for samples stored at $-20^{O}C$ (a typical situation), additional room temperature samples are analyzed using a modified extracting solvent (e.g., adding 0.5% HCl) and recovery is again compared to that for samples stored at $-20^{O}C$. As long as recovery for the samples stored at room temperature remains statistically less than that of the $-20^{O}C$ samples, extracting solvent modifications are tried until all options are exhausted. If the recovery for samples stored at room temperature is equivalent to that for the samples stored at $-20^{O}C$, then extraction time is optimized.

When the optimization procedure is complete, a report presenting all the data and decision criteria for the optimization, with final extraction parameters is generated.

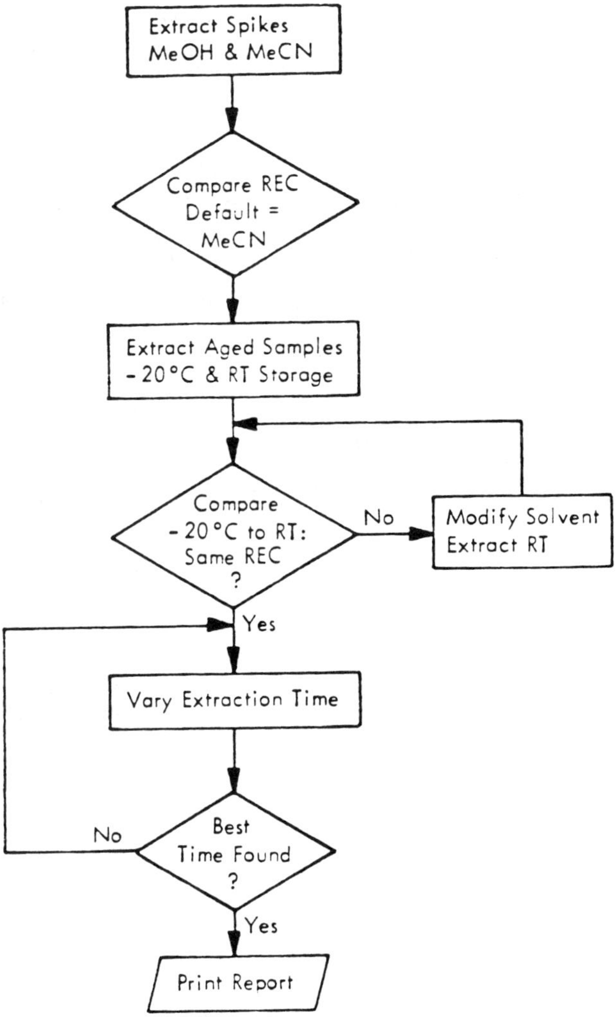

FIGURE 4. Flow diagram for methods optimization.

SUMMARY

Methods Optimization

The methods optimization system is now in reliability testing using chemicals which have been previously studied at Midwest Research Institute. The results of the automated methods optimization will be compared to those obtained previously by manual optimization procedures. If comparable extraction solvent compositions are found by the automated method optimization procedure, then a variety of current test chemicals will be studied. This will determine whether the initial goal of an automated system to successfully optimize extractions for 70% of the chemicals was achieved.

As more sophisticated computer interfacing becomes available, more refined decision-making and expanded robot control can be accomplished. In addition, an investigation of more complex optimization algorithms (e.g., Simplex optimization) for upgrading the current BASIC program could likely result in improved efficiency of this process.

Zymark and Zymate are registered trademarks of Zymark Corporation.
IBM and IBM PC/XT are trademarks of International Business Machine Corporation.

DEVELOPMENT OF A ROBOTIC CAFFEINE ANALYSIS

Marilyn Dulitzky
Thomas J. Lipton
800 Sylvan Avenue
Englewood Cliffs, NJ 07632

ABSTRACT

A Zymate Laboratory Automation System was designed and constructed to perform caffeine analyses on a variety of sample types.

The robot was set up to carry the procedure through all of its phases with a minimum of human assistance. To this end the robot goes through an initial sample preparation consisting of weighing the sample and adding the necessary reagents. It then heats the sample for an hour, adding water when necessary. It then replaces water lost during digestion, mixes and cools the sample. At this point it prepares the sample for filtration, filters it, takes an aliquot and mixes it with internal HPLC standard and finally, injects it into an HPLC.

The output from the HPLC is interfaced with a Perkin-Elmer LIMS system to facilitate data management and remove the tedious calculations from technician time.

The robot programming is serial in nature rather than being in a batch mode. This makes de-bugging simpler as this approach to programming is more logical and easier to follow. It also allows the robot to handle each sample identically.

INTRODUCTION

The Zymate System is used for the implementation of a routine method for the determination of caffeine in various solid and liquid samples. It is a direct descendent of the original A.O.A.C. procedure[1] which handles samples on a "one at a time" basis. In order to increase sample throughput when the need to perform a hundred samples per week arose, the A.O.A.C. procedure was modified into a batch mode accommodating 20 samples per batch.[2] With the need to increase productivity still further, while at the same time freeing technical personnel for more complex tasks, a robotic system became the logical next step in the progression. It is interesting to note, that while the first step in the evolution of the robotic system was to a batch mode, the robotic system itself is most efficient using a serialized procedure.

Since the manual method in use had been validated, the robotics system was designed to match the method whenever possible.

This paper discusses how the robot performs the manual procedures, where modifications have been made, and compares results obtained by the manual and robotic methods. A discussion of the difficulties encountered during start up and their resolution is included.

MANUAL METHOD

The technician weighs a set of samples (20) into 200 mL straight Tecator tubes. An appropriate amount of MgO is added to each tube and 100 mL water transferred to each tube using an automatic pipetor. Each tube and its contents is then weighed on a top loader balance and its total weight recorded. The rack of tubes is placed into a preheated Tecator digestion block for 70 minutes (including 10 minutes for the samples to come to a boil). When the digestion is complete, the rack is removed from the block and the tubes allowed to cool. Each tube is brought back to original weight and hand shaken to mix well. Each sample is then first filtered through fast filter paper and then through a .45 micron filter unit. The final steps are taking an aliquot of sample, making an appropriate dilution while adding an internal standard (8-chlorotheophylline) and injecting into an HPLC equipped with an HP85 data handler.

ROBOTIC METHOD

In order to more easily and logically program the robot for the complex series of manipulations needed to perform this analysis, the operations were organized into groups which could be timed individually. This organization is illustrated in the top level of the robots' programming:

```
     Init
 10  Initial.Sample.Prep
     Heat.Sample
     Add.Water.To.Digestion.Tube
     Prepare.Sample.For.Filtration
     Filter.Sample
 20  Inject.Samples
     Terminate.Program
     Goto Next.Step
100  I
```

Total robotic time to perform one analysis is twenty-five minutes. For an overall view of the robot, see Figures 1 and 2.

Manual operations to prepare for the robotic method include:

- Load samples into 16 mm sample tubes in approximate amount.
- Load racks with empty 25 mm tubes for digestion and filtration.

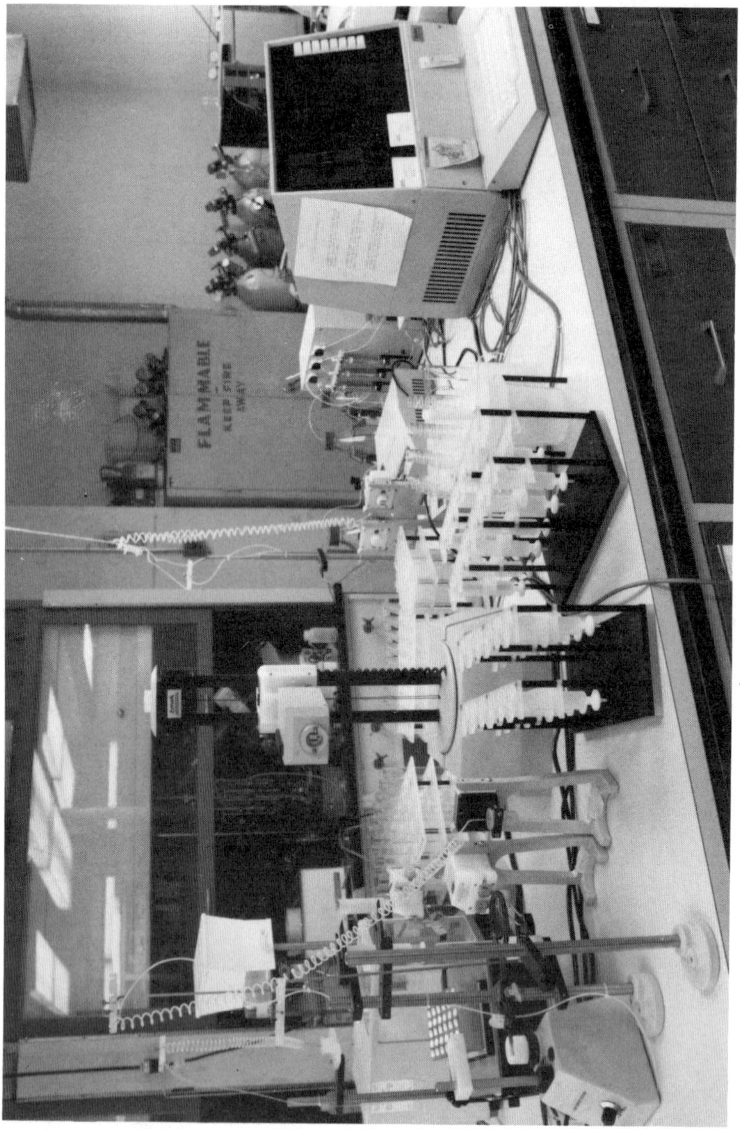

FIGURE 1. Zymate System for caffeine analysis.

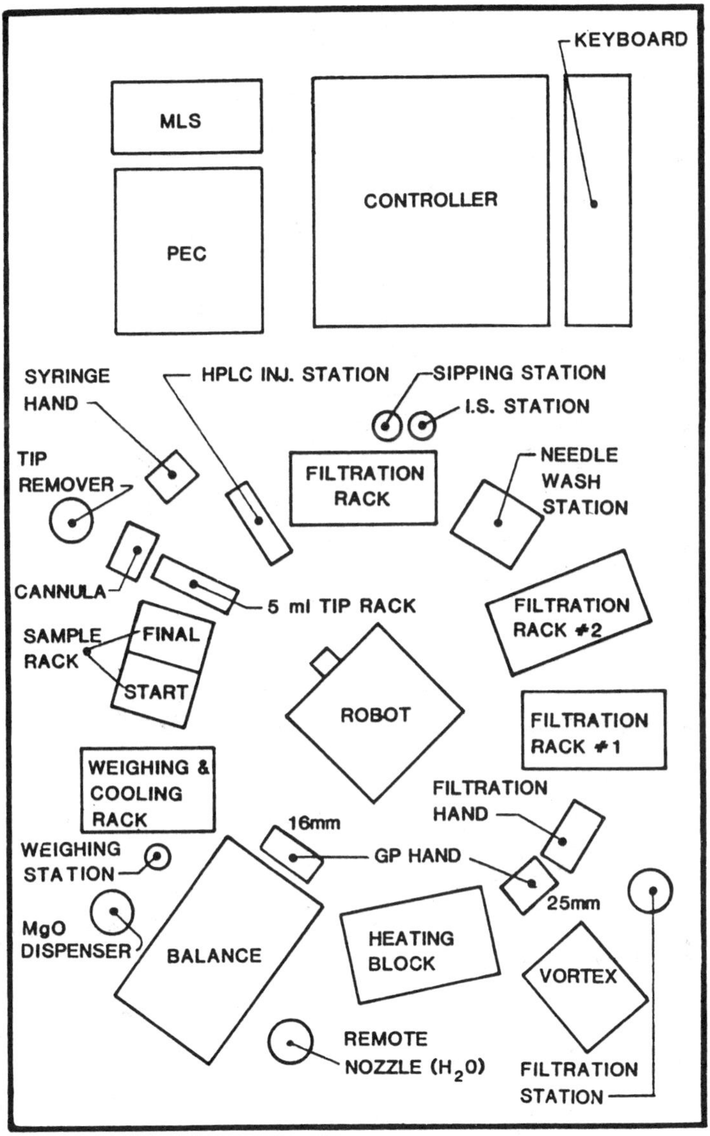

FIGURE 2. Zymate System layout for caffeine determination.

- Spray anti-foam into digestion tubes.
- Load rack with 16 mm tubes for final sample aliquots.
- Load 5 mL tips into pipet rack.
- Load syringes into holders. Preset these with .45 micron filter units.
- Load internal standard station.
- Turn on heating block.
- Fill water container and MgO dispenser.
- Fill HPLC mobile phase reservoir and check waste container.
- Turn on UV detector, pumps and recorder.
- Tare balance.
- Enter sample type and number into controller.

Table 1 indicates the timing for this serialized method.

Initial.Sample.Prep

This phase of the operation takes the robot 8 minutes to perform.

1. Tare 25 mm test tube and bring to sample transfer station.
2. Transfer sample to 25 mm test tube.
3. Weigh sample and record value in memory.
4. Calculate and dispense MgO - at MgO dispenser.
5. Dispense water - at remote nozzle and via master lab station (MLS).
6. Weigh sample, MgO, and water and record values in memory.

It should be pointed out here that many moves and hand switchings are implied in the operations rather than stated.

TABLE 1. Robot Timing and Cycling Operation for Caffeine Determination

Operation	Robotic Time in minutes	n passes through top level program									
		1	2	3	4	5	6	7	8	9	10
Initial Sample Prep	8.0	1	2	3	4	5	6				
Heat Sample	0.3	1	2	3	4	5	6				
Add water	0.7		1	2	3	4	5	6			
Prepare for Filtration	3.0				1	2	3	4	5	6	
Filter	7.0					1	2	3	4	5	6
Inject	6.2					1	2	3	4	5	6
Terminate	--	-	-	-	-	-	-	-	-	-	yes
	25.2										
		Start up			Equil.			Shut down			

Where n = the number of passes through the top level program.

number = indicates which operation is actually performed during that pass and on which sample - the robot makes 10 passes through the program for 6 samples.

It takes 125 minutes before the first sample is injected and then 25.2 minutes per sample thereafter.

Heat.Sample

The robot places the 25 mm tube into a preheated block and at this point sets a timer for 75 minutes. This allows for 1 1/4 hours heating time and ~15 minutes for the contents to reach boiling stage.

Add.Water.to.Digestion.Tube

In this phase of the analysis, which takes place part way into the digestion cycle, 5 mL of water is added by bringing the remote nozzle to the heating block and dispensing the water via the MLS.

Prepare.Sample.For.Filtration

As the program cycles through each phase of the operation, it checks timers that have been set for the various operations. If the heating timer has run out, the program instructs the robot to proceed to the next step.

Do.Sample.Prep.and.Cool

1. Remove tube from heating block.
2. Weigh tube and calculate water to be added (to correct for water lost during digestion).
3. Add needed water at remote dispenser.
4. Vortex sample.
5. Cool sample (for 25 minutes in original rack).

Filter.Sample

Here the instructions are:

1. Get filtered sample tube.
2. Place filtered sample tube in holder.
3. Get syringe filter - these are 30 mL syringes pre-loaded with .45 micron filter units.
4. Park syringe filter in fork - this is positioned over the tube holder.
5. Get sample to filter.
    a. Get tube from cooling rack.
    b. Bring to sipping station - one place for all samples to be brought to in order to withdraw a sample.
    c. Attach tip - 5 mL tip on pipet hand.
    d. Sip sample - 5 mL of supernatant is withdrawn via MLS.
    e. Bring to filter and empty.
    f. Remove tip.

6. Push sample through filter - here the filtration hand is used and air pressure applied through a valve in the MLS - pressure is also released through this valve.

7. Remove used filter assemble.

8. Remove filtered sample tube from holder.

9. Return filtered sample tube (to filtrate rack).

10. Return tube from sipping station to cooling rack.

Inject.Samples

At the proper time the robot is instructed to transfer sample aliquot (using 6" cannula) as follows:

1. Get aliquot of internal standard (IS) from internal standard station.

2. Get aliquot of sample (amount determined by sample type).

3. Put aliquot of sample in final tube (This includes the IS - both of which are now in coiled tubing above the cannula and attached to a 10 mL syringe at the MLS).

4. Add water and triturate. This results in obtaining a particular dilution for any pre-determined sample type and contains an exact and pre-calculated amount of I.S. diluted from the stock.

5. Get final sample with syringe hand.

6. Bring sample to HPLC injector.

7. Run HPLC. The output from the HPLC is sent to the LIMS system which calculates the results based on standards run, and sample weights and dilutions given to it. (These are printed out at the Zymark printer.)

8. Clean syringe hand.

9. Clean cannula hand.

PROBLEMS/SOLUTIONS

It is not expected that a precise, custom made piece of equipment should operate perfectly immediately. This is especially true of a robotic system requiring intricate programming.

The nature of inanimate matter being what it is, our unit had certain problems which needed to be addressed to make the system operational.

1. **The problem:** Pouring samples into tubes for weighing. Liquids could be made to pour, but certain powders would not pour even when the containers were turned upside down.

   The solution: Installation of a small vibrator on the fingers of the pouring hand.

2. **The problem:** Dispensing MgO. MgO is a powder which is difficult to dispense. The system came with a glass Kontes powder funnel. The design of this funnel incorporates a narrow neck over a horizontal auger which is made of plastic and has shallow groves. The MgO would stick to the glass walls, bridge over the neck and fill the grooves of the auger. A vibrator motor did little to make the powder flow, neither did the addition of a stirrer. It appeared that another design was needed.

   The solution: The use of a trough (eliminating bridging over a narrow neck) through which a wire coil rotates and carries the powder through an orifice from which it drops into the waiting tube. The MgO does not pack into the coil, which has two strips of metals attached to it that serve to scrape the sides of the trough as the coil rotates. (see Figure 3)

3. **The problem:** Slippage of tubes through 25 mm fingers. Some difficulty was encountered in removing hot tubes from the digestion block.

   The solution: The fingers were lined with silicone pads.

4. **The problem:** The robot dropped used syringes instead of removing them at the proper station. This was due to air pressure used in filtering being retained in the syringes.

   The solution: Installation of a pressure release valve in the MLS.

5. **The problem:** Filtering samples. The original program called for pouring the cooled and mixed sample into the syringe and using air pressure to filter. The filters rapidly plugged and only a few drops of liquid were filtered before the back pressure caused the assembly to bounce.

   The solution: Samples were brought back to original weight and mixed before cooling. This allowed the sludge to settle and the supernatant liquid to be pipeted and brought to the syringe for filtering.

6. **The problem:** Coiled tubing. The coiled tubing attaching the cannula to MLS and that connecting filtration hand to MLS would sag and become entangled when parking its respective hand.

FIGURE 3. Exterior and interior views of MgO dispenser.

The solution: In one case the tubing was held out of the way by means of a post extender. In the other case elastic cord attached to a light fixture did the trick.

MODIFICATIONS

Very few modifications were made in order to adapt the original method to a robotic procedure. Modifications included:

1. All quantities were quartered so that the robot could use smaller equipment.
2. Heating time was increased to 1 1/4 hours in order to allow the robot time to cycle through the program.
3. Some dilution factors were changed so that the MLS could handle all necessary dilutions with the same syringe.
4. The samples were adjusted for weight loss before cooling rather than after as in the manual method.
5. Only mixed supernatant solutions are filtered by the robot. Manually, the entire sample would be filtered.
6. The mobile phase was adjusted to accommodate it to a single rather than a dual pump system.

None of these modifications affects the accuracy of the method (See Table 4).

VALIDATION

Master Laboratory Station (Aliquot dilution and transfer)

The precision and accuracy of aliquot dilution and transfer were measured by presenting the robot with a standard solution of caffeine and having it make replicate dilutions. The measured concentration was compared to the calcualted values (Table 2).

TABLE 2. Replicates of the MLS Diluting a Standard Caffeine Solution.

Calculated Value ug/mL	6.67	20.00	50.00
Measured Value ug/mL	6.60	18.94	47.02
	6.46	19.28	47.78
	6.48	19.31	47.68
	6.64	19.15	47.20
	6.52	19.30	47.27
	6.52		
Average	6.54	19.20	47.39
s	±.07	±.16	±.33

Any observed deviation from the calculated values are of little concern because they have no effect on the final validation.

HPLC and LIMS System (Laboratory Management System)

Samples done completely by the robot and calculated by the LIMS were compared with subsamples taken from the robot after dilution and run on our Beckman Unit (Table 3).

TABLE 3. Comparison of HPLC and Data Handling Systems - Percent Caffeine

Beckman-Nelson System	LIMS
0.19	0.20
0.19	0.19
0.20	0.23
0.20	0.20
0.21	0.22
0.22	0.22
0.14	0.15
0.14	0.15
0.32	0.34
0.32	0.34
1.50	1.55
0.99	1.00
0.42	0.43
0.02	0.02
3.37	3.22
3.00	2.92
3.02	2.95
2.94	3.01

The results of the LIMS method correlates very well with our standard Beckman method and show it to be accurate to with +0.1% on a 95% confidence interval.

Sample Preparation and Dilution

The robot analyzed replicate samples of a standard tea leaf that had been set aside as a control for both the manual and robotic procedures. The robotic samples were taken after sample dilution and run on an HPLC system dedicated to caffeine analysis by the manual method (a Beckman system with Nelson Analytical software). Table 4 gives the results.

TABLE 4. Comparison of Robotic and Manual Procedures - Percent Caffeine.

n		Robotic Method	Manual Method
1		2.98	2.76
2		2.81	2.80
3		2.71	2.68
4		2.78	2.76
5		2.92	2.75
6		2.95	2.68
7		2.79	2.78
8		2.89	2.84
9		2.92	2.85
	average	2.86	2.77
	s	±.09	±.06

SUMMARY

The Zymate Laboratory Automation System has demonstrated sufficient accuracy, precision and reliability to perform caffeine analysis unattended. Statisical evaluation found no significant difference between the mean caffeine levels obtained by each procedure.

Using the "Measured Workday" figure of 9 minutes per sample, it would take a person 3.6 hours to manually run 24 samples. It would take that person 1.5 hours to prepare the robot to run the same 24 samples. Therefore, a person using the robot to run caffeine analyses can have 60% more time for other tasks. Since setting up the robot to run 10 samples is not proportionately less than for 24, the percent time saved by running 34 samples is less than that saved by running 24.

At present the robot analyzes powder (solid), aqueous or alcohol dissolved samples. When plans for a hood are completed, the next step will be to attempt to run samples in more volatile solvents. The robot will also be programmed to run analyses which do not require the digestion step.

ACKNOWLEDGEMENT

I would like to acknowledge the expertise of Kevin Tucker of Zymark Corporation in assisting with the serialized programming of the robot and for his suggestions on technical problems.

REFERENCES

1. Official Methods of Analysis, Association of Official Analytical Chemists, Washington, D.C., 13th ed., 1980, p. 236, Method No. 15.051.

2. Dulitzky, M., De La Teja, E., Lewis, H.F., Journal of Chromatography, 317, 403 (1984).

THE ROLE OF ROBOTICS IN THE AUTOMATED DETERMINATION OF THE NUTRITIONAL COMPOSITION OF FOODS -- A PROGRESS REPORT

Harry G. Lento, Ph.D., Michael D. Grady
and Harvey J. Hastings
Campbell Institute for Research & Technology
Campbell Soup Company
Campbell Place
Camden, New Jersey 08101

ABSTRACT

The relationship between health and nutrition has prompted the need to develop and obtain more and more information regarding the nutritional composition of foods. Of the many nutrients present in foods those of most concern and consequently most often requiring analysis include solids, fat, protein, carbohydrates, thiamine, riboflavin, niacin, vitamins A and C, as well as calcium, iron, sodium and potassium.

As part of our efforts to obtain information regarding the nutritional composition of our products and ingredients, we have undertaken a comprehensive program to develop an integrated system for totally automating the analysis of these nutrients. With the advent of modern robotics and the Zymate System, this objective now seems attainable. In this paper we will follow the development of our technology through auto-analysis and computer processing of the data to our present work in the direct application of robotics for sample preparation and sample weighing. Finally, a proposed scheme for the systematic transport of the sample through the analytical train will be presented. Included in this system will be a number of robotically controlled stations for conducting sample preparation, weighing, cleanup and analysis.

INTRODUCTION

Within recent years there has developed an ever increasing awareness and concern regarding the nutritional composition of foods and its relationship to good health. One need only cite as examples the relationship of cholestrol to coronary heart disease; sodium to hypertension; and fat, protein and fiber to certain cancers. It is no wonder, therefore, that these concerns have prompted the need to develop more and more information regarding the nutritional composition of foods - and with this the development and application of a variety of rather sophisticated technologies to generate this information in the most cost effective manner possible. For the most part, these analytical efforts have been directed toward determining those nutrients for which recommended daily allowances (R.D.A.) have been established (Table 1). It is also interesting to note that these same nutrients are also those requiring analysis under the nutritional labeling regulations of 1973.[1]

TABLE 1. Nutrients with Established RDA's.

	U.S. RDA
Calories	2800
Protein	65/45 gm
Vitamin A	5000 IU
Vitamin C	60 mg
Thiamine	1.5 mg
Riboflavin	1.7 mg
Niacin	20.0 mg
Calcium	1.0 gm
Iron	18.0 mg

As part of our efforts to obtain information regarding the nutritional composition of products and ingredients, we have undertaken a comprehensive program to develop an integrated system for totally automating the analysis of each of these nutrients. With the advent of modern robotics and the Zymate System this objective now, more than ever before, seems attainable.

In this paper we describe the progress that we have made to date in this area and particularly in the direct application of robotics to sample preparation and sample weighing. Finally, a proposed scheme for the systematic transport of the sample through the analytical train will be presented. Included in this system will be a number of robotically controlled stations for sample preparation, sample weighing, cleanup and analysis.

EQUIPMENT

1. The standard Zymark asynchronous computer communication package was used to connect to the Hewlett Packard 1000 3357 Lab Data System via an 8-channel asynchronous multiplexer.

2. The sample holder assembly is a customized unit designed for automated Karl Fisher titration and is available from the Zymark Corporation.

3. Tekmar SDT Series Tissumizer with 1810 motor, 182EN shaft and generator, stand and clamp available through Tekmar Company, 10 Knollcrest Drive, P.O. Box 371856, Cincinnati, Ohio 45222-1856.

SAMPLE PREPARATION

Twelve Unit composites of samples representative of a day's production are homogenized under conditions suitable to prevent the loss of any unstable nutrients. A sub-sample of this composite is transferred to an 8 ounce sample jar, and the sample diluted with water such that the total solids content is adjusted to approximately 15%. The sub-sample is kept frozen and then thawed overnight in the refrigerator prior to use.

SAMPLE WEIGHING

Figure 1 shows a schematic representation of the robotic system under development in our laboratory. One to 20 sample jars are loaded into the sample rack (station 8). To assure that a representative sample is then taken during the weighing operation, the sample jar with its contents is transferred to the tissumiser (station 9) while being supported on the sample holder. After blending at high speed for approximately one minute, the blending speed is reduced and an aliquot of sample removed by affixing a pipette to the robotic hand from station 1. The sample aliquot is then transferred to the electronic balance (station 6) containing a previously tared test tube (tare weight for each tube is established prior to the sample weighing operation) or other suitable receptacle. Pipetting and weighing of sample into the test tube is continued until a predetermined sample weight is obtained. The sample weight was established from the normal estimated level of particular nutrient normally expected to be present and this amount previously programmed together with the sample identification code at the time the samples are first composited. This algorithm not only

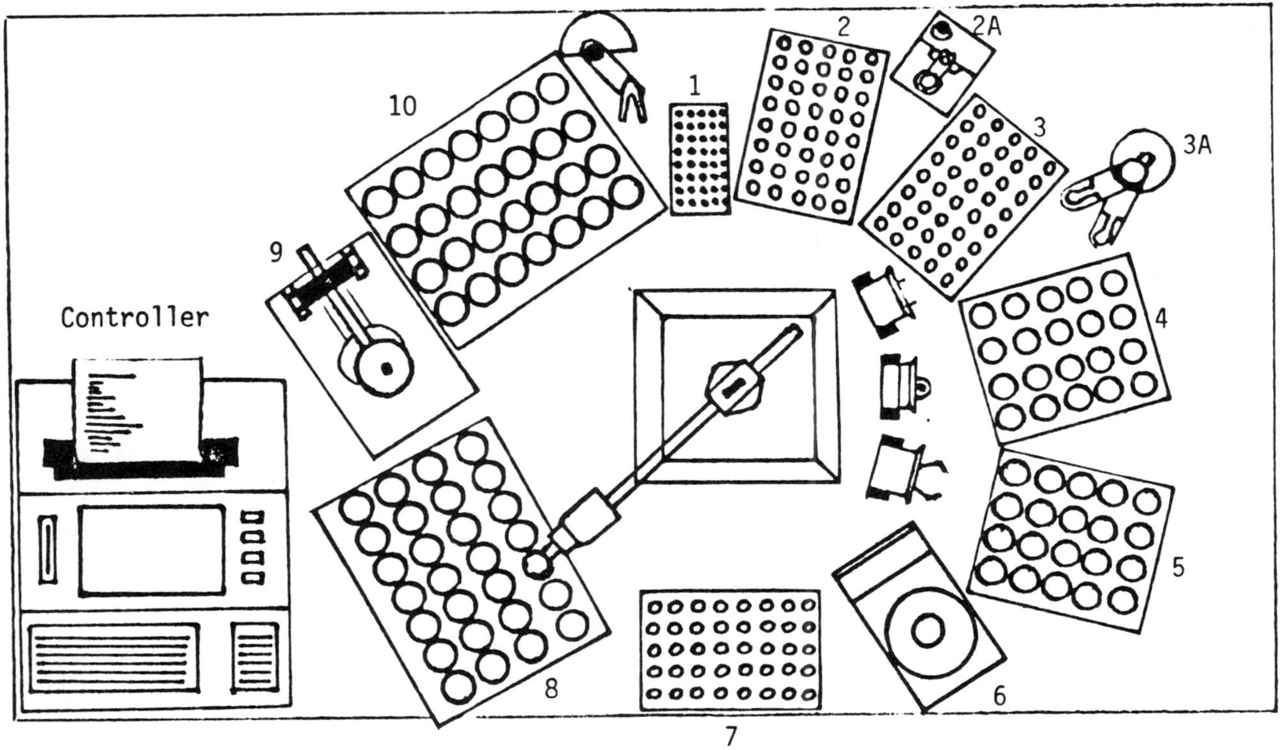

FIGURE 1. Robotic sample blending-weighing module (1 = pipette tip rack; 2 = test tube rack; 2A = vortex; 3 = test tube rack; 3A = diluting station; 4 = protein rack I; 5 = protein rack II; 6 = balance; 7 = transfer rack; 8 = sample rack; 9 = tissumizer; 10 = fat sample

assures optimal sample size but also includes certain safeguards to prevent malfunction or "sorcerer's apprentice effect" should something go awry during this period.

Solids

Following the determination of sample weight described, the sample tray containing the test samples are transferred to a vacuum oven, and dried at 70°C in a slightly inclined position overnight or to constant weight. The sample rack containing the dried samples is returned to lab station 3 and the weight of solid material remaining determined by reweighing the tubes robotically.

Ash

Following this operation the test tube containing the dry solids are placed in a high temperature furnace (525 C). Ashing is usually complete after approximately 18 hours. After cooling, the tubes are returned to station 3 for reweighing to determine the ash content. Following this operation, 2 mL of a 1:1 mixture of conc. HCl and conc. nitric acid and 8 mL of water are added manually. The ash is then solubilized by heating on a steam bath for approximately one hour. The samples are then returned to station 3 and approximately 40 mL of water added from dispenser station 3A. The total dilution of the sample solution so prepared is then determined gravimetrically by robotically reweighing at station 6. Following manual mixing and filtering of the sample solutions the concentration of sodium, potassium, calcium and iron are determined on these solutions by atomic absorption (AA) spectroscopy in accordance with the method described in the AOAC for the determination of these elements.[2] The percentages of these elements present in the original sample are then calculated on the Hewlett-Packard computer from the original sample weight as were the ash and total solids content.

Other Nutrients (Sugars, Protein and B-Vitamins)

Sample weighing as well as gravimetric dilutions are handled much in the same manner as those for solids and ash. However, due to the wide variation in the relative amounts of these individual constituents in foods, the optimum sample size that is robotically weighed is consistent with that concentration of component necessary to provide optimum range when quantitated by automated colorimetric determination of total and reducing sugars.[3]

Protein, as well as the B-Vitamins, thiamine, riboflavin and niacin are determined by official AOAC methods or automated modification of these procedures.

RESULTS AND DISCUSSION

Since most nutritional compositional data are usually expressed on a percentage basis, it becomes quite clear in examining the cost effectiveness in the automation of the nutritional analysis of foods that one of the most labor intensive phases is in the weighing of the sample prior to the analysis. This fact, together with the routine nature of the operation makes this facet of the work especially suitable to robotics. Because of the tendency of samples to settle, it is critical in automating the weighing step that a representative sample be taken for analysis.

Our early attempts to homogenize the sub-sample prior to weighing involved the blender assembly shown in Figure 2. Repeated difficulties in robotically aligning the cup sprocket with the blending rod as well as the stress caused on the robotic arm by the physical weight of the unit lead us to abandon this technique in favor of the Tissumizer-plastic holder arrangement shown in Figure 3. This unit which was developed as part of a sample holder for automated titrations works quite well for the purpose of allowing free mobility of the arm around the Tissuermizer while at the same time providing adequate support for the sample jar.

FIGURE 2. Blender and blender jar.

Aside from the above, we were also confronted with the problem of designing a suitable sample holder that would handle the variety of receptacles into which samples had to be weighed, such as plastic bottles, test tubes, crucibles, digestion tubes, weighing dishes, etc. (Figure 4). Figures 5A and 5B show the sample holder designed for this purpose and presently in use in our laboratories.

Handling the aluminum weighing dish used in the determination of solids and the porcelain crucibles used in the assay of ash and metal was a more formidable problem. Investigation soon established, however, that these assays could be conducted simply and conveniently in ordinary test tubes (23 x 45) without undue loss of solids or the pickup of contaminants from the glass. Tables 2-5 show the typical data which support the validity of this approach.

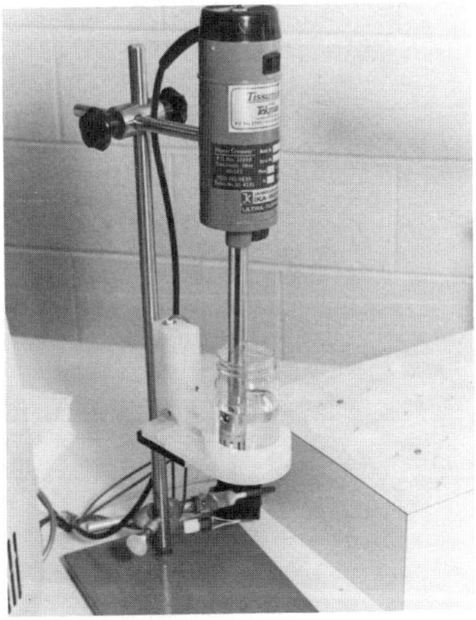

FIGURE 3. Tissumizer with sample jar and holder.

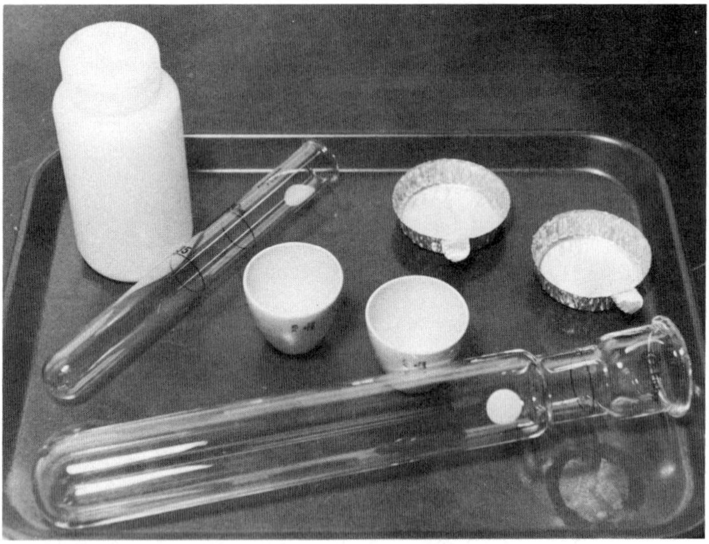

FIGURE 4. Various containers used for holding samples during manual weighing.

In the case of ash (Table 2), some accuracy may be lost in the tube-ashing technique. This is probably related to the amount of residue (ash) being weighed and the accuracy of the balance used. The error has only a minimal effect on the calculation of calories, and as shown in Table 3 does not affect the loss of essential metals or the accuraacy of the determination of these metals by AA.

TABLE 2. Comparison of Solids and Ash.

	Solids (%)		Ash (%)	
Sample	Manual	Robot	Manual	Robot
1	17.9	18.2	0.7	0.7
2	19.4	19.2	1.2	1.2
3	21.1	21.4	0.9	1.0
4	15.8	15.8	1.1	1.2
5	19.8	20.2	0.7	0.8
6	25.6	26.1	1.2	1.4

FIGURE 5A. Multipurpose sample holder for robotic weighing - top view.

FIGURE 5B. Multipurpose sample holder for robotic weighing - side view.

TABLE 3. Comparison of Metals.

	Iron (mg%)		Sodium (mg%)		Calcium (mg%)	
Sample	Manual	Robot	Manual	Robot	Manual	Robot
1	1.4	1.6	512	560	25	23
2	0.3	0.3	374	358	9	10
3	1.5	1.6	369	402	15	15
4	0.5	0.4	515	519	35	38
5	1.2	1.2	556	675	20	18
6	2.4	2.5	686	676	7	5

In Tables 4 and 5 are shown the results obtained in the application of robotic technology in sugars and protein; and in B-vitamins analysis. As can be seen, good to excellent results are obtained in using this approach as compared to conducting the blending, weighing and dilution steps in these assays by more conventional techniques.

TABLE 4. Comparison of Other Nutrients.

	Sugar (%)		Protein (%)	
Sample	Manual	Robot	Manual	Robot
1	7.4	7.2	1.9	1.9
2	0.9	0.7	1.9	1.8
3	0.8	0.8	1.9	1.8
4	3.4	3.4	1.9	1.9
5	0.6	0.7	1.9	1.9
6	7.2	8.0	1.9	1.9

In 1984 a method was officially adopted for the determination of fat for use in nutritional compositional analysis.[2] Preliminary findings to date suggest that the method, which is based on chloroform-methanol extraction of the fat, may be amenable to automated analysis. Such a

TABLE 5. Comparison of B-Vitamins.

Sample	Thiamine (mg%)		Riboflavin (mg%)		Niacin (mg%)	
	Manual	Robot	Manual	Robot	Manual	Robot
1	0.22	0.26	0.11	0.10	0.2	0.2
2	0.11	0.09	0.04	0.03	1.0	0.8
3	0.05	0.04	0.04	0.03	0.6	0.6
4	0.10	0.10	0.09	0.07	1.3	1.2
5	0.06	0.05	0.07	0.05	1.3	1.9
6	0.08	0.05	0.20	0.19	2.4	2.9

system would certainly lend itself more readily to the robotic addition of solvent, extraction, separation, evaporaton and gravimetric determination of the crude fat residue. Automation should easily permit the quantitation of provitamin A (β-Carotene) which is extracted with $CHCl_3$/methanol. We also believe an entirely integrated system for the complete automation of the nutrients present in this paper is equally feasible, and we have already taken our first step in this in realizing this objective. For example, Figure 6 shows a preliminary diagrammatic representation of such a system. Here one can envision an analytical train consisting of several robotically controlled stations for the transport of samples through the system. Module 1, as presented in this paper, has been essentially developed. It comprises a series of sub-stations for blending, homogenizing and weighing. Module 2 is envisioned as a location at which a variety of cleanup operations such as extraction, filtration, evaporation, centrifugation, etc. are performed. In all probability, a dual hand or another robotic unit may be required to perform these operations as well as to physically transport the sample to the analytical module (3) to perform the actual analysis desired. Recent developments in the area of HPLC suggest that

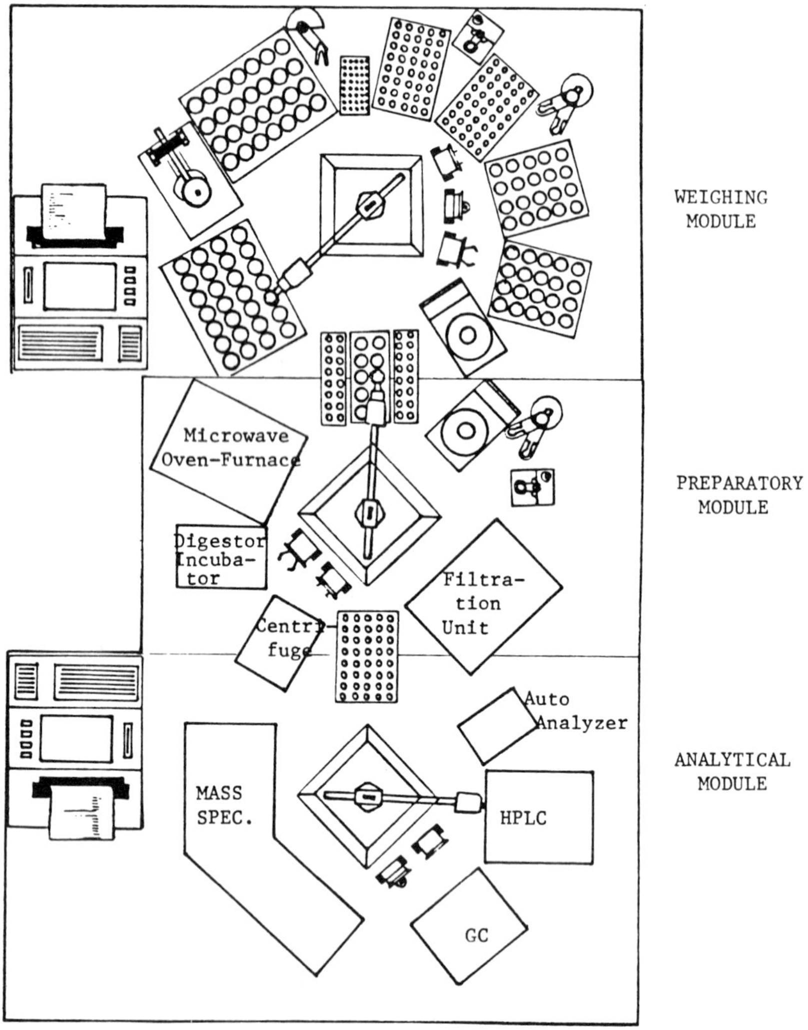

FIGURE 6. Robotic system for automated nutritional composition analysis of foods.

this technique, in combination with UV, florescence and electrochemical detection would allow the simultaneous determination of the vitamins, and possibly cation and anions (ion chromatography). Microwave drying and ashing in combination with inductive coupled plasma-atomic absorption spectroscopy should also prove readily adaptable for this purpose.

REFERENCES

1. Federal Register 1973, Vol. 38, pp. 2124-2164.

2. Book of Official Methods - Association of Official Analytical Chemists, 14th ed.

3. Technicon Industrial Methods Manual Nos. 142-71A.

AUTOMATED ROBOTIC EXTRACTION AND SUBSEQUENT ANALYSIS OF VITAMINS IN FOOD SAMPLES

Darla J. Higgs, Joseph T. Vanderslice and Mei-Hsia A. Huang
Nutrient Composition Laboratory
Beltsville Human Nutrition Center
Agricultural Research Service, USDA
Beltsville, MD 20705

ABSTRACT

A robotic extraction procedure, previously developed for the analysis of vitamin C in foods, has been modified so as to make it more applicable for the chemical analysis of a wide range of water soluble vitamins. In contrast to the extraction of vitamin C, most sample extracts containing the nutrient of interest are passed through a cleanup column prior to injection into a high-performance liquid chromatographic analytical system. The modified procedure is illustrated for the analysis of thiamine content in several samples. This procedure can also be used for the analysis of vitamin B_6 in biological samples.

INTRODUCTION

In recent years, nutrition in relation to health has become a very important issue and thus the levels of nutrient content in foods, as eaten, is of prime importance.[1] Because of this interest, large numbers of food samples need to be analyzed to provide data for dietary information and food product labeling. Also, in recent years, automation of high performance liquid chromatography (HPLC) has become a viable means to do analysis of vitamins.[2] However, HPLC methods require nutrients to be extracted from complex food matrices so as to provide "clean" samples for analysis.[3] Since such extractions normally demand tedious manual operations, it is apparent that the automation of such procedures would be advantageous. Automation of the extraction of nutrients from food samples had been a real "stumbling block" until the advent of robotics which has provided the necessary capability. Now it is possible to completely automate the extraction/analytical methods for a given nutrient. For example, an automated method of extraction and analysis of Vitamin C has been developed and demonstrated to be an efficient and reliable avenue for collecting nutrient data from different types of foods.[4,5] The next step has been to adapt this method and expand its capabilities for use with other methods of vitamin analyses.

Modular laboratory robotics provides flexibility, which is a prime consideration when making modifications in existing extraction procedures. Since the vitamin C extraction method has been developed, it is possible to apply similar protocols to the analysis of thiamine and vitamin B_6 contained in food. Because both of these vitamins require additional steps in the extraction procedure over those used with vitamin C,[6,7] some additional modifications and extensions to fully automate the analysis of these vitamins in food samples are required. Such modifications will be discussed in this paper.

EXPERIMENTAL

Reagents

Sodium hydroxide solution (50% w/w), sulfosalicylic acid (SSA) sodium phosphate monobasic, phosphoric acid, and hexane (Fisher Scientific, Silver Spring, MD) were used as received. Thiamine hydrochloride (T), thiamine monophosphate chloride (TMP), and thiamine diphosphate chloride (TDP) (Sigma Chemical Co., St. Louis, MO) were used as standards. Amprolium (AMP) (Merck & Co., Rahway, NJ) was used as an internal standard in the thiamine method.[7] All aqueous solutions were prepared with water that had first been passed through a reverse-osmosis purification system (Millipore, Bedford, MA) and then through a Milli-Q system (Millipore, Bedford, MA).

Robotic System

The Zymate Laboratory Automation System used for the thiamine analysis is shown in Figure 1. This contains modifications of the system originally developed for vitamin C. The original system consists of all modules shown in Figure 1 that are not shaded.

This system has been fully described in the literature.[4,5] The added components are those in the shaded area. These added components include a second injector station, and a three-way valve operated by the power and event control station. The second injector station is used to introduce an aliquot of sample into the cleanup column for the HPLC

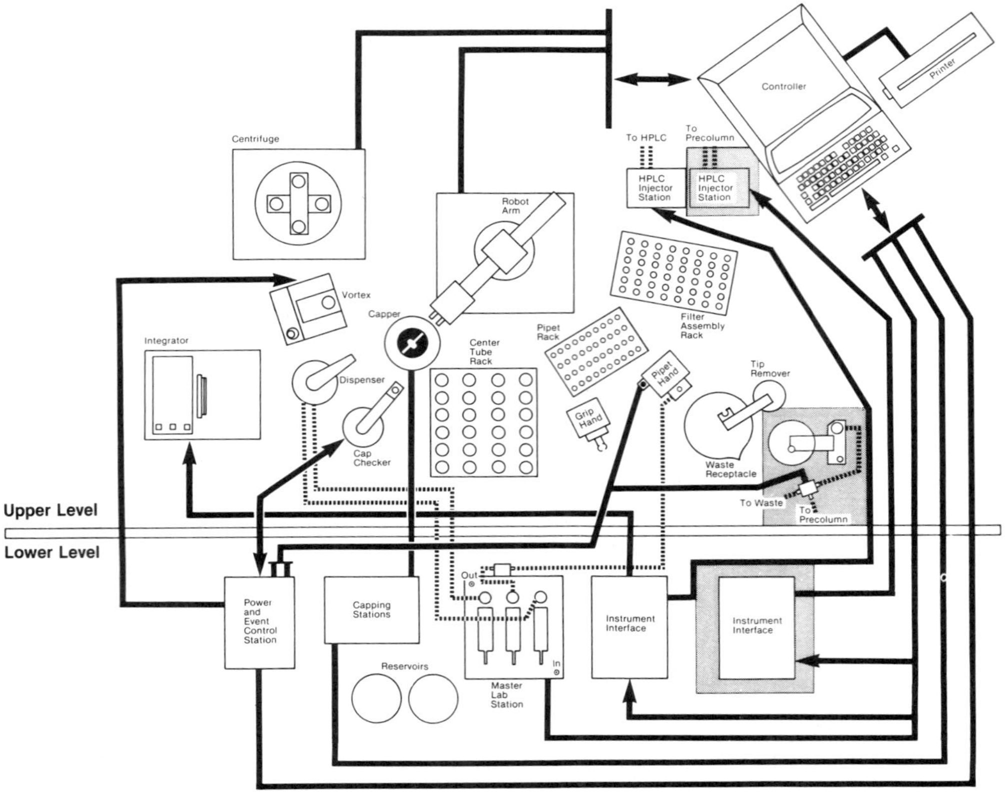

FIGURE 1. Schematic diagram of the robotic layout for thiamine. (Modules not shaded are those used for previously reported extraction of vitamin C. Additional modules, shaded, were added for the thiamine determination.)

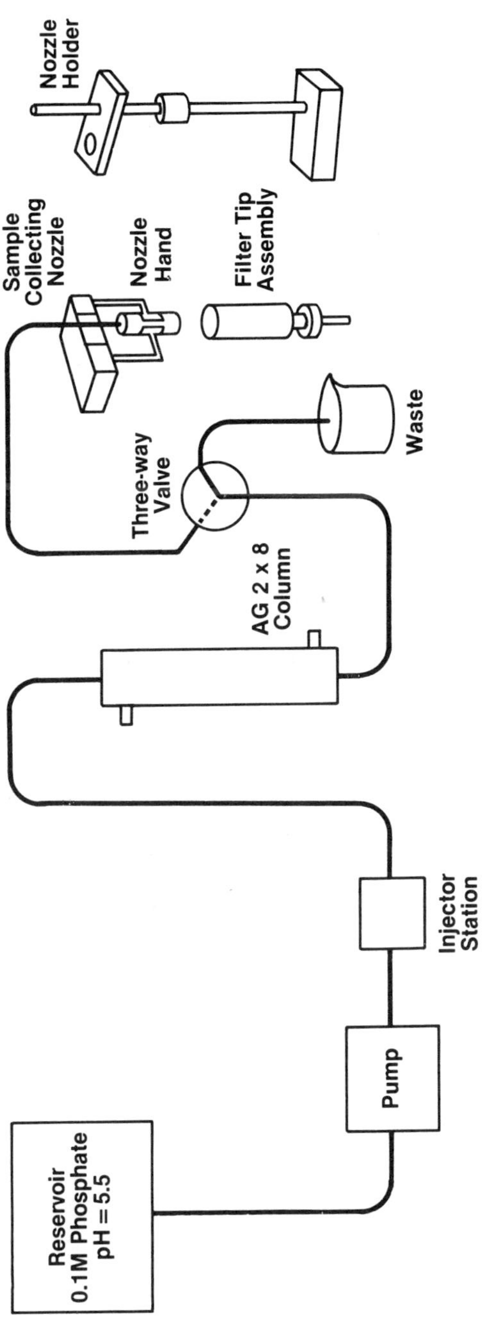

FIGURE 2. Diagram of cleanup column set up for thiamine with sample collection nozzle.

system, while the three-way valve is used to direct the eluant from the cleanup column either to waste or to a filter assembly through a movable nozzle. This movable nozzle, along with its holder, was constructed from small plastic scraps available in the laboratory. The nozzle was moved by attaching a set of fingers to the pipet hand so as to form a "nozzle hand". Details of this nozzle arrangement are shown in Fig. 2.

Sample Preparation

Solid food samples, a whole box of cereal for example, is homogenized in a food processor; meats are homogenized using entire serving portions. No pretreatment is required for pollen samples. Weighted samples with known amounts of amprolium, the internal standard, are placed into centrifuge sample tubes to be stored in the robotic sample rack. Ten mL of 5% SSA and ten mL of hexane are added to the sample by the robot and replaced in the rack until further treatment.

Thiamine HPLC Cleanup System

The cleanup system is outlined in Figure 2. The eluting solution which is 0.1M sodium phosphate, pH=5.5, is passed through a Glenco glass column (Howe Scientific, Cut and Shoot, TX) packed with AG 2-X8 anion exchange resin (Bio-Rad, Richmond, CA). A 500 uL sample is introduced through the injector station and a 4.3 mL of eluant collected as the vitamers elute (from 1 to 2 min after injection depending on the column used). For more details see reference 7.

HPLC Analytical System

This HPLC system is a straightforward gradient system which separates the vitamers from one another (T, TMP, TDP) and from the internal standard (AMP). Post-column chemistry is used to convert the eluting vitamers to their thiochrome derivatives which are detected fluorometrically. Amprolium, the internal standard, also produces a thiochrome derivative. More details are given in reference 7.

Extraction Procedures

Figure 3 shows a block diagram of the manual procedure for the extraction of thiamine from foods.[7] The robotic extraction procedure is a modified version of the manual method (Figure 4). The procedure is similar to the vitamin C extraction method except 5% SSA is used in place of citric acid and a cleanup column is required.

A typical procedure for thiamine consists of the initial addition of acid and hexane to the sample containing the internal standard, mixing, centrifuging, removal of sample aliquot, filtering and the injection of the sample into the cleanup column. This procedure is identical to that already described for vitamin C except that in the latter case, the injection of sample is directly into the analytical system. The cleanup column carrier goes directly to waste until it is time to collect the vitamin fraction. Shortly before the vitamin fraction emerges from the column, the nozzle hand moves the nozzle from the holder to the filter assembly and the three-way valve switches to dispense the fraction into the filter assembly. Once a full fraction is collected, the three-way valve is switched back to waste and the nozzle hand parks the nozzle.

The hand is then switched from nozzle to pipet hand which picks up the

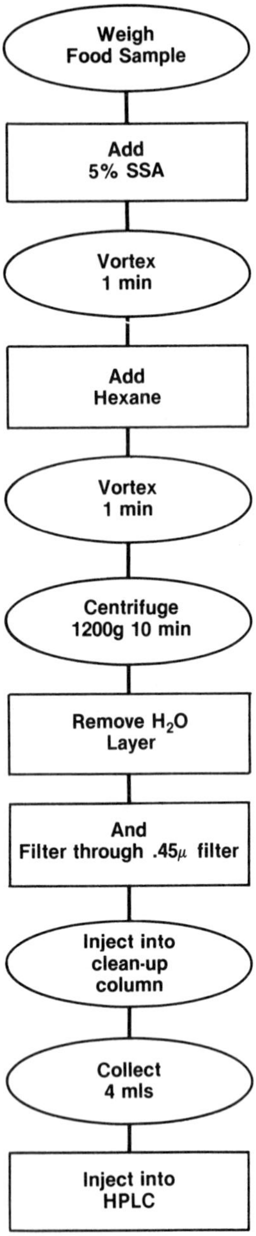

FIGURE 3. Block diagram for the manual extraction of thiamine from food samples.

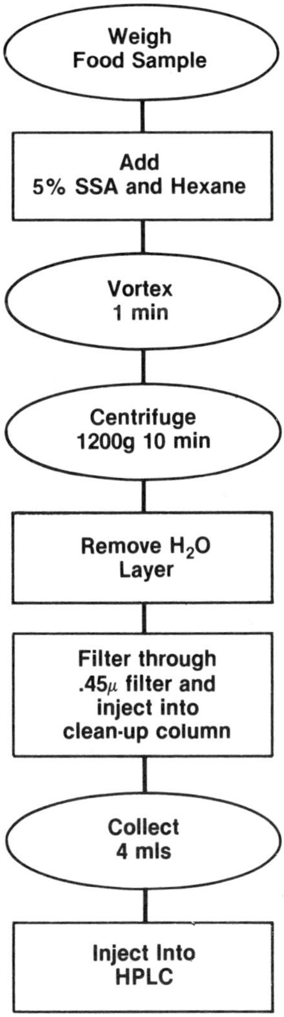

FIGURE 4. Block diagram for the robotic extraction of thiamine from food samples.

filter assembly and "rocks" the latter gently back and forth two times before injecting the vitamin fraction into the analytical column. If vitamin B_6 is to be determined rather than thiamine, an additional pH adjustment is required prior to injection into the analytical column.

The steps needed for pH adjustment in the vitamin B_6 samples are as follows. Once the vitamin fraction is collected from the cleanup column, the hand switches from nozzle to pipet hand. Then the robot picks up a pipet, obtains an aliquot of NaOH, and dispenses it into the filter assembly which contains the sample. The robot discards the pipet, then picks up the filter assembly and "rocks" the filter assembly back and forth twice before injecting the sample into the analytical column.

The extraction and analytical procedures for all vitamins are carried out under yellow light.

RESULTS AND CONCLUSIONS

Chromatograms obtained with the manual and robotic methods of extraction with subsequent analysis for thiamine are shown in Figure 5 for a sample of standards and for a breakfast cereal. In Table 1, data is given for three different types of samples analyzed. The results are comparable (agreement to within 10%), with coefficients of variations less than 20% even for samples low in thiamine content.

From the preliminary results reported here, it should be evident that the introduction of laboratory robotics has made it possible to fully

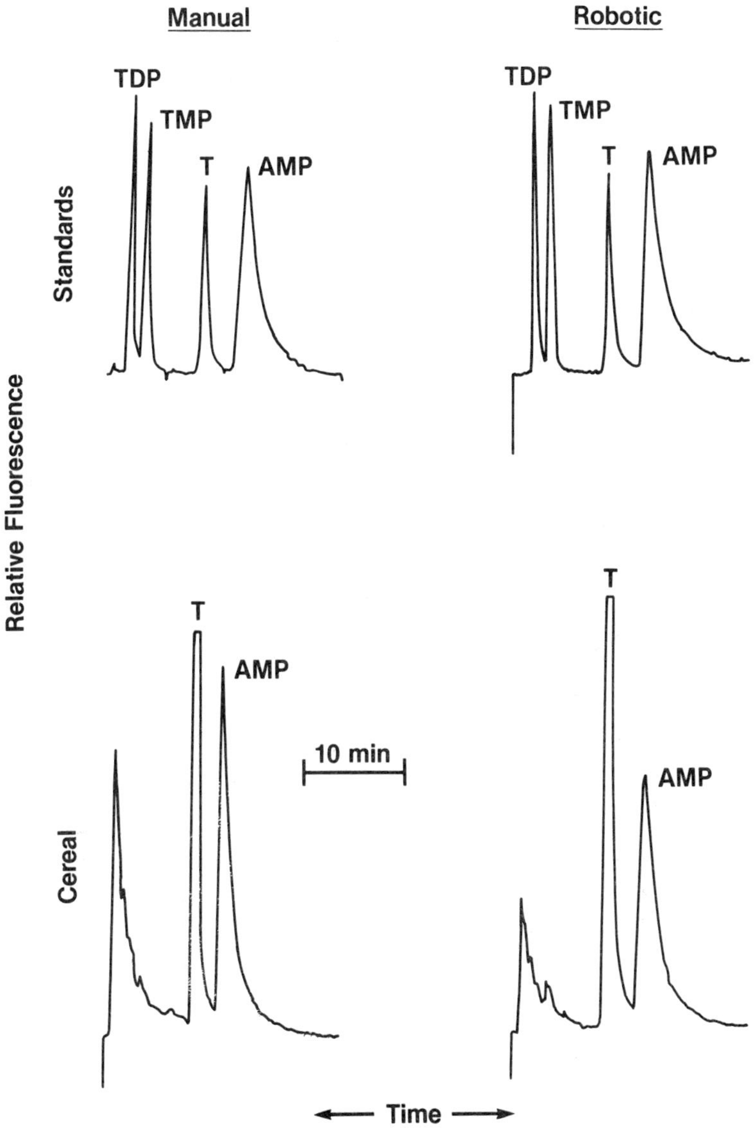

FIGURE 5. Chromatogram obtained with two different methods of extraction, manual and robotic. The upper two are for standard thiamine samples while the lower are for a fortified breakfast cereal. TDP (thiamine diphosphate ester), TMP (thiamine monophosphateester), T (thiamine), AMP (amprolium).

TABLE 1. Comparison of Manual and Robotic Extraction Procedures for Thiamine

	MANUAL		ROBOTIC
		mg/100g	
CHICKEN MEAT, skinned			
Breast	.049 ±.004		.053 ±.002
Leg	.062 ±.009		.061 ±.012
Thigh	.050 ±.004		.053 ±.013
CEREAL, bran	2.02 ±.127		2.55
POLLEN	.763 ±.067		.701 ±.060

automate the extraction and analytical procedures for biological samples for at least three of the water soluble vitamins. There is no reason to suspect that these procedures could not be extended or modified to cover analytical methods for other vitamins of future interest.

Of particular interest is the fact that extraction and cleanup procedures for the analyses of vitamin B_6 and thiamine in foods are essentially identical so one extracted food sample can form the basis for the simultaneous analysis of both vitamins. The robot can be programmed to inject an aliquot of sample into each of the two HPLC analytical systems. Since both the vitamin B_6 and thiamine procedures use internal standards which are added to the sample prior to analysis, it is not necessary to monitor the various volume dilutions that may occur during the analytical steps.[3]

Vitamin C could be analysed simultaneously with thiamine or vitamin B_6

or both. However, since the extraction procedures for this vitamin differ from that of the lattter two, it would be necessary to deal with different food samples. The robot can be programmed to run one extraction procedure and then switch to the other, since the analytical time period, at present, is longer than the time it takes to extract a sample.

DISCLAIMER

Mention of a trademark or proprietary product does not constitute a guarantee or warranty of the product by the U.S. Department of Agriculture and does not imply its approval to the exclusion of other products that may also be suitable.

REFERENCES

1. Higgs, D. J. and Vanderslice, J. T., Pittsburgh Conference and Exposition on Analytical Chemistry and Applied Spectroscopy, Feb.25-March 2, 1985, Abstract No. 582A.

2. Vanderslice, J. T. and Higgs, D. J., J. Micronutr. Anal., 1, Sept. 1985.

3. Vanderslice, J. T., Maire, C. E., Doherty, R. F. and Beedher, G. R., J. Agric. Food Chem., 28, 1145 (1980).

4. Vanderslice, J. T. and Huang, M-H. A., (manuscript in preparation).

5. Diet, Nutrition and Cancer, Report of the Committee on Diet, Nutrition and Cancer, Assembly of Life Science, National Research Council, National Academy Press, Wash., D.C., 1982.

6. De Leenheer, A. P., Lambert, W. E. and De Ruyter, M. G. M., Modern Chromatographic Analysis of the Vitamins, Marcel Dekker, Inc., New York, 1985.

7. Vanderslice, J. T., Brownlee, S. G., Cortissoz, M. E. and Maire, C. E., in Modern Chromatographic Analysis of the Vitamins, De Leeheer, A. P., Lambert, W. E. and De Ruyter, M. G. M. (eds.), Marcel Dekker, N.Y., 1985, p. 435.

Zymate is a registered trademark of Zymark Corporation.

AUTOMATION OF MULTIPLE ANALYTICAL PROCEDURES IN AN INDUSTRIAL LABORATORY (TRACE ORGANICS IN WATER & SOIL, RESIDUAL MONOMERS, ANIONIC SURFACTANTS, PREPARATION OF STANDARDS FOR GC, LC & IC, ETC.)

M. Markelov, M. Antloga, S.A. Schmidt
The Standard Oil Company (Ohio)
Sohio Research Center
4440 Warrensville Center Road
Cleveland, Ohio 44128

ABSTRACT

This paper describes applications of a commercially available Zymate System in a diversified analytical laboratory with fast changing scope of analytical tasks. The enviornment of most non-QC industrial research laboratories rarely calls for processing more than 10-20 samples of the same nature. Rather, it requires rapid switching from one analytical technique to another. Several of the analytical procedures most frequently used in our laboratory were identified as "robotizable". The wide application of pneumatic cylinders, controlled via solenoid valves connected with Zymark power and event controllers, permitted us to incorporate a variety of sample processing stations in one robotic system. These stations are located on a pegboard which economizes space in the valuable area of the robot's reach. The large number of different stations within the robot work area creates a flexible environment for fast method development and modification. Over ten different procedures are set up within a single robotic arm work area. Moreover, the robot's flexibility was further enhanced by placing it on a mobile cart and constructing a special chemical bench along a pegboard covered wall. This bench is equipped with fixed receptacles to accomodate the mobile cart in preprogrammed areas. Several analytical procedures such as headspace gas chromatography, colorimetry, standard preparation, etc., and work area designs will be presented and discussed.

INTRODUCTION

The introduction of the Zymate Laboratory Automation System by Zymark Corporation was enthusiastically welcomed by analytical chemists involved in quality control and other analyses that require continuous repetitive sets of actions. However, questions were raised about the applicability of robotics in laboratories involved in research or analytical testing where analytical procedures were to be modified constantly and the need for one type of sample analysis was not substantial. This paper will demonstrate the application of robotics in a fast-changing analytical enviornment.

All automated procedures described below are set up on single benchtop, within a single robot's reach (see Figure 1).

The ease of adopting the robot to a variety of specific analytical tasks is achieved via the following:

1. Creating a collection of specific and generic stations such as those shown in Figures 3, 4, and 5.
2. Expanding the robot work area using a cart, bench, and pegboard combination (Figure 1). Most of the generic stations (e.g., balances, colorimeter, standard preparation) are located on the robot cart. The relatively specific stations (e.g., headspace, environmental and pipette racks, pipette cleaning and waste stations) are located on the bench. The pegboard, which is connected to the bench with hinges, permits easy connection and disconnection of lightweight stations (e.g., salt dispenser. selective-electrode and colorimetric cells).

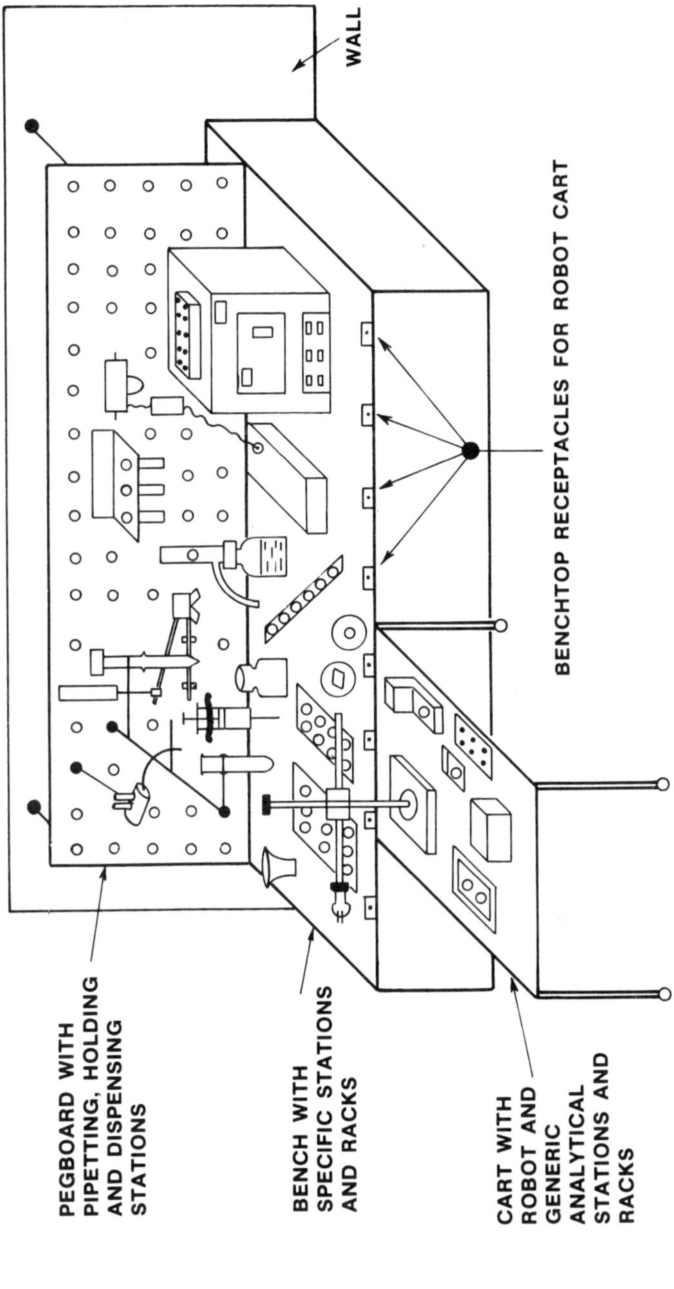

FIGURE 1. Expansion of robot useful area for multiple analytical procedures.

3. Programming every station separately. Accumulation of a substantial number of such modular programs creates a specific dictionary in the robot's language. This might consist of the names of programs to perform such functions as pick up a vial, dispense salt, dilute 10 to 10, crimp, etc. If the langauge is rich enough, an operator can easily modify or create a new program to meet the requirements of many different analyses. As an illustration, it took us about four months to set up the first analytical method on the robot. The last one was operational in about an hour after the request for robozation automation.

4. Extensively using linear or rotational actuators (pressure-actuated cylinders) in the design of robotic stations. These actuators save valuable space around the robot by bringing vials to and accepting vials from the sample processing stations. Actuators capable of circular and linear motion permit simultaneous manipulation of several objects by the robot.

ANALYSES OF VOLATILE COMPOUNDS IN WATER, SOIL, AND PLASTICS

These anlalytical procedures are grouped together because all of them use headspace methodology,[1] which makes use of the thermodynamic equilibrium between gaseous and condensed phases (liquid or solid) in a heated, sealed vial. A typical headspace vial is shown on the upper left corner of Figure 2.

Once equilibrium is established in the vial, the concentrations of volatile compounds in the vapor phase uniquely correspond to their original concentrations in the sample. A portion of the gaseous phase is introduced into an anlytical instrument (usually a gas chromatograph). All chromatographic quantification techniques, such as external standard, internal standard, and standard additions, are applicable to headspace methodology. Moreover, techniques specific to headspace sampling (such as variable loading equilibration[2,3] and

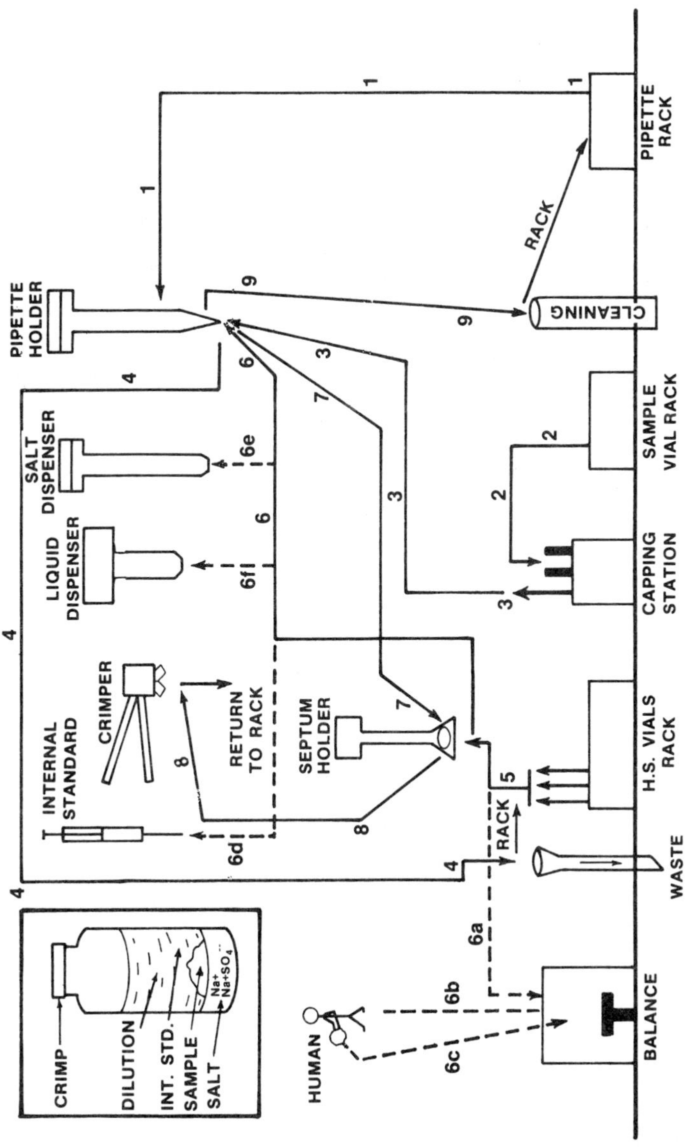

FIGURE 2. Robotization of headspace analysis of water, soils and plastics for ppb level volatile organics.

multiple headspace extraction[4]) eliminate the need for standard preparations in the sample matrices. Automation of headspace methods is attractive for several reasons. The wide variety of samples that can be handled by the same robotic method increases sample loading, thus justifying automation as a cost-effective approach. The gas chromatograph is set up for analyses of an air matrix, independent of the actual sample matrix. This minimizes variability of instrumental conditions (e.g., columns, sample sizes, temperatures) for various samples. Highly automated headspace analyzers are available which only require the user to place a prepared vial in the heated equilibration bath and to press the "start" button.

Sample preparation is the most labor-intensive step in headspace analyses. This is the only step amenable to direct instrumental automation. Principles of manual sample preparation and automation via robotics are described below.

<u>Manual Sample Preparation for Headspace Analyses</u>

Water Samples: An aliquot (5-10 mL) of sample is transferred into a headspace vial (~20 mL). An internal standard can be added (2-100 uL). If necessary, the sample is diluted with water and saturated with salt. The vial is then crimped with a Teflon-lined septum and placed in a headspace analyzer.

To allow statistical evaluation of the results, three headspace vials are usually prepared from every water sample. The salt is added for three reasons:

1. To take advantage of the salting out effect, which increases the sensitivity

2. To eliminate variations in solubility of organics in natural waters (fresh and seawater) containing different levels of inorganic salts

3. To destroy living organisms that might effect the solubility and stability of the volatile compounds.

Soil and Sludge Samples: The internal standard, salt, and distilled water are added to the headspace vial, as described previously. The weight of the vial with the cap is then recorded. One to five grams of the sample is added; the vial is then crimped and reweighed.

Polymer Samples: Sample preparation for headspace analysis of plastics is similar to that for soils except salt usually is not added and an organic solvent is used for sample digestion.

The presence of residual monomers and solvents (often those that are used in manufacturing the plastic) is highly regulated for plastics used for food packaging. The FDA (Food and Drug Adminstration) considers these residuals food additives.

Robotic Sample Preparation for Headspace Analyses

We have developed a robotic procedure that automates the sample preparation as described. A schematic of the robot's movements is shown in Figure 2. The robot essentially handles only three objects, all of comparable diameters - a pipette, a headspace vial, and a sample or environmental bottle. This design avoids changing robotic hands and speeds up the method. The dashed lines in Figure 2 represent optional movements specified by the user prior to running the program. For example if "salt option = 0" is specified, the robot will omit the salt dispensing step.

In the description of the following robotic methods, the path numbers in parantheses are depicted in Figure 2.

Water Samples: The robot picks up the pipette from the pipette rack and introduces it into the pipette holder (Path 1), where it is held pneumatically. The pipette holder is connected to a computer-controlled syringe/3-way valve assembly that fills the pipette and dispenses preprogrammed volumes under vacuum control.

The robot picks up a 44 mL environmental bottle containing the sample (Path 2) and unscrews the cap using the capping station. After disposing of the screw cap, the vial is brought to the pipette (Path 3) where 34 mL of sample are sucked into the pipette. The robot dumps the remaining sample into the waste funnel (Path 4) and returns the used bottle back to the environmental sample rack.

The robot picks up a headspace vial covered with both a Teflon-lined septum and an aluminum cap and carries it to a septum holder (Path 5) where the septum with the cap is temporarily held by a vacuum. The vial is then brought to the salt dispensing station (Path 6), which is operated by computer-controlled pneumatic cylinders. Approximately 6 gm of salt is dispensed from a suction tube into the vial by releasing the vacuum that held the salt in the tube. The vial is delivered to the pipette (Path 6e) and 10 mL of sample are dispensed into the vial. Depending upon the requirements, the vial is either first delivered to the internal-standard dispensing nozzle (Path 6d) or directly to the septum holder (Path 7). The septum with an aluminum cap is then put back onto the vial. The robot delivers the vial to the crimper (Path 8), which is actuated by a computer-controlled pneumatic cylinder. The crimped vial, ready for headspace analaysis, is returned to the headspace vial rack.

Paths 5 through 8 are repeated three times to generate replicates of the sample for statistical evaluation. The pipette is then removed from the holding station and inserted into a cleaning station, (Path 9) where the pipette is subjected to heat and vacuum. The whole process is repeated until the preprogrammed number of samples is exhausted.

Soil and Sludge Samples: The robot picks up a headspace vial, leaves the septum at the septum holder (described previously), and delivers the vial to the balance (Path 6d) where the weight of the vial is recorded and stored in the computer. The robotic arm brings the vial to a human (Path 6b) who adds a sample specimen and presses a switch to let the robot know that the human part of the sample preparation is finished. Without human assistance, the present state of technology does not permit any machinery to take representative samples of soil, which may contain worms, leaves, etc.

The robot returns the vial to the balance (Path 6c) where the weight is recorded again. The vial is then delivered to a solvent (water) dispensing nozzle (Path 6f) or, if an increase of sensitivity is described to the salt dispensing tube (described above) to take advantage of the slating-out effect.

Depending upon the specifics of the analysis, the amount of water dispensed into the vial can be kept constant or can be varied to maintain a constant ratio of sample to water. This is acheived using the computer-controlled syringe/3-way valve assembly (master lab station, described previously). Again, there is an option of internal-standard addition. The program also permits sample addition to follow addition of salt, water, and internal standard.

The vial is then brought to the septum holder, to the crimper, and finally back to the headspace sample rack, ready for analysis.

Polymer Samples: This procedure can be run in the same fashion as the soil analysis or can be completely automated if the polymeric materials to be tested are in powder or pellet form.

In the fully automated mode, the robot picks up the vial and brings it to the balance where the vial is tared. While the vial is on the balance, the robotic arm, equipped with a vacuum suction tube, moves to the polymer sampling rack. Sample pellets are vacuumed into the tube. The arm returns to the balance and dispenses the pellets into the vial by substituting the vacuum with a flow of air via a computer-controlled/ 3-way solenoid valve.

After the weight of the polymer is recorded, the vial is transferred to the solvent dispensing nozzle where options for constant solvent volume or constant sample/solvent ratio are available. The vial containing the sample and the solvent is capped and crimped and can be put into a shaker. The vials are removed from the shaking rack at the anlayst's discretion.

COLORIMETRIC AND SELECTIVE ELECTRODE MEASUREMENTS

These procedures, sketched in Figure 3, are grouped together, in spite of the dissimilarity of their applications and principles of analytical measurements, because

a. The same peripheral robotic devices (i.e., master laboratory station, power and event controller, printer) are used

b. The workcells are small, similar in design, and have flexible connections to analytical equipment or robotic peripheral devices allowing them to be removed from the robot's reach area

c. The robotic movements and glassware handled are practically identical.

The operations described below do not require human assistance. Flexibility is achieved in three ways:

- By manipulating computer-controlled syringes equipped with 3-way line selection valves using the master laboratory station
- By appropriately positioning the robotic arm carrying a specific vial
- By feeding back the results of measurement to the robot, which then makes the choice of the next movement based on the analytical instrument readings.

The design of the workcells (Figure 3) permits -

- Introducing the desired volume of liquid sample into the cells with the accuracy of the master laboratory station and without contaminating the master laboratory syringes, valves, or connecting lines by the sample.

- Adding the necessary amount of diluting solvent, buffer, electrolyte solution, or reagent with the sample already present in the cell.

- Mixing the sample vigorously with reagents, solvents, electrolytes, etc., by forcing air through the cell.

- Dispensing a portion of the sample to a waste bottle and diluting the remaining solution with the new portion of solvent. These operations can be repeated until the sample concentration in the cell is within the linear portion of the calibration curve stored in the computer/controller memory. The same approach is used for generating a calibration curve from a single concentration.

- Measuring either layer of a sample in an extraction vial or extracting within the cell itself, followed by measuring either or both sample layers.

- Washing the cells after each measurement with the appropriate cleaning solution, either by bringing the vials containing the cleaner to the cells or by purging the cells from an independant solvent supply line.

The colometric cell is a plastic, solvent-resistant tube, which is positioned vertically. The tube has a hole on the side to accomodate the color probe, which is inserted horizontally. Horizontal positioning avoids trapping air bubbles in the fiber optic lens or on the optically active area of the color probe.

The electrochemical cell is positioned at a 45° angle to avoid loss of contact between the electrode surface and its filling solution and to avoid bubble entrapment during mixing with air.

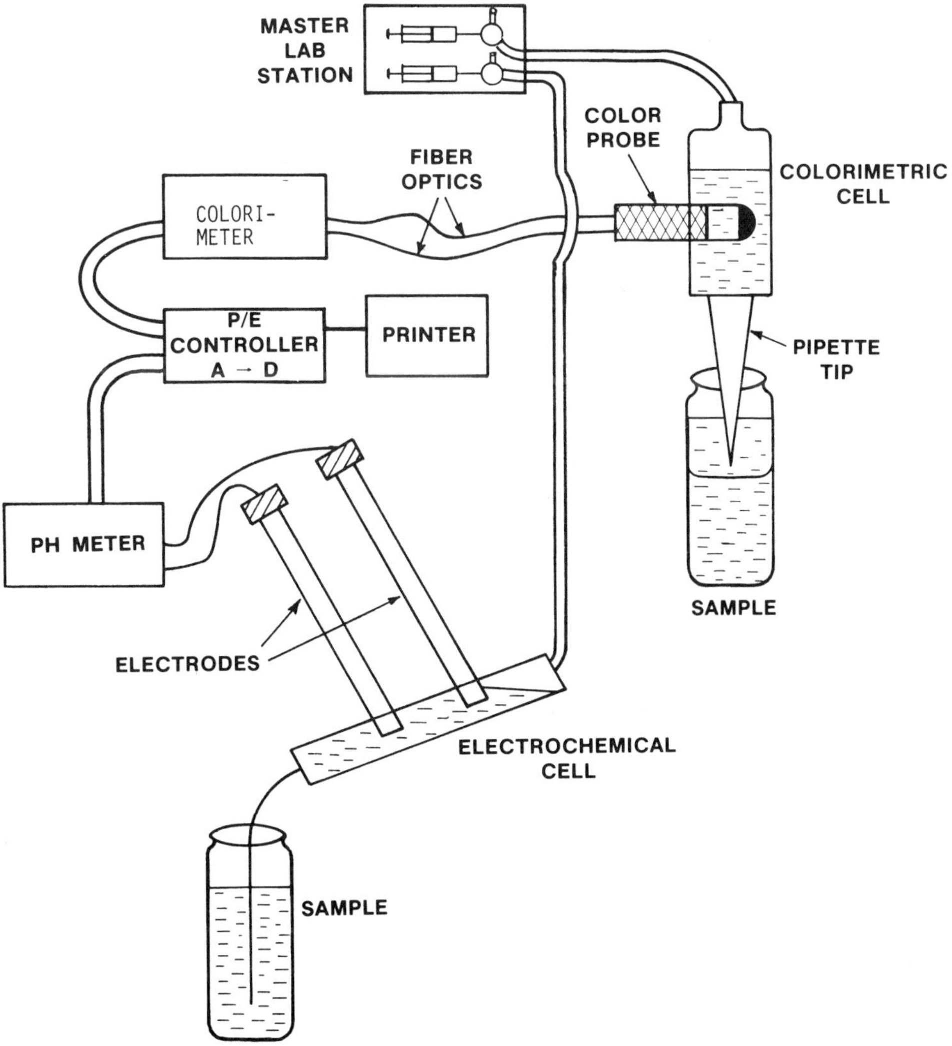

FIGURE 3. Robotic stations for colorimetric and electrochemical measurement.

These simple and inexpensive cells can be used in conjunction with the robot and master laboratory station for a variety of analytical procedures, one of which is in the following section.

Analysis of Anionic Surfactants in Seawater

Principles of the Procedure: The method is based on the reaction of anionic surfactants with a dye. The pink-colored product is soluble in chloroform ($CHCl_3$), which is used to extract it from the water matrix. The method calls for titration of the resulting two-layer system with a chemical that forms a stronger chloroform in soluble complex. Titration and shaking is continued until complete discoloration of the bottom chloroform layer occurs.

Automation of the Procedure: Titration with vigorous mixing of partially emulsified phases is a complicated task for a robot to perform and requires quite sophisticated and expensive equipment. Also, the procedure can be lengthy if complete emulsion breakdown is required. Therefore, for automation, the procedure is converted from a two-phase titration to a colorimetric method, which involves determination of the absorbance of the pink surfactant/dye complex.

The procedure consists of two steps: preparation of reagents and colormetric measurement of the organic layer at 520 nm. The sample can be added to the bottle either before or after it is filled with reagents. In the former mode, the water solution of dye, the chloroform, and the sample are simultaneously dispensed into the bottle, thus providing good mixing. These liquids are dispensed using computer-controlled, pneumatically actuated dispensing bottles (Figure 4).

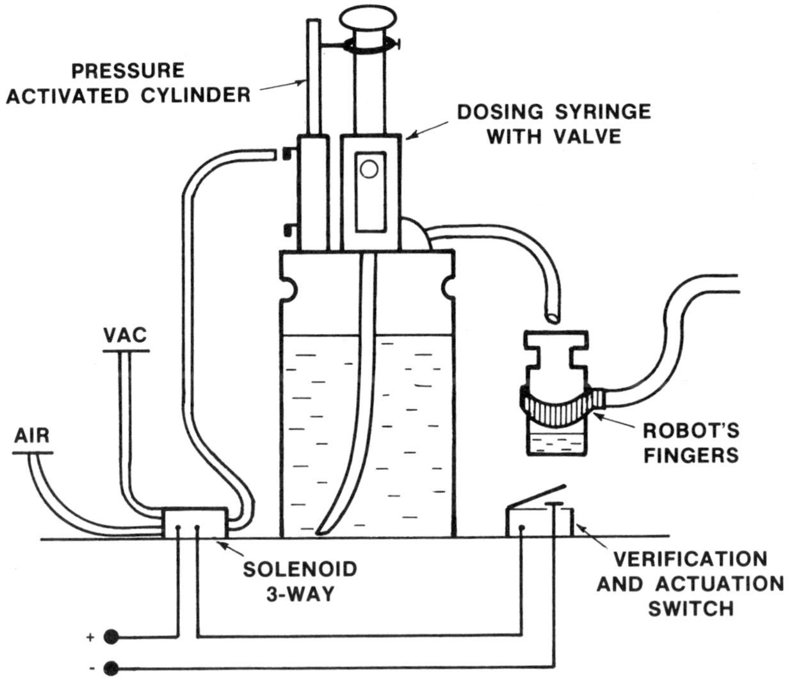

FIGURE 4. Adjustable constant volume liquid dosing system.

The robot picks up a sample bottle and brings it to a cylinderical sample cell equipped with a suction nozzle and connected to a computer-controlled syringe/3-way valve asembly. (Figure 3). The cell contains a fiberoptics color probe inserted horizontally through the side wall. The bottom (pink organic) layer is pulled into the cell, and after allowing one minute to stabilize the colorimeter, the reading is taken and processed by the computer, and the concentration is recorded. The sample is then dispensed back into the sample bottle, which is then returned to the sample rack.

A SIMPLE INEXPENSIVE LIQUID-DELIVERY SYSTEM

The major differences between colorimetric and ion-selective-electrode procedures are variations in the reagents (i.e., dyes versus electrolyte solutions). Therefore, in order to be flexible in switching from one procedure to another, one should have a large number of liquid-delivery systems. The Zymate master laboratory station has only three syringes and is quite expensive. The liquid-dosing system shown in Figure 4 is designed and constructed to meet this challenge. This system is essentially a liquid-dispensing bottle whose syringe is connected to a pressure-actuated cylinder. The cylinder is powered from a solenoid that supplies either vacuum or pressure to one end of the cylinder or pressure to both sides of the cylinder.

The solenoid can be controlled via outputs on a power/event controller or through an independent power supply. We found that an independent power supply has a number of advantages.

- A solution will never be dispensed unless the robot possesses a bottle, and the bottle is in an upright position, directly under the nozzle. When these conditions are met, the bottom of the vial will trigger a switch, and the solution will fill the syringe. The robot then moves the vial straight up, turning the switch off, and the reagent in the syringe is dispensed into the vial.
- The power and event controller of the Zymate System has only eight inputs and outputs, which are used extensively for method development. This dosing system does not use any of these I/O devices. The switch supplies and interrupts power to the solenoid and at the same time verifies the presence of the vial under the dispensing nozzle.
- The system is not dedicated to the robot--manually pressing the switch will also cause the system to dispense.

The dispensing bottle does not need to be located within the robot's reach. Only the dispensing nozzle (connected to the bottle with a flexible capillary) and the switch occupy the robot's valuable work area. This setup is quite flexible, and the volume dispensed can be easily varied by adjusting the stopper knob on the syringe.

MULTICOMPONENT STANDARD PREPARATION AND SAMPLE SPIKING SYSTEM

Preparation of standards, especially multicomponent standards, is the most common, time-consuming activity in any analytical laboratory. This activity is usually considered by lab personnel as a noncreative and unfortunate assignment. On the other hand, it is the most crucial part of any analysis - if calibration is incorrect, the analysis results are worthless. Automation of the standard-preparation process should improve cost effectiveness, moral, and reliability of results.

System Description

A schematic of the system is presented in Figure 5. The heart of the system is a single-sample GC injector with its controller. The rack on the right of the injector contains pure chemicals or their concentrated solutions for use in preparing standards. The pure chemicals can be injected into a bottle that has been prefilled with solvent or into a T-tube that is plugged with a septum at one end. The bottle is placed in a holder that can swing in and out under the GC injector needle. The swing motion is controlled via a pressure-activated cylinder. The side arm of the T-tube is attached to a solvent bottle containing the diluent. The remaining end with the septum serves as a dispensing nozzle for introducing diluted standard into a sample receptacle or, if

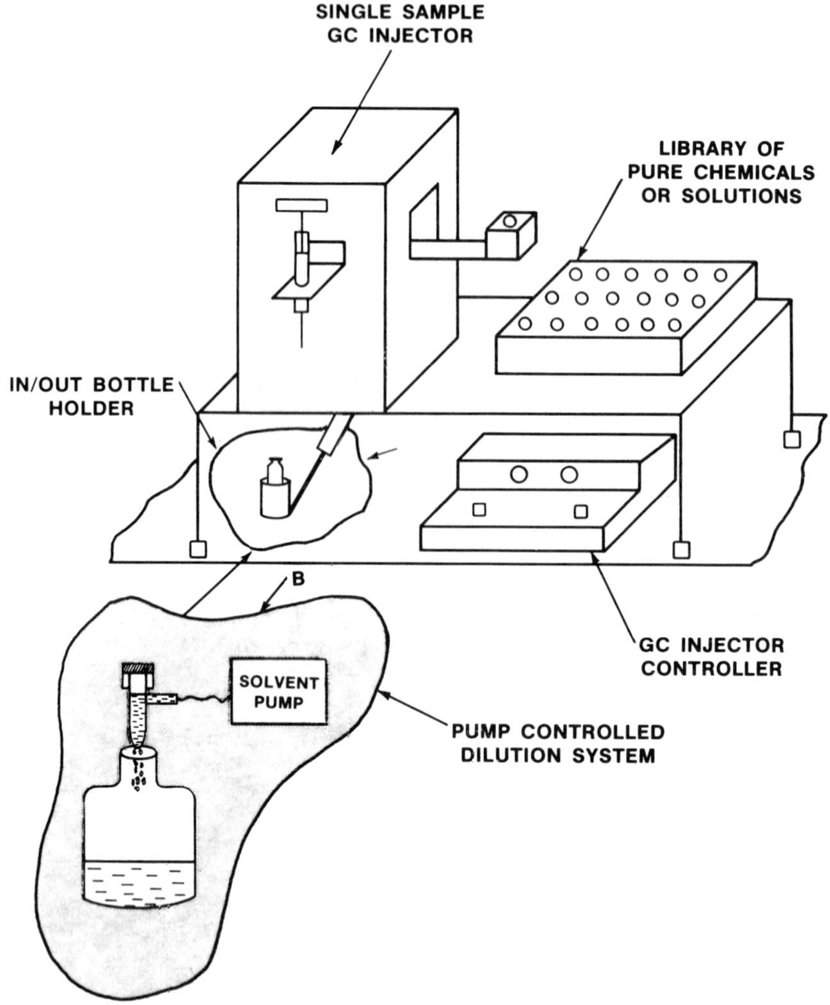

FIGURE 5. Standard preparation and sample spiking systems.

the standard or diluent is volatile and there is concern about their loss, into a large syringe.

Automation of Standard Preparation

The robot carries a bottle (~20 mL) prefilled with solvent to the liquid dosing system (described in Figure 4). Depending upon requirements, the bottle can be capped and crimped (as described in the headspace section of this paper). The robot places the bottle into the in/out bottle holder (see Figure 5A) and presses the switch to swing the bottle under the injector needle. The robotic arm then moves to the chemical rack and picks up a vial containing a standard compound. The vial is placed into a receptacle on the right side of the GC sampler. The robot then pressed the inject button on the GC sampler controller. The sampler flushes the dosing syringe with the chemical and automatically injects an aliquot (1-10 uL) into the bottle.

The robot then changes the vial for another one or repeats the injection if a larger concentration of the chemical is desired. This process continues until all the chemicals specified for the standard are injected. When it is finished, the robot presses the bottle-holder actuating switch, causing the bottle to swing away from the injector needle, and picks up the ready-to-use standard. The process can be repeated with a new solvent-containing bottle.

The same methodology can be used for spiking samples with internal standards. The system is also useful for standard addition methods especially when several components of the sample are to be quantified.

OTHER USEFUL GENERIC LABORATORY PROCEDURES

Washing Glassware

The robot picks up a sample bottle from the bottle rack. It inverts the sample bottle over the waste bottle to dispose of the bulk of the sample, and brings the sample bottle to the cleaning station.

The inverted bottle is placed over the cleaning-station nozzle. While the bottle is moved up and down over the nozzle, a mist of air and washing liquid is dispensed. After several washings the liquid supply is interrupted, and the continued stream of air drys the bottle.

Preparing Binary Mixtures of Variable Composition

The robot picks up a vial, and brings it to a liquid-dispensing nozzle where two liquids are dispensed simultaneously in preprogrammed proporations via computer-controlled syringes (master laboratory station).

The concentration of the components is varied by setting the dispensing volume of the syringes in such a way that the total dispensing volume remains constant. When this function is programmed into a loop with the desired concentration increment, a full range of concentrations is obtained.

CONCLUSIONS

All analytical procedures, glassware and instrumentation are designed for use by humans who have a total of over one hundred degrees of freedom in their joint movements. Even if we fully understood how a human performs a task, the cost of duplicating these motions in a single robot would be prohibitive. In order to transfer any procedure to a robot which has only 4-1/2 degrees of freedom, a combination of two approaches must be used:

1. Construction of automated sample processing station inside of the robot's work area.
2. Modification of the procedures, glassware and instrumentation to fit the robot's requirements.

It is relatively simple to construct a robotic system that performs a repetitive operation (e.g., industrial robots on assembly lines or laboratory robot in a QA/QC lab analyzing one type of sample). However, the sample load in a typical industrial laboratory rarely involves the analysis of hundreds of samples of the same nature. Rather, it consists of requests for 10-20 different analyses on quite different samples submitted in batches from 1 to 50. Automation in this environment call for:

1. Identifying tests which are performed relatively frequently and appear to be automatable.
2. Grouping tests with similar sample processing requirements (dilution, digestion, dissolution, cleaning, measuring techniques, etc.).
3. Constructing flexible automated sample processing stations.

4. Optimizing the position of work stations within robotic handreach in a non-conflicting fashion.

In our opinion, an efficiently run laboratory should consist of a combination of robotics, computers, direct automation and human common sense.

ACKNOWLEDGMENT

The authors would like to thank Milan Radjenovich for his invaluable help in electronic designs and electrical hardware maintentance. The authors also appreciate the contributions from L. Pagliano, M.A. Cinq-Mars, S. Ivy, and A. Rynaski in implementing specific analytical procedures. This work would never have been accomplished without support and encouragement from Analytical Sciences Laboratory management (L.E. Wolfram and J.G. Grasselli).

REFERENCES

1. Kolb, B., Applied Headspace Gas Chromatography, Heyden and Son, Inc., Philadelphia, PA, 1980.

2. Markelov, M., Mendel, D., Talanker, L.E., Method for Determination of a Wide Range of Organic Compounds at ppb Levels in Industrial Waters, Pittsburgh Conference Abstracts, Paper 206, (1983).

3. Markelov, M., Devolatilization of Polymers, Biesenberger, T., ed., Hanser Verlag, 1984.

4. Kolb, B., Multiple Headspace Extraction, Chromatographic, 15 (9), (1982).

DUAL FUNCTION ROBOTICS SYSTEM: AUTOSAMPLER FOR THERMAL ANALYSIS AND APPLICATIONS IN CORROSION STUDIES

Frank M. Prozonic
Air Products and Chemicals, Inc.
Allentown, PA 18105

ABSTRACT

A single Zymark Laboratory Automation System has been installed to perform two distinct applications on an alternating basis. In one of these applications, the robot functions as an autosampler for the determinations of polymer glass transition temperatures by Differential Scanning Calorimetry (DSC). The Zymate controller is interfaced to a DuPont 1090 Thermal Analyzer via an IBM-PC, which serves as a command interpreter. Productivity is further improved by use of the DuPont 912 Dual Sample DSC module, which analyzes two samples simultaneously.

The alternate application is involved with the determination of corrosion rates in process hardware at various plant sites. Here, the robot is used to weigh metal coupons before and after exposure to the process environment. The weight data is then transferred to an external computer where corrosion rate calculations are performed. In this application, errors are reduced by relieving operator tedium in the weighing step and by eliminating manual transcription of numerical data.

INTRODUCTION

Like many other companies in today's economic environment, Air Products has recognized the benefits of automation to improve productivity. In the Corporate Analytical Department, automation has typically implied the use of autosampling devices and computerized data reduction for improving the productivity of instrumentation. When this department decided to extend its use of automation into the area of robotics, possible applications were screened on the basis of two major criteria: high volume of usage and high probability of success. The latter of these implied that the applications be relatively simple, or ones for which previous experience existed. Within these criteria, a single system was designed to function in two distinct applications:

1) as an autosampler for a Differential Scanning Calorimeter (DSC)
2) to weigh metal coupons used in corrosion testing.

Autosampler for DSC

In supporting the research efforts of the company, one of the tasks of the Corporate Analytical Department is the analysis of samples which are submitted by the various R&D centers. From those groups involved in the synthesis of new polymers, one of the most frequently requested analyses is the determination of the glass transition temperature, or Tg. This is a key parameter used in characterizing the behavior of a polymeric material in its end use.

The glass transition of a polymer is manifested by changes in a number of physical properties of the material; e.g., heat capacity, specific volume, and elastic modulus. Each of these properties is determined by a specific analytical technique; e.g., Differential Scanning Calorimetry (DSC), Thermomechanical Analysis (TMA), Dynamic Mechanical Analysis

(DMA), respectively. Therefore, several options are available for the determination of Tg.

Because of its simplicity, one of the most widely used methods for Tg determination is DSC.[1] Until very recently, however, no manufacturer of Thermal Analysis insturmentation offered commercial equipment for automated loading and unloading of samples in routine DSC procedures. It was the objective of this program to design and implement a robotics system which would serve as an autosampler for the routine determination of polymer Tg by a DSC procedure.

Corrosion Studies

In the operation of full scale chemical plants, one of the concerns of the plant manager is the long term integrity of the process hardware (piping, reaction vessels, heat exchangers, etc.) in the presence of corrosive chemicals. To predict the usable lifetime of this hardware, corrosion rates are commonly determined at the critical points in the process stream. This testing is generally done in accordance with standard ASTM test methods.[2,3,4]

In ASTM D 2688-83,[2] for example, the corrosion rate in a cooling tower is determined from the weight loss observed for a metal coupon, after it has been in contact with the flowing cooling water stream for an extended period of time. The procedure involves accurate weight determinations on the coupon before and after exposure to the corrosive environment.

When a large number of plants and sampling points are included in an ongoing testing program, the number of accurate weighings required can be very large. A robotic system was designed to automate the weighing operation which is an integral part of the corrosion testing program. Automation not only frees an operator from the tedium of repetitive weighing of samples, but also eliminates human transcription errors by computerizing the data transfer.

Both the corrosion testing and DSC applications were designed into a single robotics system, in order to utilize the automation to its fullest extent.

SYSTEM DESCRIPTION

Overview

As was indicated previously, the overall robotics system was designed to perform two unrelated functions on an alternating basis. The hardware for each function occupies about half of a 3' x 6' counter top. It will simplify the description of the system to consider each half separately in subsequent sections. On the left side of Figure 1, the DSC module and sample racks are shown. Completed samples are dropped into a container for disposal. The right side of Figure 1 shows the slotted racks for coupons to be weightd, the Mettler balance, and the receiving spindle for finished coupons.

In each application, the Zymate controller serves as the time sequencer for the overall operation. The overall electrical diagram is shown in Figure 2. For the DSC application, the Zymate controller not only

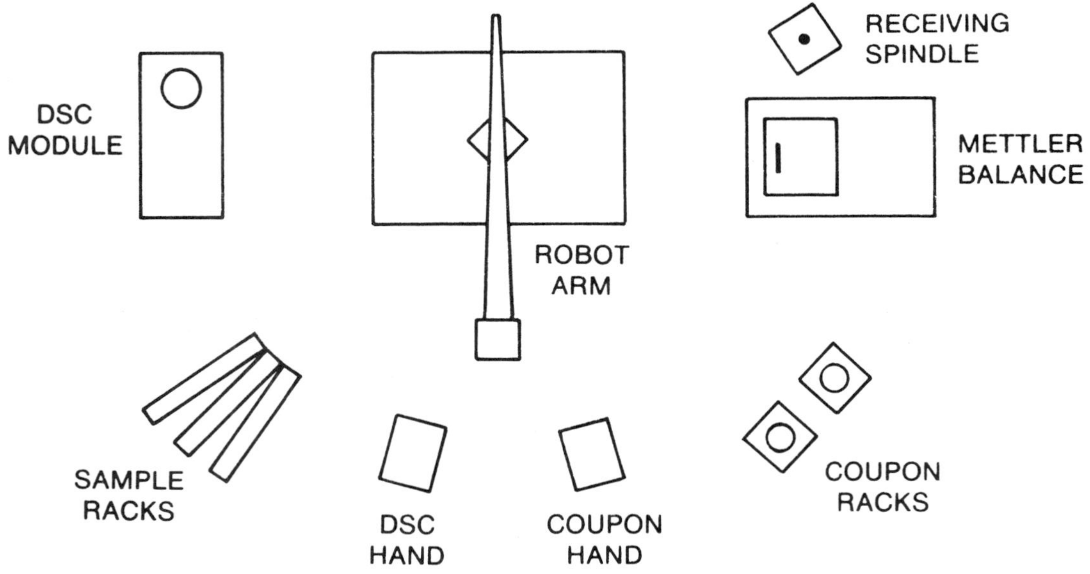

FIGURE 1. Zymate System - mechanical schematic diagram for DSC and corrosion studies.

instructs the robotic arm for movement of the samples, but also sends commands to the Thermal Analyzer, via the IBM-PC, for the selection of temperature program methods and storage of the raw data. In the corrosion application, the Zymate controller permits movement of the metal coupons to and from the balance and sends the weight data to a printer for the final report. A detailed description of each application is given in the sections that follow.

Autosampler for DSC

Using the DSC technique, the glass transitions region for a polymer is exhibited by a step change in the specific heat vs. temperature curve. Generally, the sample is conditioned at a temperature above Tg to remove residual moisture or solvent and eliminate the thermal history. After the sample is quench cooled to a temperature below Tg, heat flow data is taken while the sample is heated at a specified rate. Specifically, the steps involved in the procedure include sample preparation, sample loading into the DSC, conditioning, cooling, temperature programming and data acquistion, sample removal, and data analysis. All but the first and last of these steps have been automated with the use of the robotics system.

DSC Hardware

To perform the procedure outlined above, the following instrumentation was assmebled:

> Zymate Laboratory Automation System: The Zymate controller directs the robotic arm for removal of lids on the DSC and movement of sample pans in and out of the DSC cell. It also functions as the timekeeper for the overall sequence of events in the procedure - i.e., after loading samples into the DSC, commands are sent to the Thermal Analyzer to activate one

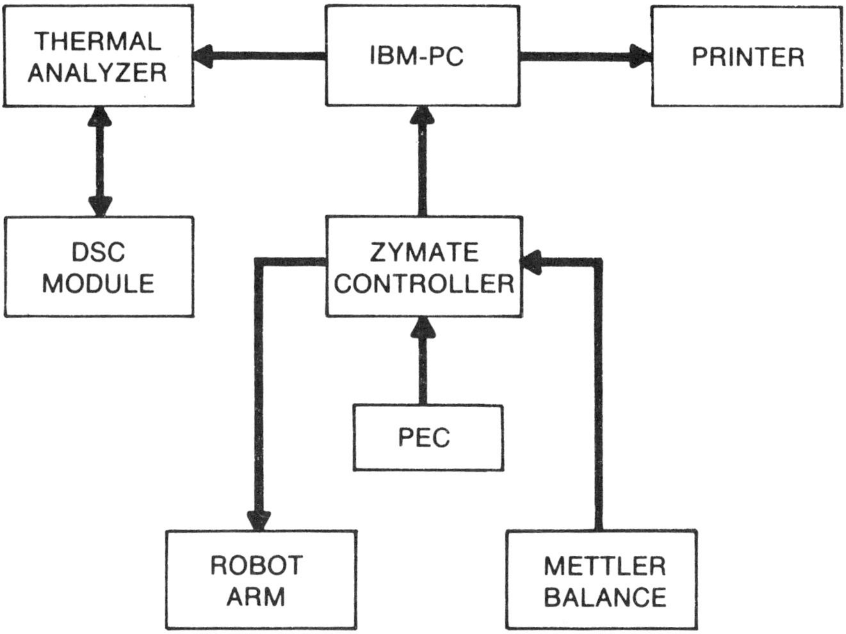

FIGURE 2. Zymate System - electrical schematic diagram for DSC and corrosion studies.

of several temperature programs.

DuPont 1090 Thermal Analysis System: The 1090 System controls the operation of the DSC module and acquires the heat flow data for storage on its integral 8" floppy disks. Data analysis is performed with standard DuPont DSC software. Communication with the 1090 System is accomplished through DuPont's RS-232 option.

A DuPont module 912 Dual Sample DSC is used in this system. This cell has the capability to analyzed two samples simultaneously for a further increase in productivity.

DuPont Mechanical Cooling Asscessory (MCA): For this application, the DuPont MCA is operated in the temperature range of -70 to +100 C, which is adequate for most of the polymers analyzed in our laboratory. With the MCA, access to the DSC cell requires the removal of two small lids. The use of liquid nitrogen as a coolant offers a wider temperature range but has the disadvantage of requiring the movement of relatively large bell jar and cooling vessel, in addition to regulating the flow of coolant.

IBM-Personal Computer: Because the 1090 System is not able to accept commands directly from the Zymate controller, an IBM-PC was interposed as a translator. Custom software was written to enable the IBM-PC to interpret the Zymate System commands before transmitting them to the Thermal Analyzer.

Sample Preparation

In our laboratory, emulsion polymers (polymer latexes) are prepared for DSC analysis by casting a thin film from the latex and allowing the film to air dry. Residual moisture is removed by further drying in an oven at 100°C. The film is then placed in a standard aluminum pan which is closed by crimping. At this point, the prepared samples are placed into the sample racks.

Operation of the Robotic System

When the sample racks are filled, the robot arm is activated to start its autosampler sequence. An overall flowchart of the operation sequence is given in Figure 3.

to perform the various manipulations which are required, a dual function robot hand was designed at Zymark. The fingers of this hand are used to remove the two small lids from the MCA and the DSC cell. When the wrist is rotated by 180 , access is gained to a vacuum head, which is used to pick up and move the small DSC pans. The vacuum head is machined to precisely match the inside configuration of the DSC cell.

As shown in Figure 3, the sequence begins with removal of the two lids which provide access to the DSC cell. The vacuum head then picks up a pair of sample pans and deposits them into the DSC cell. Rotating the hand by 180 again allows the lids to be replaced. Throughout this procedure, microswitches are used to verify proper movement of the samples. If a pan is not in its intended position, the run is terminated, and an error message is displayed on the Zymate System screen.

When the samples are in place in the DSC cell, the Zymate controller signals the Thermal Analyzer to perform the appropriate temperature programs. To do this, a numeric code is sent to the IBM-PC. This code activates a series of ASCII commands which are sent to the Thermal Analyzer via its RS-232 port. A typical command sequence is: turn data storage 'on', activate temperature program #1, print a status message on the plasma display.

When the temperature programs have been completed, the robot arm is instructed to remove the sample pans from the DSC cell, dispose of

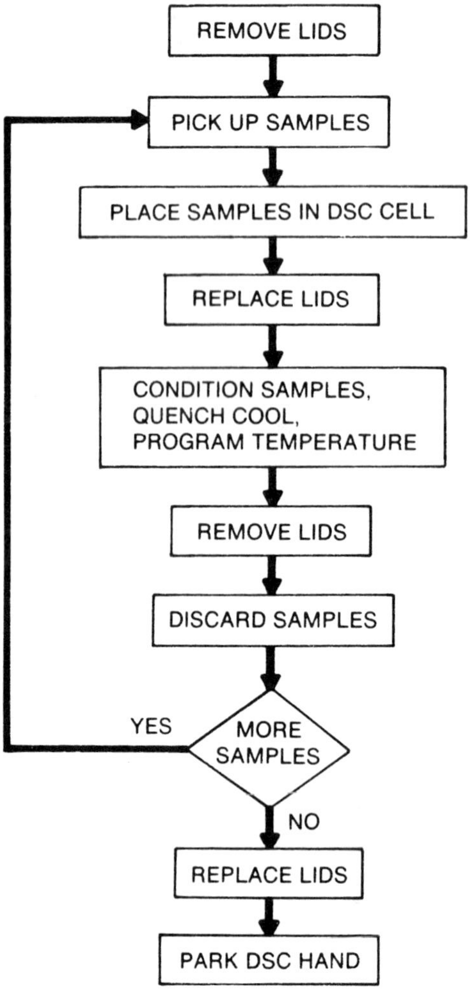

FIGURE 3. Program flowchart for DSC application.

the pans, and load the next pair of samples into the DSC for analysis.

Heat flow data is automatically taken from the DSC cell during the test and stored on a floppy disk. Data analysis is performed at the operator's convenience, using standard DuPont software. An example of the DSC curve for determination of polymer Tg is given in Figure 4.

Corrosion Sutdies

As explained in the introduction, the robotic system is used in corrosion studies in our laboratories to minimize the possibility for operator error in performing a large number of routine weighings. Through the use of electronic data acquistion, manual transcription errors are eliminated.

Coupon Weighing Hardware

In addition to the Zymate controller and robot arm, a Mettler AE-163 electronic analytical balance is used in the coupon weighing procedure. Communication between the balance and the controller is accomplished with the balance interface option. A standard robot gripping hand is used for manipulating the coupons and for opening and closing the balance door. The fingertips of the hand are made of a polymeric material to prevent scratching of the coupons, and they are slotted to accomodatae the coupon thickness.

Prior to weighing, the metal coupons are placed into two slotted

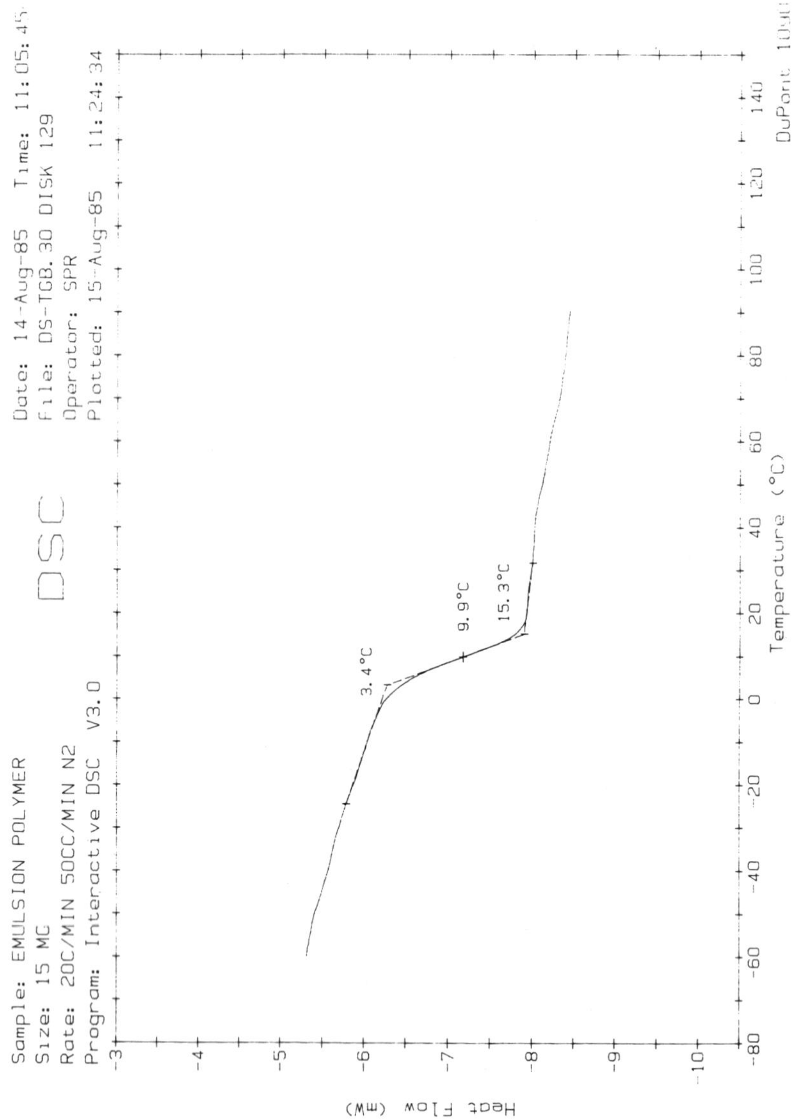

FIGURE 4. Typical DSC curve for determination of polymer T_g.

racks, which stand vertically for ease of access. A raised platform is attached to the balance pan to facilitate access to the balance through the top door. After weighing, the coupons are stacked onto a spindle in order to maintain the proper sequencing of the samples. The spindle is made in two pieces so that the weighed coupons can be removed without affecting the spindle alignment on the counter top.

Operation of the Coupon Weighing System

The robotic procedure for weighing corrosion coupons is relatively simple: pick up a coupon, put it on the balance, weigh it, remove it from the balance, and stack it on the receiving spindle. During the procedure, proper movement of the balance door is verified throuth the use of microswitches, which are connected to the power and event controller of the Zymate System.

The weight data which is obtained is stored in an array in the Zymate controller. Individual coupons are identified by matching the array index number to a code which is stamped into the coupon. At the completion of a series of coupons, a hardcopy of the indexed weight data is obtained from the system printer.

CONCLUSIONS

By judicious choice of initial applications for robotics in the Corporate Analytical Department, the time required for implementation of the above system was minimized. About two months elapsed from the date of installation of the completion of the entire system. Since that time, significant improvements in productivity have been realized. The turnaround time for polymer samples requiring Tg determination has been shortened substantially with a concurrent reduction in the cost per analysis. Statistical testing of a cross section of emulsion polymer samples has demonstrated that consistency of Tg results has been maintained in transferring to an automated procedure. In the corrosion studies application, the goals of eliminating operator tedium and data transcription errors have been attained.

Improvements to the present system have also been planned, as part of an ongoing effort in automation. The use of liquid nitrogen as a coolant in the automated DSC procedure is under consideration; this approach would increase the useful temperature range of the method and extend its applicability to a larger variety of sample. A magazine-type sample rack has been designed for the coupon weighing application, in order to accomodate more samples per batch, and simplify the operation of the system. Further, software will be written for the IBM-PC so that the final corrosion rate calculations can be done without additional data transfer.

Finally, the Corporate Analytical Department at Air Products is evaluating the use of robotics as a sample preparation tool in several of its laboratories.

ACKNOWLEDGEMENTS

The author gratefully acknowledges the efforts of the Zymark Corporation staff, especially M. Nallen, in the design and implementation of this system. At Air Products, special thanks are due to W. D. Hanagan for writing all the software for the IBM-PC, and to S. P. Ritter for her expert assistance in the laboratory.

REFERENCES

1. ASTM D 3418-75, "Standard Test Method for Transition Temperatures of Polymers by Thermal Analysis".

2. ASTM D 2688-83, "Standard Test Methods for Corrosivity of Water in the Absence of Heat Transfer (Weight Loss Methods)".

3. ASTM G1-81, "Standard Practice for Preparing, Cleaning, and Evaluating Corrosion Test Specimens".

4. ASTM G31-72, "Standard Recommended Practice for Laboratory Immersion Corrosion Testing of Metals".

AUTOMATED SAMPLE PREPARATION PROCEDURES FOR LIQUID AND GAS CHROMATOGRAPHIC ANALYSIS OF POLYMERIC MATERIALS

Kenneth A. Klinger
Borg-Warner Chemicals, Inc.
Technical Centre
Washington, West Virginia 26181

ABSTRACT

The first automated sample preparation procedure involves the ambient temeprature dissolution of a polymeric material in an appropriate organic solvent at a concentration of two-tenths of one percent. Filtered solutions at this concentration are used to determine the molecular weight(s) and polydispersity of a polymer sample by an independent automated analytical measurement technique known as Size Exclusion Chromatography. Knowledge of the molecular weight distribution of a polymer is essential for predicting the physical, processing, and end-use properties of any subsequent plastic product.

The second automated sample preparation procedure involves the dispersion or dissolution of a polymeric material in a suitable organic solvent at a concentration of five percent. Mixtures at this concentration (which include an appropriate internal standard) are used to measure the unreacted monomers or other non-aqueous volatiles by an independent automated analytical measurement technique, gas chromatography. Information concerning the concentration of these volatile organic components is useful in identifying end-use applications for any finished plastic product.

Collectively, these two automated sample preparation procedures illustrate a variety of robot-compatible operations including electronic weighing, liquid diluting, syringe hand and remote dispenser pipeting, crimp and screw top capping, linear shaker conditioning, confirming, vertical tube cap racking, and disposable assembly membrane filtering. Data is presented comparing the precision and accuracy of these robotic sample preparation procedures to the previously employed manual methods.

INTRODUCTION

Future generations may well refer to the time in which we live as the age of plastics or polymers. Perhaps no other class of man-made materials has had a more pronounced impact on today's society.

The properties and uses of plastics are influenced by a number of polymer parameters including:

- Chemical Composition
- Molecular Weight Distribution
- Stereochemistry
- Topology
- Morphology
- Additives

The molecular weight distribution (MWD) is perhaps the most fundamental characteristic, since it directly affects solution, melt flow, and final mechanical properties. Although several techniques exist for evaluating the MWD of any soluble polymeric material, probably the most versatile and widely used method is a liquid chromatographic procedure known as Size Exclusion Chromatography (SEC).

Among the additives effecting both the processing and the end-use performance of a plastic product are the non-aqueous volatiles (NAV), generally comprised of residual monomers and other non-polymerizable volatiles. The presence and concentration of these smaller molecular weight species can be determined by another instrumental analytical technique known as Gas Chromatography (GC).

Fully automated state-of-the-art SEC and GC instrumentation is available from a number of manufacturers, permitting unattended analysis if the MWD and NAV in many plastic materials. However, before any polymeric sample can be routinely analyzed by these modern instrumental techniques, preliminary sample preparations including weighing, liquid handling, manipulating, conditioning, and filtering are required. Collectively, these multiple step procedures are typically manpower intensive and often time consuming.

In a continuing effort to automate as many routine testing procedures as possible, the Analytical Section at the Borg-Warner Chemicals (BWC) Technical Centre acquired the Zymate Laboratory Automation System during the third quarter of 1984. This flexible automation system integrates robotics with typical laboratory scale apparatus and allows the analyst to orchestrate the movement of a robot arm to the laboartory stations and other devices located within the cylinder of space it services. Our initial goal was to use this capability to automate the SEC and GC sample preparation routines. It was fully anticipated that other manpower intensive sample preparation and analysis procedures could also be developed, thus providing additional opportunities to increase productivity in our testing laboratories.

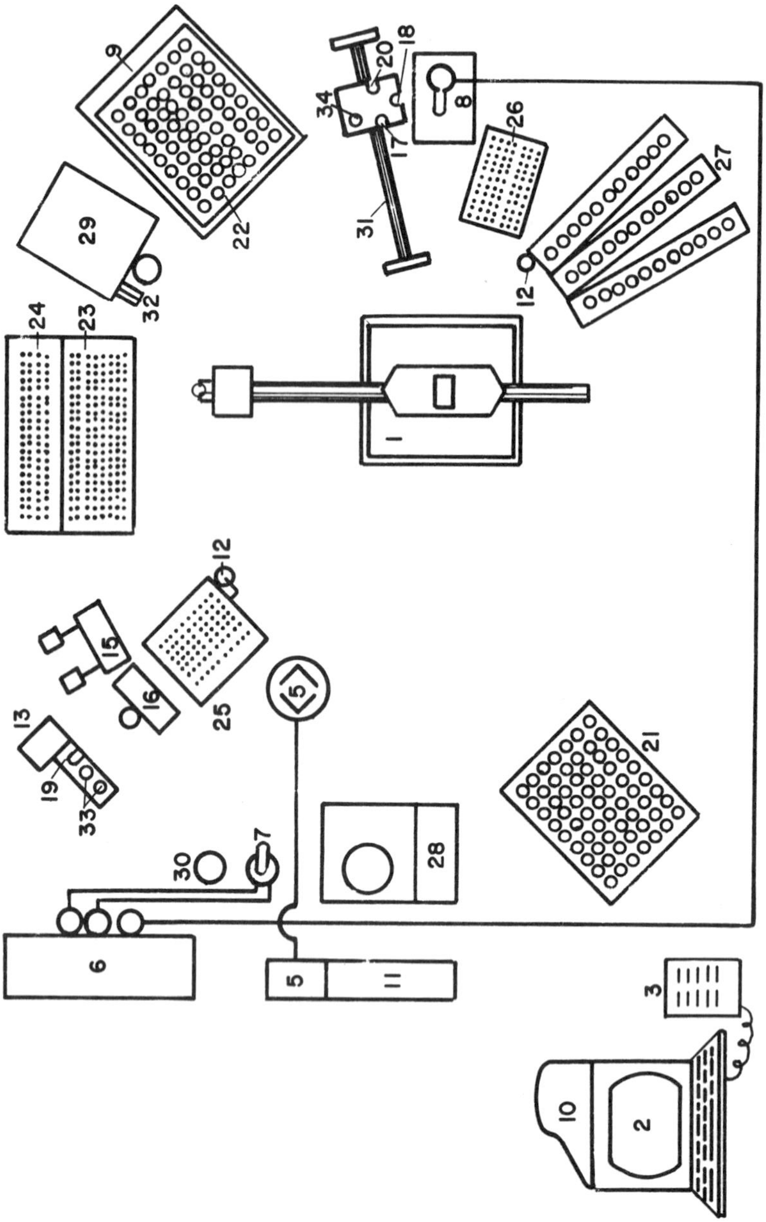

FIGURE 1. Schematic of bench layout for size-exclusion chromatography and gas chromatography.

TABLE 1. List of Equipment for GC and LC.

Zymark Corporation (Hopkinton, MA)
1. Z110 Zymate Robot
2. Z120 Zymate Controller
3. Z121 Remote Robot Teaching Module
4. Z150 Zymate Bench Top
5. Z410 Capping Station (with small gripper fingers)
6. Z510 Master Laboratory Station (with three 10 mL syringes)
7. Remote Fixed Four Tube Solvent Dispenser
8. Remote Single Tube Cannula (for 5 mL pipet tips and 10 mL filter assemblys)
9. Z611 Sample Conditioning (Linear) Station
10. Z820 Printer (with support stand)
11. Z830 Power and Event Controller
12. Electronic Verification Switch(s)
13. Optical Verification Sensor
14. Z850 Balance Interface
15. Z900HT General Purpose Hand (with standard fingers)
16. Z910 Precision Microliter Syringe Hand (with 1 mL syringe)
17. Filter Assembly Fork Holder
18. Filter Assembly Shucking Plate
19. Pipet Tip (1 mL) Shucking Plate
20. Pipet Tip (5 mL) Shucking Plate
21. Sample Vial Rack (8 x 12 inch, 24 position)
22. Shaker Vial Rack (8 x 12 inch, 24 position)
23. SEC Autosampler Vial Rack (6 x 12 inch, 50 position)
24. GC Autosampler Vial Rack (6 x 12 inch, 60 position)
25. Pipet 1 mL Tip Rack (3.5 x 6.5 inch, 50 position)
26. Pipet 5 mL Tip Rack (4 x 8 inch, 50 position)
27. Filter Assembly Rack (three 2.5 x 12 inch, 10 position each)

Mettler Instruments Corporation (Hightstown, NJ)
28. AE163 Electronic Balance (with CL data interface)

Wheaton Instruments (Millville, NJ)
29. Crimpmaster Cat. No. 224550 (with 11 mm seal crimper)

Borg-Warner Chemicals, Inc. (Washington, WV)
30. Vertical Tube Cap Rack
31. Moveable SEC Autosampler Vial Holder
32. GC Autosampler Vial Parking Platform/Cap Holder
33. Sample Vial and SEC Autosampler Vial Cap Holder(s)
34. Conditioned Sample Vial Holder (for SEC filtering)

Miscellaneous Disposable Accessories
1 mL (Oxford) Pipet Tips
5 mL (Oxford) Pipet Tips
1 mL (Wheaton Scientific) Crimp Top GC Autosampler Vials
5 mL (DuPont Co.) Screw Top SEC Autosampler Vials
10 mL (Philip Fishman Corp.) Polypropylene Syringe Barrels
0.45 micron (Gelman Sciences) PTFE Memabrane Filter Assemblys
20 mL (Kimbel Glass co.) Screw Top Scintillation Vials

FIGURE 2. Assembled bench layout for size-exclusion chromatography and gas chromatography.

INSTRUMENTATION

The manual SEC and GC sample preparation procedures were automated using a Zymate System with a bench layout illustrated in Figure 1 with the equipment listed in Table 1 and shown in Figure 2.

A unique vertical tube rack (designed and fabricated at BWC) for the 20 mL sample vial screw top caps is illustrated in Figure 3. The needle valve at the base of the rack controls the gas flow used to raise the column of stacked caps upward after each cap is removed at the slot located near the top of the tube.

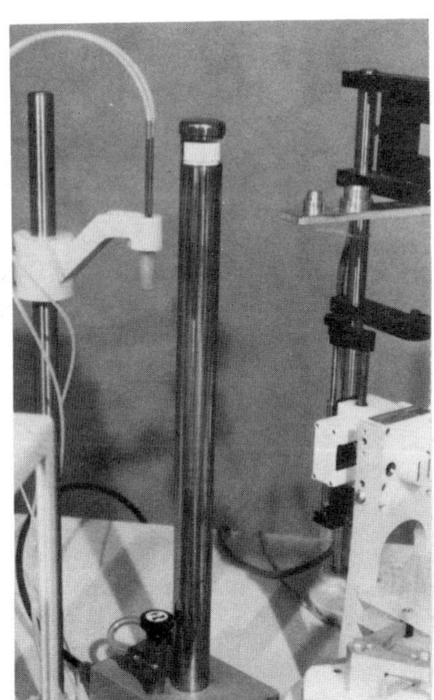

FIGURE 3. Vertical tube cap rack.

The air-powered Wheaton Instruments (Millville, NJ) 'Crimpmaster' fitted with a seal crimper (11 mm) is depicted in Figure 4. By incorporating an electronic relay and a three-way solenoid valve (per Zymark instructions) this crimping station was modified to operate through the 5 volt supply on the Z830 power and event controller (PEC). A spring-loaded mechanical cap holder was fitted with a GC autosampler vial parking platform and mounted on the left front side of this crimp capping station.

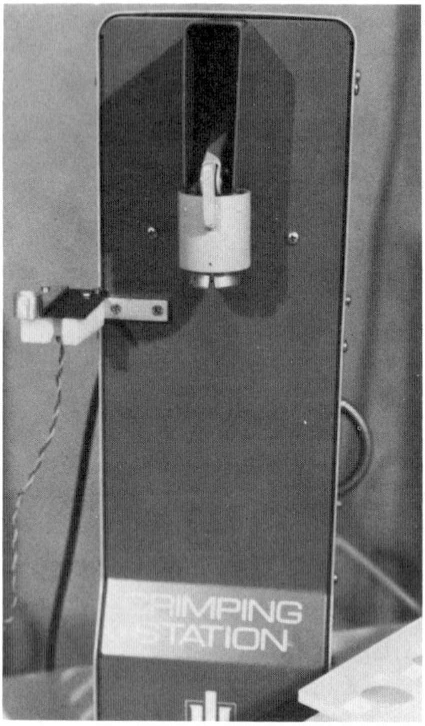

FIGURE 4. GC vial crimp capping station.

The moveable SEC autosampler vial holder (also designed and fabricated at BWC) used for locating an open, empty autosampler vial underneath a disposable filter assembly is shwon in Figure 5. After filtering is completed the air-actuated piston (operated through the 12 volt supply on the Z830 PEC) is used to push the filled autosampler vial out from underneath the filter fork for easier access by the general purpose hand.

PROCEDURES

For both the SEC and GC sample preparation procedures, the basic operations performed include:

(1) Automated batch taring of sample vials

(2) Manual sample addition to each of the tared vials

(3) Automated batch dilution of the weighed samples to a known concentration followed by sample conditioning

(4) Automated batch transferring of conditioned sample solution (by filtering or pipeting) into appropriate autosampler vials

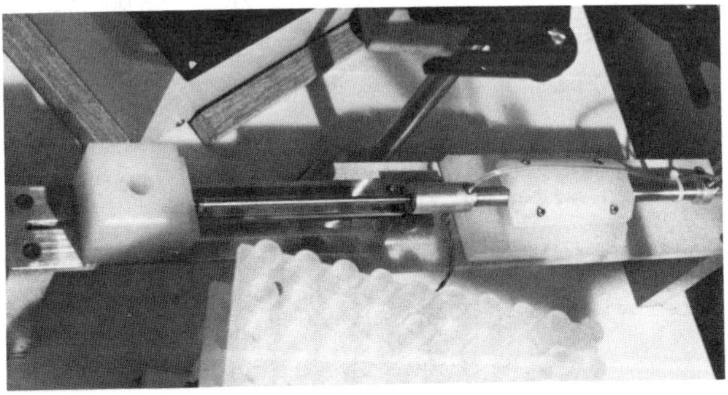

FIGURE 5. SEC filtration station.

Due to the many and varied sample geometries (including liquid latex despersions, powdered resins, compounded pellets, thin films, and sections cut from test specimens or from finished molded applications) which are routinely encountered, samples are added manually to the pre-tared vials by the analyst involved.

SEC Sample Preparation

After initializing the Zymate System by attaching the general purpose hand, identifying variable names, and creating arrays for data storage; the robot arm begins an automated batch weighing process for a specified number of uncapped empty vials to be used during sample makeup. The sequence of steps for taring each empty vial are given in Table 2.

TABLE 2. Taring Sequence for Empty SEC Vials.

Get the uncapped vial from its rack.
Move the vial to the Mettler AE163 balance station.
Read the balance zero.
Place the vial onto the balance.
Read the balance weight and correct it for the zero reading.
Store the empty vial tare weight in the computer.
Return the tared vial to its rack.
Repeat this cycle for each of the empty vials to be used.

The taring sequence typically requires one minute per each empty vial. After the specified number of vials is tared, the general purpose hand is parked, and the robot arm pauses while the analyst manually loads each open tared vial with soluble polymer sample. After all sample addition is completed, the Zymate System continues the automated SEC sample preparation by purging solvent from the Z510 master laboratory station (MLS), re-attaching the general purpose hand, testing the

TABLE 3. Sample Make-up Sequence for SEC.

Get the uncapped sample vial from its rack.
Move it back to the Mettler AE163 balance station.
Read the balance zero.
Place the open sample vial onto the balance.
Read the sample vial weight and correct it for the zero reading.
Calculate the sample weight by subtracting the previously determined empty vial tare weight from the sample vial weight.
Move the weighed sample vial to the fixed four tube dispenser.
Calculate the volume of organic solvent needed for dilutiong this sample weight to a concentration of 0.2 (weight/volume) percent.
Deliver the calculated volume of solvent form the Z510 MLS into the sample vial now located under the fixed four tube dispenser.
Place the diluted sample vial into the Z410 capping station.
Get the next cap from the vertical tube cap rack.
Confirm cap pickup on an electronic verification switch.
Cap the made-up sample vial by moving the robot arm down as the capping station turns counterclockwise.
Move the capped sample vial to the Z611 shaker station.
Start the linear shaker conditioning station.
Return to repeat this dilution cycle on the next sample vial.

operation of the Z611 linear shaker and Z410 capper modules, checking the status of the optical verification sensor, and operating the moveable SEC autosampler vial holder at the filtering station. The sequence for continuing the automated batch sample makeup process is given in Table 3.

The sample make-up sequence generally requires 3.5 minutes per each sample vial. While all the made-up samples are being conditioned on the Z611 shaker, a report giving sample numbers, empty vial tare weights, sample weights, and sample volumes is output on the Z820 printer.

Table 4 gives the final sequence of steps for completing the automated batch filtering of each of the conditioned sample solutions into the SEC autosampler vials.

TABLE 4. Filtration Sequence for SEC.

Get an empty capped SEC autosampler vial from its rack.
Uncap the autosampler vial in the Z410 capping station.
Confirm the uncapping operation in the optical sensor detector.
Park the autosampler vial cap on its nearby cap holder.
Place the uncapped vial in its moveable SEC autosampler vial holder at the filtering station.
Use this air-actuated piston-operated holder to reposition the open empty autosampler vial under the filter assembly fork holder.
Get the conditioned sample vial from the Z611 linear shaker.
Test for sample vial pickup on a verification switch.
Uncap the conditioned sample vial in the Z410 capping station.
Confirm the uncapping operation in the optical sensor.
Park the sample vial cap on yet another nearby cap holder.
Move the uncapped conditioned sample vial to its stationary holder at the filtering station.
Get the remote single tube cannula dispenser which is connected to an air syringe at the Z510 MLS.
Use this remote dispenser to get a filter assembly from its rack.
Test for filter assembly alignment on a verification switch.
Postion the filter assembly in its fork holder.
Use the remote dispenser to get a 5 mL pipet tip from its rack.
Test for pipet tip alignment on a verification switch.
Aspirate 4 mL of conditioned sample solution into the 5 mL pipet tip using the attached air syringe at the Z510 MLS.
Dispense the 4 mL of the sample solution into the filter assembly.
Remove the used pipet tip on its nearby shucking plate.
Return to force conditioned solution through the filter assembly using a fast 4 mL air displacement of the MLS syringe connected to the remote single tube cannula dispenser.
Continue the filtration with an additional slower 4 mL air displacement from the same MLS syringe.
Use the remote cannula dispenser to remove the used filter assembly from its fork holder and discard it on a shucking plate.
Return the remote cannula dispenser to its parking station.
Use the moveable holder to push the filled SEC autosampler vial out from underneath the filter fork holder.
Place the filled autosampler vial into the Z410 capping station.
Retrieve the parked autosampler vial cap from its holder and confirm pickup in the optical sensor detector.
Recap the filled autosampler vial.
Move the filled SEC autosampler vial back to its rack.
Get the sample vial from its holder at the filtering station.
Check for vial pickup on a verification switch.
Place the open sample vial into the Z410 capping station.
Retrieve the parked sample vial cap from its holder and confirm pickup in the optical sensor detector.
Recap the sample vial.
Move the recapped sample vial back to its stationary rack.
Return to repeat this cycle on the next conditioned sample.

The sample filtering sequence routinely requires 10.5 minutes per sample. After all the conditioned solutions are filtered into autosampler vials, the automated SEC sample preparation is concluded by parking the general purpose hand and moving the robot arm to a rest position.

GC Sample Preparation

After initializing the Zymate System by attaching the general purpose hand, identifying variable names, and creating arrays for data storage, the robot arm begins an automated batch weighing process for a specified number of capped, empty vials to be used during sample makeup. The sequence of steps for taring each empty vial are the same as given for the SEC sample preparation (Table 2).

The taring sequence requires one minute per each empty vial. After the specified number of vials is tared, the general purpose hand is parked, and the robot arm pauses while the analyst manually uncaps, loads, and caps each tared vial with polymeric sample.

After all smaple addition is completed, the Zymate System continues the automated GC sample preparation by purging solvent from the Z510 MLS, re-attaching the general purpose hand, testing the operation of the Z611 linear shaker and Z410 capper modules, and checking the status of both the optical verification sensor and the GC autosampler vial mechanical cap holder at the Wheaton crimping station. The sequence of steps for continuing the automated batch sample makeup are given in Table 5.

TABLE 5. Sample Makeup Sequence for GC.

Get the capped sample vial from its rack.
Move it back to the Mettler AE163 balance station.
Read the balance zero.
Place the capped sample vial onto the balance.
Read the sample vial weight and correct it for the zero reading.
Calculate the sample weight by subtracting the previously determined empty vial tare weight from the sample vial weight.
Place the weighed capped sample vial into the Z410 capper.
Uncap the vial by moving the robot arm up as the capping station turns clockwise.
Confirm the uncapping operation in the optical sensor.
Park the sample vial cap on its nearby cap holder.
Move the weighed uncapped sample vial to the four tube dispenser.
Calculate the volume of organic solvent needed for diluting this sample weight to a concentration of five (weight/volume) percent.
Deliver the calculated volume of solvent from the Z510 MLS into the sample vial now located under the fixed four tube dispenser.
Return the diluted sample vial to the Z410 capping station.
Retrieve the parked sample vial cap and confirm pickup in the optical sensor detector.
Recap the made-up sample vial by moving the robot arm down as the Z410 capping station turns counterclockwise.
Move the recapped made-up sample vial to the Z611 shaker station.
Start the linear shaker conditioning station.
Return to repeat this dilution cycle on the next sample vial.

The sample makeup sequence requires 3.5 minutes per sample vial. While all the made-up samples are being conditioned on the Z611 shaker, a report giving sample numbers, empty vial tare weights, sample weights, and sample volumes is output on the Z820 printer.

Table 6 gives the final sequence of steps for completing the automated batch pipeting of each of the conditioned sample solutions into the GC autosampler vials.

TABLE 6. Pipetting Sequence for GC.

Get an empty loose-capped GC autosampler vial from its rack.
Move the autosampler vial to the parking platform at the Wheaton crimping station and reposition the finger grip to permit removal of the uncrimped cap.
Uncap the autosampler vial by inserting it into the specially designed mechanical cap holder at this crimping station.
Replace the uncapped autosampler vial onto the parking platform.
Get the conditioned sample vial from the Z611 linear shaker.
Test for sample vial pickup on a verification switch.
Uncap the conditioned sample vial in the Z410 capping station.
Confirm the uncapping operation in the optical sensor.
Park the sample vial cap on its nearby cap holder.
Park the general purpose hand.
Attach the syringe hand.
Get a 1 mL pipet tip from its rack with the syringe hand.
Test for pipet tip pickup on a verification switch.
Aspirate 1 mL of conditioned solution from the open sample vial with the syringe hand.
Dispense the 1 mL sample into the open GC autosampler vial.
Remove the used 1 mL pipet tip on its nearby shucking plate.
Park the syringe hand.
Get the filled GC autosampler vial at the parking platform and reposition the fingers on the vial.
Replace the uncrimped cap held in the mechanical cap holder onto the GC autosampler vial.
Seal the cap tightly onto the vial by using the Wheaton crimper.
Reposition the vial in the fingers at the parking platform.
Move the filled, crimped autosampler vial back to its rack.
Retrieve and confirm pickup of the parked sample vial cap.
Recap the sample vial in the Z410 capping station.
Move the recapped sample vial to its stationary rack.
Return to repeat this cycle on the next conditioned sample.

The micropipeting sequence requires 7.5 minutes per sample solution. After all the sample solutions are pipeted into autosampler vials, the automated GC sample preparation is concluded by parking the general purpose hand and moving the robot arm to a rest position.

RESULTS

Statistically designed tests were conducted to evaluate differences in precision between (1) the previously employed manual and (2) the newly developed robotic SEC and GC sample preparation procedures. Since a predetermined concentration is being sought in both of these sample preparation routines, chromatography peak areas were used to quantify precision aspects. Consequently, the precision of each independent analytical measurement technique (SEC for MW and GC for NAV) had to first be established. This was accomplished by loading appropriate autosampler vials (24 for SEC and 18 for GC) with aliquots from an accurately prepared master solution of a well-characterized polymeric sample. The manual and automated sample make-up procedures were next compared by preparing two additional sets (with like numbers of individually madeup samples of the same polymer and the most qualified laboratory analyst versus the Zyamte System.

To evaluate the SEC sample make-up procedure, a typical polystryrene (PS) with a weight average molecular weight (Mw = 250,000) and polydispersity (Mw/Mn = 2.5) was selected. The SEC instrumental operating parameters for analyzing this polymer along with a typical chromatogram showing its MWD are given Figure 6.

The area under the broad polymer peak was integrated and the results are reported in Table 7.

Data for the single master solution showed the analytical SEC measurement technique to have a relative standard deviation or coefficient of variation (C.V.) of 0.7%. Meanwhile, data for the individual analyst prepared samples showed a C.V. of 2.7% and finally, data for the individual robot prepared samples exhibited a C.V. of 1.2%.

TABLE 7. Precision of SEC Sample Preparation Procedures.

Autosampler Vial Number	Single Master Solution	Individual Analyst Prepared	Individual Robot Prepared
1	6268722	6046423	6185122
2	6149889	5736456	6204019
3	6216656	6161406	6174092
4	6217949	5643106	6235229
5	6238351	6002643	6078786
6	6192560	5967837	6181850
7	6248978	6110184	6295481
8	6261249	6019405	6312532
9	6232278	5950841	6239766
10	6245957	6021112	6129080
11	6130128	6147072	6189439
12	6196419	6064400	6120315
13	6167603	5747900	6061738
14	6212808	5964464	6244487
15	6251851	6303961	6236012
16	6153409	6133559	6244590
17	6204324	5899903	6156511
18	6187714	5984285	6156018
19	6154370	5998073	6047415
20	6155822	6068795	6106739
21	6162133	6151056	6011993
22	6169464	5849034	6166541
23	6169115	6357655	6108912
24	6138396	5971181	6147558
Mean	6196922	6012531	6168092
Std Dev	42078	165223	76695
C.V. (%)	0.7	2.7	1.2

The robot procedure shows improved precision over the manual method.

After calibration vs narrow MWD polystryene standards,[1,3] the molecular weight average(s) and polydispersity were calculated for each sample analysis. Next, the mean weight-average molecular weight (Mw) was computed for each of the three SEC sample preparation procedures. The percent relative error (for these mean Mw) ranged from 2 to 3%, suggesting the robot procedure to have accuracy equal to the previously

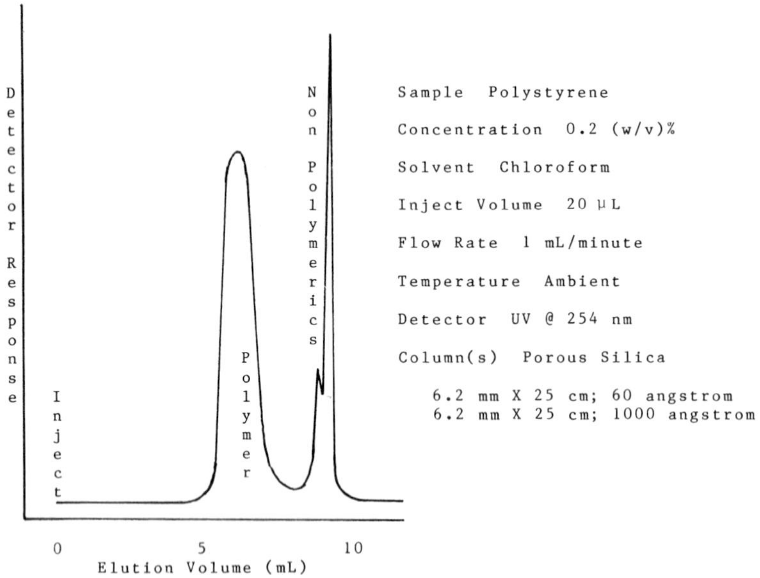

FIGURE 6. SEC evaluation of polystyrene.

used manual method.

To evaluate the GC sample make-up procedure the same PS with a NAV content of 0.65% (40% being styrene monomer) was again used. The GC operating parameters for analyzing this polymer along with a typical chromatogram showing its NAV content are given in Figure 7.

The chromatographic peak areas were integrated and the styrene monomer peak was area normalized against the internal standard peak. The results are reported in Table 8.

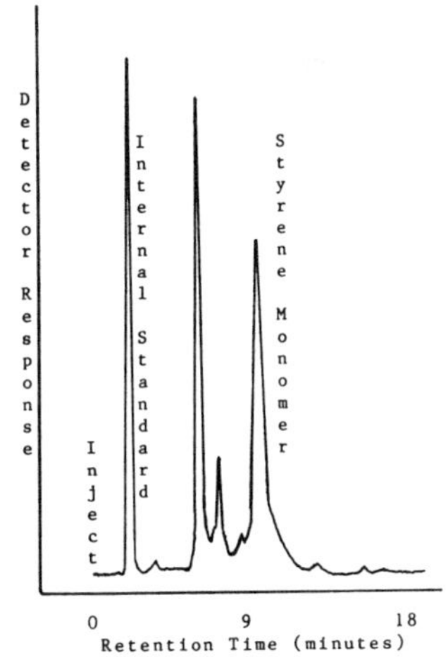

Sample Polystyrene

Concentration 5.0 (w/v)%

Solvent Dimethyl Acetamide

Inject Volume 1 μL

Carrier Gas Nitrogen

Flow Rate 24 mL/minute

Temperature 125 Centigrade

Detector FID

Column 7% (Amine AT220/ Carbowax 20M) on Chromosorb WHP

12 feet X 1/8 inch stainless steel

FIGURE 7. GC evaluation of polystyrene.

TABLE 8. Precision of GC Sample Preparation Procedures.

Autosampler Vial Number	Single Master Solution	Individual Analyst Prepared	Individual Robot Prepared
1	64697	62119	63262
2	62683	62642	62612
3	62353	62875	62961
4	63081	60453	61705
5	62156	61406	59721
6	61897	61365	62830
7	62910	60535	61765
8	61711	61297	60178
9	61119	61412	61850
10	61664	63258	62620
11	61550	61565	61425
12	61287	60487	60924
13	60866	62188	61485
14	60756	60098	60431
15	60548	61433	61096
16	61353	57892	63704
17	60700	61333	64526
18	60715	60948	62230
Mean	61780	61295	61963
Std Dev	1069	1209	1260
C.V. (%)	1.7	2.0	2.0

Data for the single master solution showed the analytical GC measurement technique to have a C.V. of 1.7%. Meanawhile, data for the individual analyst prepared samples showed a C.V. of 2.0% and finally, data for the individual robot prepared samples exhibited a C.V. of 2.0%. The robot procedure shows precision equal to the manual method.

After calibration using the same internal standard,[2,3] the concentration of unreacted styrene monomer was calculated for each sample analysis. Next, the mean styrene monomer concentration was computed for each of the three GC sample preparation procedures. The percent relative error (for these mean styrene levels) ranged from 1 to 2%, suggesting the robot procedure to have accuracy equal to the previously employed manual method.

CONCLUSION

In summary, the precision of the SEC and GC sample preparation procedures as performed by the Zymate Laboratory Automation System is equal to or better than the previously employed manual methods. The accuracy of both the MWD and NAV test results for these automated sample preparation methods is equal to the prior manual techniques. The performance of these automated sample preparation procedures has been sufficiently reliable to permit unattended overnight operation.

The use of flexible automation (robotic sample preparation) coupled with dedicated automation (chromatography autosamplers) has proven most effective in maximizing efficiency in the analysis of the MWD and NAV of polymeric samples.

ACKNOWLEDGEMENTS

I would like to thank Mr. Mel N. Jameson of BWC for his many valuable ideas and technical assistance in programming and testing these automated SEC and GC sample preparation procedures. I would also like to extend thanks to everyone at Zymark for all of their help.

REFERENCES

1. ASTM D3536 Test Method for Molecular Weight Averages and Molecular Weight Distribution of Polystyrene by Liquid Exclusion Chromatography, 1984 Annual book of ASTM Standards, Vol. 08.03, ASTM, Philadelphia, PA, 1984.

2. G. Zweig and J. Sherma, CRC Handbook of Chromatography: General Data and Principles, Vol. II, CRC Press, Cleveland, OH 1977.

3. C. G. Smith, N. E. Skelly, R. A. Solomon and C. D. Chow, CRC Handbook of Chromatography: Polymers, CRC Press, Boca Raton, FL, 1982.

LABORATORY ROBOTICS: AN APPLICATION IN AUTOMATED TITRATIONS

Larry A. Simonson*
Zymark Corporation
Zymark Center
Hopkinton, MA 01748

ABSTRACT

This paper describes two stations for totally automated titrations which integrate the instruments and apparatus for sample preparation, titration, and data collection into a unified system capable of unattended and flexible operation. This level of automation is made possible by the use of a laboratory controller which coordinates the actions of its modules, not the least of which is a robotic arm with interchangeable hands. Besides its supervision of the robotics, the controller interfaces with an analytical balance, commercial titrators, and various other apparatus. The controller also acquires the analytical data, prints a final report, and senses proper operation of the system.

To test the performance of the system, a series of titrations were run using different sampling methods for acid/base and Karl Fischer determinations. Data from these titrations show the system to be very satisfactory with respect to precision, accuracy, reliability, and throughput.

* On leave from Framingham State College, Framingham, MA 01701 (1984-1985).

INTRODUCTION

The continued need for improvements in laboratory productivity has been partially answered by more and more laboratory automation. Significant improvements in sample throughput, precision, accuracy, and sensitivity have been realized by use of modern computerized analytical instruments and powerful data systems. However, complete automation of a given procedure requires not only analytical measurement and data reduction, but also sample preparation and manipulation. The recent introduction of laboratory robotics technology has addressed these issues with an ultimate goal of the "TOTAL" automation of a wide variety of laboratory procedures. The successful operation of several hundred laboratory robot installations attest to the viability of robotic technology as a useful, perhaps necessary, component in the total automation picture.

The extensive use of titrimetric methods of analysis in the various chemical, pharmaceutical and biological industries is strong evidence of the performance of these methods. While the performance is notable, it is often at the expense of careful and labor-intensive attention to detail. Commercially available automatic titrators have eased some of this tedium, but regular technician attention is still required for sample preparation and loading. Based on these considerations, routine titrimetric methods are a logical candidate for complete automation. The pioneering work of Owens and Eckstein[1] showed that the complete automation of titrimetric procedures was feasible.

The incorporation of laboratory robotics into an automated analytical procedure requires some level of inter-module communication to synchronize the flow of samples and information. An issue of current interest is the integration of sample preparation, analytical measurement, and data handling into a closed loop procedure under the control of a single high level computer/controller. In such a system, not only would the flow of samples and data be synchronized, but the feedback loops channel the analytical information itself back to the controller, giving it intelligence beyond the typical binary data obtained from switches. The controller can now use this information to effect the course of the experiment, recognize a malfunction and call for a remedy, or perhaps best of all, report that everything is working correctly. In this report, we describe two integrated systems for totally automated titrations. The main instruments used in these systems were a Mettler DL40RC "MemoTitrator", and a Mettler DL18 "Karl Fischer Titrator", each interfaced to a Zymate Laboratory Automation System. The objective was to demonstrate the satisfactory operation of fully automated titration stations with regard to precision, accuracy, and throughput.

ZYMATE/DL40RC GENERAL PURPOSE TITRATION STATION

This system was programmed to first standardize a dilute aqueous solution of NaOH, and then to use this standardized solution to determine the purity of commercially available unknown acid. The standardization was done against primary standard grade potassium hydrogen phthalate (KHP), (Mallinckrodt, St. Louis, IL). The unknown acids were impure KHP samples (Thorn Smith, Beulah, MI). Since any error introduced in the standardization would be transferred to the unknown measurement, this combined method offers a reasonably good test of overall performance.

Experimental

The first system tested involved the Zymate Laboratory Automation System (Controller and Robotic Arm) interfaced to a Mettler DL40RC general purpose titrator. This system was further equipped with a Mettler AE-160 analytical balance, a power and event controller, racks for capped test tubes and titration cups, two peristaltic pumps, a compressed air tank, and assorted valves, tubing, and air cylinders. Figure 1 illustrates the block diagram of the instruments and modules used in this system.

The general operation of this titration station is explained by the EasyLab program shown in Figure 2. The sample was prepared and introduced to the titrator, the titration was started, and preparation of the next sample was begun. After completion of the preparation of this sample, the controller obtained the results of the previous titration, washed the electrodes and stirrer, disposed of the used cell, and introduced the next sample to the titrator. This simple serialization increases throughput by perhaps a factor of two as compared with strict batch processing.

The sample preparation included obtaining a new titration cup from a custom designed cup dispenser (Figure 3), placing the cup on the balance, closing the balance door pneumatically and weighing the empty cup. All weighings were done to the nearest 0.1 mg. After a change of hands, the balance door was opened and the capped crystalline KHP was

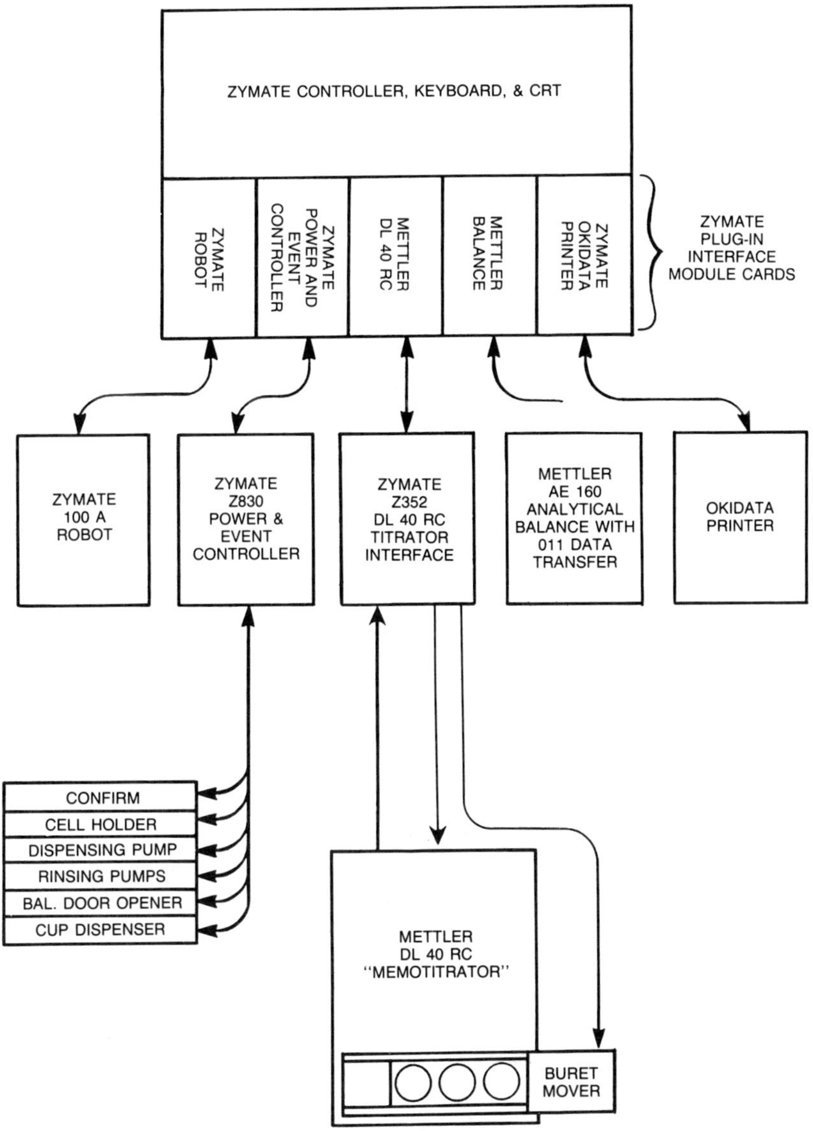

FIGURE 1. Block diagram of a fully automated general purpose titration station.

```
   INITIALIZATION
   PREPARE.FIRST.SAMPLE
5  TAKE.SAMPLE.TO.TITRATOR
   START.TITRATION
   GET.NEXT.SAMPLE.READY
   FINISH.CURRENT.TITRATION
   CALCULATE.AND PRINT RESULTS
   GOTO 5
```

FIGURE 2. Listing of general titration program.

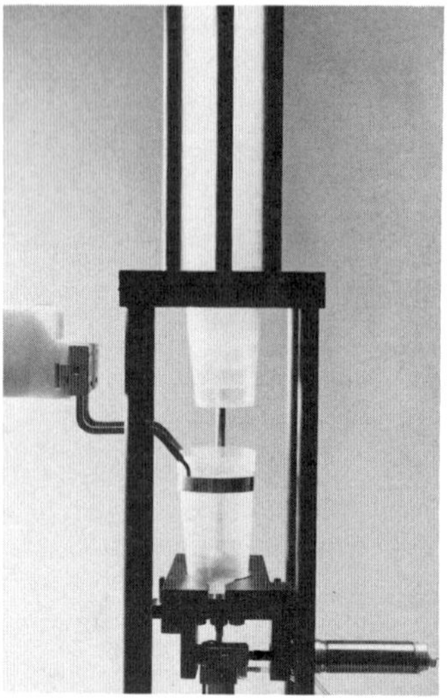

FIGURE 3. Titration cup dispenser.

removed from a test tube rack. The slip cap was removed and held while the KHP was poured into the cup using a vibrating general purpose hand. The pouring, with continuous weighing, was done until 300 mg was transferred. After pouring was complete, the balance door was closed for an accurate weighing of the sample. The tube was recapped and returned to its rack, and the original hand was reattached. The titration cell, now containing the weighed sample, was removed from the balance to a solvent addition station where approximately 50 mL of distilled water was added. The water addition was done using a peristaltic pump controlled through the power and event controller (PEC). An EasyLab timer determined the time the pump was on and thus the amount of water added.

The sample was introduced to the titrator (Figure 4) and held in position by a pneumatically controlled platform operated by a positive

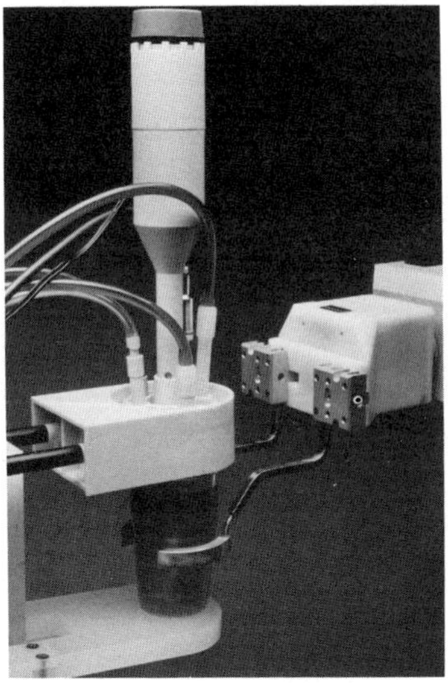

FIGURE 4. General purpose titration station.

displacement gas valve through the PEC. The titration was started from the EasyLab program through the Zymate System/DL40RC interface. First, the desired buret was moved into the active position. Then a previously stored titration method in the DL40RC was called and initiated. This particular method included a 60 second pre-stir and titration to an end point at pH 9.00. The results of the titration were obtained in mL of base added, and the calculation of the normality of the base was done by the Easylab program taking into account the mass of the KHP. The second experiment to determine the percent purity of some KHP unknowns was done in the same way as the standardization.

RESULTS AND DISCUSSION

Tables 1 and 2 show the results of the standardization and unknown analysis.

TABLE 1. Results of Standardization of Two NaOH Solutions.

NaOH Solution	n	Normality	RSD
1	10	0.0901	0.27%
2	10	0.0895	0.24%

TABLE 2. Results of Analysis of Thorn Smith Unknown KHP Samples. Trials 1 and 3 were done using NaOH solution 1. Trials 2 and 4 were done using NaOH solution 2. The values %KHP (known) were obtained from the manufacturer.

Trial	Sample#	n	%KHP(found)	%KHP(known)	RSD	%Error
1	357	6	64.94%	64.79	.24%	+.23%
2	349	6	46.32	46.38	.13	-.12
3	349	6	46.40	46.38	.21	+.04
4	313	6	49.83	49.72	.19	+.22

These results indicate an excellent performance of the system in terms of precision and accuracy, and compare very favorably with that expected from strictly manual methods. The throughput was 10 to 12 titrations per hour for this particular procedure. Perhaps a 5 to 10 percent increase would be expected if this station were run using Concurrent EasyLab, such that all the DL40RC functions would operate under the second interpreter. In regard to reliability, this station ran hundreds of titrations and endured several moves around the country without showing any significant problems.

Although not specifically shown in the program in Figure 2, error detection checks are imbedded in these routines. Because the controller not only controls the robotics, but also acts as a data aquisition-report writing system, it can use the data from the balance, titrator, and other sensors to detect and in some cases remedy errors. The closed-loop nature of this type of system aids in increasing the reliability of the automation. Based on balance data, one can check if a sample of appropriate mass was transferred. Time limits can also be imposed on the weighing and titrating steps to guard against clogged or empty sample tubes and titrations where end points are not reached. Based on titration volume data samples requiring too small (large) a volume can be repeated. If the calculated results are out of some specified range, EasyLab can easily branch to rerun the sample or implement other actions.

ZYMATE/DL18 KARL FISCHER TITRATION STATION

This system was programmed to determine the titer of a commericial Karl

Fischer reagent in mg water per mL titrant. The reagent, Hydranol Titrant-5 (Cresent Chemical) was standardized against tartaric acid disodium salt dihydrate (TDD), (Baker Chemical) in Hydranol solvent. The purpose of this experiment was to determine the overall precision of the fully automated station for a rather demanding multi-step procedure.

Experimental

This demonstration system was equipped essentially like the first station with the controller, robot, balance, and power and event controller. This station used a Mettler DL18 Karl Fischer Titrator and a Zymark solvent exchange kit specifically designed for these titrations. The general operation of this station is explained by the EasyLab program shown in Figure 5. For this station, serialization as

```
   DRAIN.AND.FILL.CELL
11 START.PRETITRATION
   TAKE.SAMPLE.TO.BALANCE
   WEIGH.SAMPLE.AND.BRING.TO.TITRATOR
22 IF PRETITRATION.STATUS< > THEN 22
   REMOVE.CAP.AND.POUR.SAMPLE.INTO.TITRATION.CELL
   START.TITRATION
   RECAP.AND.RETURN.EMPTY.TUBE.TO.BALANCE
   REWEIGH.TUBE.AND.CALCULATE.SAMPLE.WEIGHT
   GET.RESULTS.FROM.TITRATOR
   CALCULATE.AND.PRINT.TITER
   CHANGE.CELL.SOLVENT.IF.SCHEDULED
   GOTO 11
```

FIGURE 5. Listing of Karl Fischer titration program.

in the previous procedure was not done since the non-robotic titration time was very large in comparison to the robotic sample handling time. In these cases, the robot could be put to work on other tasks, such as serving a second titrator or other instrument.

The titration cell was drained and refilled with about 40 mL of Hydranol Karl Fischer solvent using the PEC and solvent exchange apparatus. A pre-titration was started to remove traces of water in the cell. A test tube rack was loaded with 16 x 150 mm test tubes containing about 250-350 mg of reagent grade TDD and covered with plastic slip-caps. The robot transferred the loaded tube to the balance and the gross sample weight was acquired by the controller. The weighed capped tube containing the TDD was then brought to the titrator and held there until the titrator had signaled the controller that the pre-titration was complete. On completion of the pre-titration, the cell port was opened using a pneumatic cylinder, the cap was slipped off and held in a holder, and the crystalline sample was poured into the cell (Figure 6). This last sequence must be done quickly and accurately to prevent excessive air-borne water from entering the cell and to assure that no TDD was scattered on the lip of the port or other places inacessible to the titrant. The pouring was done using the vibration hand. After the pouring was complete, the cell port was closed and the slip-cap replaced on the tube. The controller now started the titration which included a 60 second pre-stir to dissolve the crystals. The capped tube was now returned to the balance for a second weighing, thus yielding the weight of the sample transferred to the cell. As above, the balance door was pneumatically opened and closed as needed for accurate weighing. The

FIGURE 6. Karl Fischer Titration Cell Port.

controller next polled the titrator for results and once they were obtained, the titer of the Karl Fischer reagent was calculated and printed. The titer is based on the weight of the TDD and volume of the titrant required for the end point. TDD has a water composition of 15.66 percent. The next sample was run in the same solvent; however, the solvent was changed after every two samples.

RESULTS AND DISCUSSION

Table 3 shows the results of several series of titrations for the determination of the titer of the KF titrant. Generally, the results are quite satisfactory in terms of precision with relative standard deviations (RSD) below 0.5%. In trial 4 the titrant was standardized manually against pure water. The RSD of this trial is in the same range as the robotic method. Satisfactory accuracy is implied based on comparison of the results of the manual standardization, since water is an independent standard.

TABLE 3. Results of fully automated (trials 1-3) and manual (trial 4) standardization of Karl Fischer titrant.

Trial	n	Titer (ave)	RSD
1	28	4.954(mg H_2O/mL)	0.48%
2	3	4.941	0.082%
3	4	4.947	0.15%
4	5	4.984	0.38%

In regard to throughput, approximately 15 minutes was required per titration. Of course, this is very dependent on the type of sample and the precision or accuracy expected. In regard to reliability of the method, no problems were noted with the equipment. However, it was found that Karl Fischer titrations in general are more demanding than normal acid-base types. Occasionally, one gets results that deviate from the expected that elude simple explanations. For example, in trial 1, two titrations gave very high results that were easily rejected based on their wide deviance but are not so easily explained. Fortunately, EasyLab can be programmed to reject values that lie outside specified limits. Furthermore, the sample can be automatically rerun if so desired.

CONCLUSION

The precision and accuracy of this system was comparible to that typically observed for manual methods. These results are not unexpected because the automated method emulates the manual method in general. The accuracy to a large degree is determined, not by the robot, but by the balance and the titrator, both of which are proven components. Finally, the reports[2] of many other laboratories have shown that the robot can equal, in fact exceed, humans in quality of work for repetitive samples.

ACKNOWLEDGEMENTS

The author thanks Framingham State College for granting him sabbatical leave, and the Zymark Corporation for hosting him during this leave. The author expresses gratitude to Drs. Jim Little and Rick Brown of the Zymark Corporation for their advice and encouragement. Finally, he thanks members of the Zymark's R & D department, and especially Mr. Arthur Martin, for their invaluable technical support.

REFERENCES

1. Owens, G. D. and Eckstein, R. J., Anal. Chem., 54, 2347-2351, (1982).

2. Advances in Laboratory Automation - Robotics 1984, G. L. Hawk and J. R. Strimaitis, eds., Zymark Corporation, Hopkinton, MA, 1984, pp. 31, 113, 133, 171, 216, 250, 269, 304.

ROBOT-ASSISTED SAMPLE PREPARATION FOR PLUTONIUM AND AMERICIUM RADIOCHEMICAL ANALYSIS

T. J. Beugelsdijk, MST-9; D. W. Knobeloch, MST-9;
A. A. Thurston, CHM-1; and N. D. Stalnaker, CHM-1
Los Alamos National Laboratory
Los Alamos, New Mexico 87545

ABSTRACT

An automated robotic sample preparation scheme for the radiochemical assay of plutonium (Pu) and americium (Am) has been developed at Los Alamos National Laboratory. Alpha counting plates are prepared for plutonium analysis and gamma-counting tubes are prepared for americium determination. Depending on the americium content of the samples, the sample preparation for plutonium involves either a weight dilution or a liquid-liquid extraction. Results to date indicate a precision equal to or slightly better than the manual proceadure and with data comparable to existing methods.

INTRODUCTION

Much of the analytical chemistry performed at Los Alamos National Laboratory (LANL) is on the actinide elements and Special Nuclear Materials in particular. Since most of these samples occur as solids, careful sampling and dissolution chemistry is first performed.

Subsequent sample preparation steps often include separation or suppression of interferences, matrix modification and appropriate preconcentration or dilution. As these procedures are often labor intensive and routine, they are an inefficient use of human talent and are subject to error. With the introduction of laboratory robotics systems, automation of such operations becomes feasible. Added benefits include improved and consistent quality, reduced and error-free record keeping, elimination of the need to run duplicates and reduced radiation exposure. A study was performed at Los Alamos National Laboratory that indicated, taking all these factors into consideration along with reducing personnel accessibility to fissile material, automating sample preparation via robotics is a cost-effective and desirable activity.

The radiochemical section of the analytical chemistry group receives many hundreds of samples per year of plutonium metal and oxide for analysis. The samples are routinely analyzed for plutonium content and the beta-decay product of plutonium-241, americium-241. These metal and oxide samples originate in the Plutonium Production Facility at LANL and result from various production processes, new process research programs, and scrap and waste recovery efforts.

As noted above, the majority of samples submitted for this analysis are solids and require dissolution prior to further processing. This step is currently performed manually either in our group's trace analysis section or by the technicians in the radiochemical section. For plutonium metal, dissolution is straightforward and proceeds quite rapidly using 8N HCl. Plutonium oxides are dissolved under pressure at

155°C using a mixture of concentrated nitric acid, concentrated hydrochloric acid, and 1.3M hydrofluoric acid. For either metal or oxide, a solution containing approximately 10 to 20 mg of Pu per mL results.

For the radiochemical analysis of plutonium and americium, all subsequent preparation steps take place on these solutions and consist of diluting, pipetting, liquid-liquid extraction, centrifugation, weighing, etc.- all accompanied by extensive record keeping. The procedure is tedious and, due to the many steps involved, susceptible to error. We have therefore developed a robotic system to execute these tasks. Human variability, once an important factor requiring all samples to be run in duplicate by different technicians, has now been eliminated.

ROBOTIC SYSTEM

The robotic system chosen to implement this sample preparation scheme is manufactured by Zymark Corporation of Hopkinton, Massachusetts. Table 1 lists all the components obtained from Zymark.

Two physical preparations result from each sample submitted to the robot - one is a 25 mm square glass cover slip which is submitted for gross alpha activity measurement using a gas-flow proportional counter, and

TABLE 1. Robotic System Components for Plutonium and Americium Dilution Procedures.

Modules
Robot
Balance Interface (Mettler PE 160)
Master Laboratory Stations (2)
Power and Event Controller
Capping Station
Centrifuge
Computer Interface (HP-85)
Printer
Accessories
Orbital Shaker
Vortex Mixers (2)
Racks (27)
Hands (9)
Dry Bath
Hot Plate
Peristaltic Pumps (2)

the other is a polypropylene test tube containing solution for gamma activity measurement using a sodium iodide scintillation detector.

For samples having a low Am-241 content, both the plutonium and americium preparation involves a direct dilution of the original sample. This scheme is presented in Figure 1. A 25 to 100 uL aliquot of the original sample is taken and diluted with 25 to 100 mL of 1N nitiric acid. A 25 to 100 microliter aliquot of this solution is then evaporated onto a glass cover slip for gross alpha counting. Similarly, for americium-241 analysis, a 0.025 - 2.00 mL aliquot is taken of the diluted solution and placed into a test tube for gamma activity measurement.

In the event that the Am contribution to the total alpha activity is

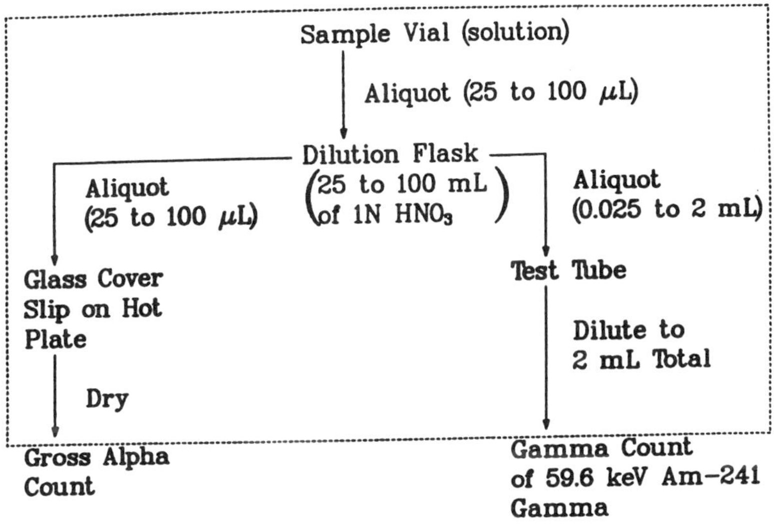

FIGURE 1. Flow diagram of the dilution procedure for plutonium and americium.

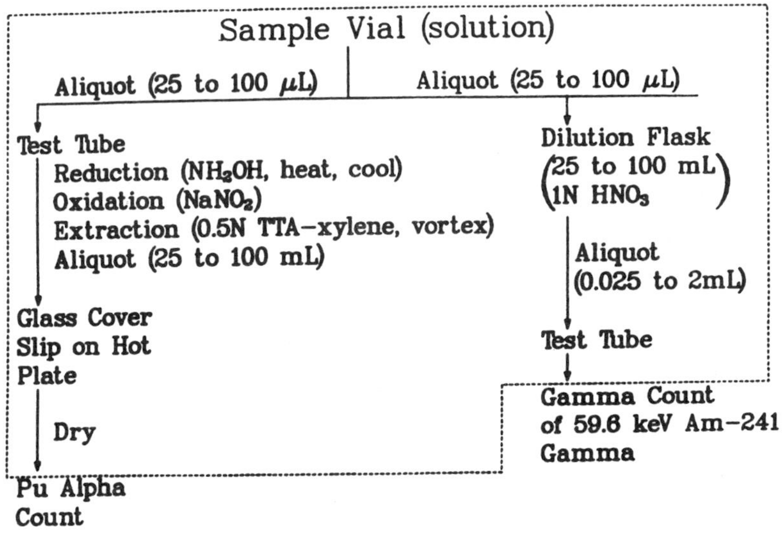

FIGURE 2. Flow diagram of the plutonium extraction procedure.

greater than 90%, the Pu must be extracted away from it. The extraction scheme is shown in Figure 2. An appropriately sized aliquot of the original sample is transferred to a clean test tube where a reducing mixture consisting of hydroxylamine hydrochloride, and nitric acid is added. The solution is then heated for ten minutes and allowed to cool for an additional five minutes. At this point, all the plutonium has been reduced to the +3 oxidation state. The plutonium is selectively re-oxidized to the +4 state using sodium nitrite, and then an extraction solution consisting of 0.5 M 2-theonyltrifluoroacetone (TTA) in xylene is added. After thorough mixing in a vortex station, the phases are separated using a centrifuge. An aliquot of the xylene layer is then deposited onto a glass cover slip for total alpha activity measurement. Due to the extraction chemistry, all the alpha activity is now due only to the plutonium content of the sample. Americium determination on this sample is done as before involving only a direct dilution of the original sample. Details of the extraction procedure are published elsewhere.[1] Not shown in Figures 1 and 2 is an additional step introduced into the robotic process, which is not part of the manual procedure. This is an initial one to ten dilution of the samples submitted to the robot. This step enables the robot to handle larger volumes throughout the process, which it can do more precisely, and also reduces the activity of the solutions in the event of a spill.

SOFTWARE FEATURES

Since the detection scheme for both americium and plutonium involves counting equipment, the sample preparation sequence must insure that the total activity presented to the counters is within their dynamic range.

As a result, allowance must be made to handle each sample differently by taking appropriately sized aliquots throughout the preparation sequence to meet the counting conditions. Robotic systems, however, are used to their greatest advantage when all samples are treated in the same way. To resolve this situation, the individual sample parameters consisting of the treatment sequence to be followed (dilution or extraction) and all the aliquot sizes are entered into a large look-up table using an interactive program written in BASIC on an HP-85 (Hewlett Packard, Palo Alto, CA) desktop computer. This program allows for entering, editing, filing, printing, reading, etc., of these parameters and also includes a software driver subroutine for communicating this information to the robot controller. At the conclusion of the robotic routine for a given set of samples, the robot controller sends the actual dilution factors for Pu and Am (determined for the aliquot sizes and actual weight readings taken form a balance) back to the HP-85. Here all the data is stored on magnetic tape for future reference, and a hardcopy is generated on an HP 82906A graphics printer. Details of the communicaitons software have been reported elsewhere.[2]

The robotic system software comprises approximately one hundred forty-five subroutines which occupy more than 99% of the 64 kilobytes of controller memory. Eighty percent of these subroutines are called by the main sample preparation program, with the balance consisting of utility and calibration routines. The utility routines include programs to flush the reagent lines on the master laboratory stations and several statistical routines used in the calibration process.

Because the entire system is located inside a hood enclosure, it was deemed desirable to have the robot calibrate itself. Routines were written so that each of the critical solution dispensing devices can be calibrated prior to use to insure their accuracy. The syringe hands are calibrated by weighing repeated aliquots of a solution of known density at 10, 30, 60 and 90 percent of the syringe volume. Least squares estimates are calculated for the slope and intercept of the calibration curve and these values are then updated in the main program and used whenever solution dispensing is called for. Experience with the syringe hands dictates that they should not be used for delivering aliquots below about 15-20% of the nominal syringe size. Since the master laboratory stations are used to dispense only fixed volumes, a single-point calibration factor is calculated at this point. The master laboratory stations deliver very precisely (RSD < 0.05%), but show a slight bias which can be corrected with a factor. The actual corrections for all the devices is often quite small, but given the large number of transfers in this routine, propagation of error considerations dictate that each transfer be made very precisely and accurately.

PRODUCTIVITY CONSIDERATIONS

The throughput of the system is limited by the speed of the robot module. Whenever possible, support modules are run concurrently so that no robot time is lost waiting. In addition, the method requires several non-robotic conditioning steps (e.g. ten minutes of heating and five minutes of cooling for each extraction sample). To prevent the robot from being idle, a fairly intelligent decision making routine has been written to assess the task at hand for any mixture of dilutions and extractions in a sample batch. This routine has the capability of changing the sequence of operations as required to keep the robot active at all times. Dilutions and extractions for Pu and dilutions for Am are thus optimally interleaved for maximum throughput.

The average cycle time for a sample requiring only a dilution for plutonium determination is 15 minutes. For a sample requiring liquid-liquid extraction, the cycle time averages 20 minutes. Hence, for a twenty sample task, the system operates continuously from five hours to six hours and forty minutes. Therefore, two such twenty sample batches can be run in an eight hour day with the second continuing unattended into the evening. Maximum throughput is thus forty samples per day. This compares favorably with the manual procedure which has a throughput of twenty samples per day (each run in duplicate).

RESULTS

Table 2 summarizes some typical results obtained with the robotic system and compares these with those from a referee spectrophotometric method for plutonium determination by the dilution procedure. Indicated differences are within the historical repeatability of the methods and show no consistent bias introduced by the robotic system. Results for plutonium by extraction and for americium show similar trends with referee methods.

TABLE 2. Comparison of Results for Plutonium by Dilution and Spectrophotometric Procedures.

Sample No.	Spectrophotometric	Robotic/Radiochemical	Difference
1	84.5	85.0	+0.5
2	87.1	85.4	-1.7
3	86.4	86.2	-0.2
4	86.9	84.1	-2.8
5	87.3	85.3	-2.0
6	86.8	89.3	+2.5
7	86.8	89.9	+3.1
8	86.9	87.1	+0.2

REFERENCES

1. J. M. Cleveland, The Chemistry of Plutonium, American Nuclear Society, Illinois, 1979, pp. 203-208.

2. T. J. Beugelsdijk, N. D. Stalnaker, L. R. Austin, "Interfacing an HP-85 with the Zymate System for the Exchange of a Data Matrix," Zymark Newsletter, Vol. 2, No. 2, 1985, pp. 4-8.

Zymark is a registered trademark of Zymark Corporation.

AUTOMATING SAMPLE PREPARATION (AND DISPOSAL) WITH A ROBOTIC WORKCELL

R.D. Jones and J.B. Cross
Phillips Petroleum Company, Research and Development,
Bartlesville, Oklahoma 74004

ABSTRACT

A Zymate Laboratory Automation System is being used by Phillips Petroleum Company for the preparation (and subsequent disposal) of samples submitted for inductively coupled plasma (ICP) and x-ray sulfur analysis. The system heats (if necessary), shakes, uncaps, pipettes, weighs and dilutes with appropriate solvents both aqueous and hydrocarbon samples. During what would otherwise be idle periods, it uncaps completed samples and dilutions, pours their contents into waste containers and disposes of the empty vials. Capable of processing up to nine different types of samples in a single, unattended operation, the robot has handled as many as 5,345 samples in a one month period and boosted productivity 50-75% in areas it supports.

The system is currently programmed to handle six different types of sample containers, ranging from four dram glass bottles to plastic test tubes with "press-on" caps. Sample weight, dilution ratio, number of dilutions and solvent used are variables easily adjusted to meet the user's specifications. Dilution weights and sample ID's are automatically printed on labels and manually affixed to the appropriate vials following preparation. Quality control studies indicate robotic dilutions are comparable to those performed manually. No special knowledge of robotics or computers is required to operate the system. The workcell's payback period is expected to be on the order of ten months despite the fact it is presently being utilized at only 50% of its potential capacity.

INTRODUCTION

In any given month, over 4,000 aqueous and hydrocarbon samples, slated for nearly 10,000 analytical tests, stream into Phillips Petroleum Company's Analysis Branch. Most are analyzed and the data returned to the submitter in less than five working days - a feat due in no small part to extensive Branch-wide efforts aimed at automation, computerization and, more recently, robotization. Three of Phillips' "islands" of laboratory robotic automation have been described elsewhere.[1,2,3] A fourth is the subject of this report.

Since November 27, 1984, a Zymate Laboratory Automation System has been diluting an average of 1,000 aqueous and hydrocarbon samples per month for quantitative, multi-element analysis by inductively coupled plasma (ICP), as well as the determination of sulfur in oils by x-ray fluorescence. More recently, the robot's duties have expanded to include "dumping" completed samples and dilutions. From a numerical standpoint, this is its most active assignment, averaging about 3,500 samples per month. The robot uncaps vials, pours their contents into a waste container slated for incineration and drops the empty vial into a separate waste bin.

SYSTEM DESIGN

Diluting and dumping samples may sound like simple, straightforward procedures, but they are not. Different analytical procedures require different sample weights, different dilution ratios, different total volumes and different solvents. Some samples (heavy hydrocarbons) must be heated; others require duplicate preparations (which, for speed and simplicity, the robot performs simultaneously). Despite efforts toward standardization, the system is still confronted by six different types of sample/dilution containers. Dumping poses the additional problem of handling vast numbers of vials in a limited work area.

The need for flexibility, user friendliness and a high degree of reliability shaped the resulting workcell design (Figure 1). As presently configured, it is capable of coping with almost all of the above variables in a single, unattended run encompassing as many as 116 samples for dilution or 720 samples for disposal (assuming, in each case, the use of four dram vials).

No special knowledge of computers or robotics is needed to operate the system, and the required data input is minimal. For example, if the operator enters the number of samples to be diluted, followed by the program name ICP.DILUTIONS, the robot will prepare a single 1:10 dilution of each, containing a minimum of 0.95 grams of sample diluted in xylene. Entering the program name TOLUENE.DILUTIONS results in duplicate preparations - diluted 1:7 in toluene, with a final weight of approximately eight grams (less total solution is required for this test). BASIC.NITROGEN produces single dilutions, containing 0.200 ($\pm$0.020) grams of sample, diluted in toluene to a total weight of five grams. An aqueous program called SHOTGUN produces dilutions in triplicate each possessing a different dilution ratio (1:10, 1:30 and 1:70).

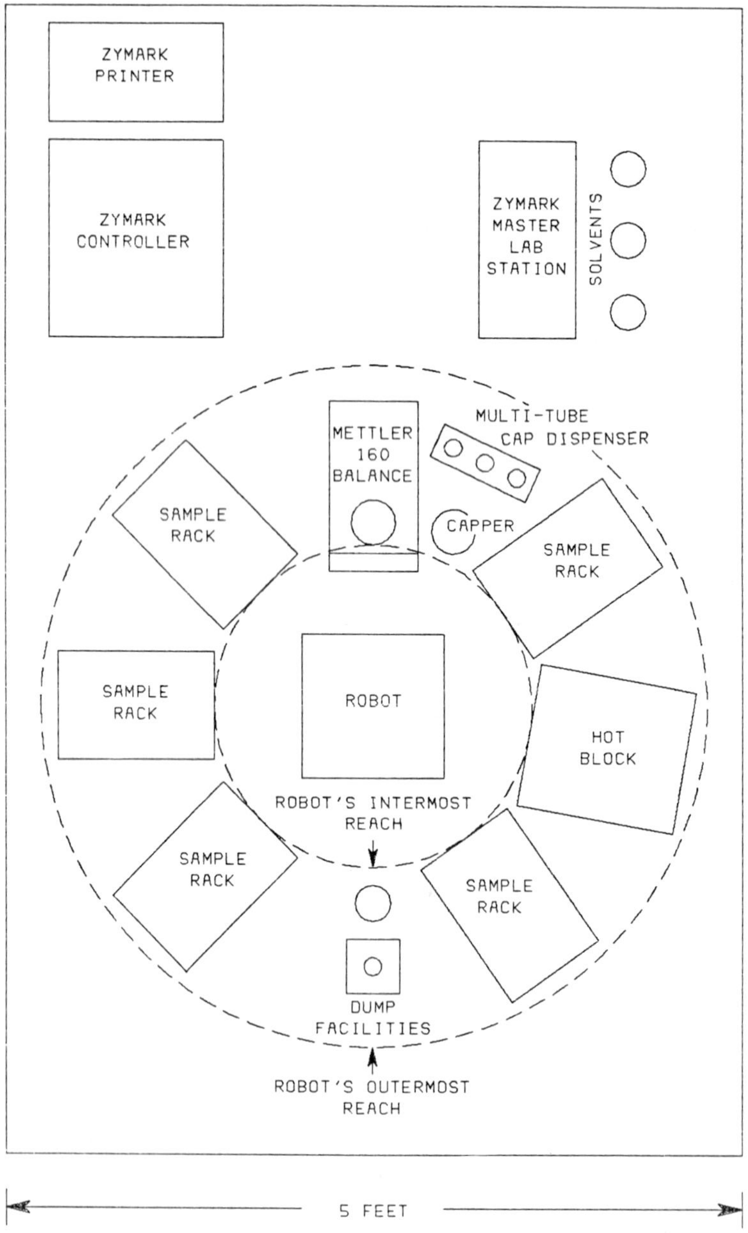

Figure 1. Layout of Robotic Workcell for Dilutions and Sample Disposal

Achieving this degree of flexibility, using a computer system of limited storage space (64K) and a programming language (Zymark's EasyLab) which does not support menu-driven routines, posed some interesting problems. Basically, our solution was to create a "menu-less menu" driven system. Instead of offering the user a choice of sample weights, final weights, solvents, container types, etc., in a menu format (which is what we would have preferred), we have made these selections for the user and "packaged" them as a series of what we call "control" programs.

An example cited earlier, ICP.DILUTIONS, is such a program. It does not directly dilute samples. Instead, it contains, in the form of a series of preset variables, the information needed by the system's "master" dilution program to execute the operation desired. Like all of the control programs, ICP.DILUTIONS is quite short, only 20 lines long. Once it initalizes the necessary variables, it calls the master dilution program as a subroutine.

Thus, the same basic program is responsible for all of the robot's dilution routines. At 14 key points, it contains statements reading GOTO some variable. For example, at program line number 73, the master lab station (MLS) must begin filling a burette with the appropriate solvent. It has three options: xylene, toluene and 10% nitric acid. Line 73 simply reads GOTO FIL (FIL being one of those variables initialized by a control program). ICP.DILUTIONS intializes FIL to a value of 320, which the robot's computer interprets as line number 320. Line number 320 reads FILL.XYLENE.BURETTE, the command which will cause the MLS to do just that. FIL will retain the value of 320 until another control program changes it. TOLUENE.DILUTION, for example, will initialize FIL to 330, at which line number the computer will find FILL.TOLUENE.BURETTE. In each case, these statements are followed by GOTO 74, which returns the computer to the next step in the dilution procedure.

While lacking the potential felxibility of a true menu-driven system, we have found this arrangement lends itself readily to modification, as well as the creation of new dilution routines (so long as they are of the same generic type - i.e., liquid samples diluted in solvent).

To perform different types of dilutions in the same run, a program called MULTIPLE.DILUTIONS links the control programs together sequentially. Samples are physically arranged by dilution type, in the same order as the robot's computer will step through the control programs, checking at the beginning of each to see if samples were entered for that particular procedure.

The software responsible for dumping samples is considerably less complex. However, by following the same strategy, we are able to accomodate different container sizes, as well as both heated and non-heated samples in the same run.

Fully loaded, the robot can dilute samples for as long as 13 hours or dump for about 18 hours. This brings up the question of reliability.

"Accidents" have happened (and no doubt will continue to happen), but it should be noted that rarely has more than one dilution been affected.

That is because numerous testing procedures continuously monitor the workcell's operation, watching for problems. Should one arise - for example, a bottle cap fails to unscrew - detailed instructions in the master program cause the robot to cease routine operation and attempt to resolve, in a satisfactory manner, the resulting situation. In the example given, the robot will: (1) make a second attempt to remove the cap, (2) test for removal, (3) resume normal operation if this attempt is successful, (4) if it fails, return the sample to its position in the rack, discard the empty dilution vial (to flag the problem sample) get the next sample and then resume normal operation. If the robot encounters a problem beyond its ability to resolve or if a problem, such as the example cited, occurs in three consecutive dilution cycles, the workcell will automatically shutdown.

Since the robot is deaf, dumb and blind, its controller relies upon sensory feedback derived from two sources: a Mettler 160 electronic balance, to which it is interfaced, and two microswitches monitored by the power and event controller. Data from the Mettler yields many clues about what is really happening in the workcell. Initially, it confirms that an empty vial has indeed been moved to the balance pan at the beginning of each dilution cycle. A weight increase, following pipetting, of less than that specified in the program sends the robot back to the sample for a second try (but not a third - if it cannot achieve its goal in two attempts, it assumes insufficient sample). A weight increase noticeably greater than specified indicated the pipette tip has fallen into the dilution vial. Zero weight increase suggests either:

1. The sample is not a liquid at room temperature
2. The pipette tip has been lost
3. The syringe mechanism is defective.

Since (1) or (2) is the most likely possibility, the robot will discard the dilution vial and return the sample to the rack uncapped. If the tip has been lost, it is most likely in the sample vial. Given the pipette tip length, part of it would be protruding above the vial's mouth, making capping impossible. If, on the other hand, the problem is the sample's viscosity, leaving it uncapped briefly poses no serious problem. As a safety precaution, the robot will also execute the steps necessary to remove the disposable pipette tip. Three consecutive occurrences of this problem triggers a system shutdown.

The microswitches yield equally important, if less richly varied, information. Touching a microswitch, immediately prior to sampling, confirms the pipette tip's presence. If this test fails, the robot returns to the tip rack, increments one position and tries to load the next tip. Three consecutive failures again shut the system down.

The second microswitch is more interesting. Diagrammed in Figure 2,

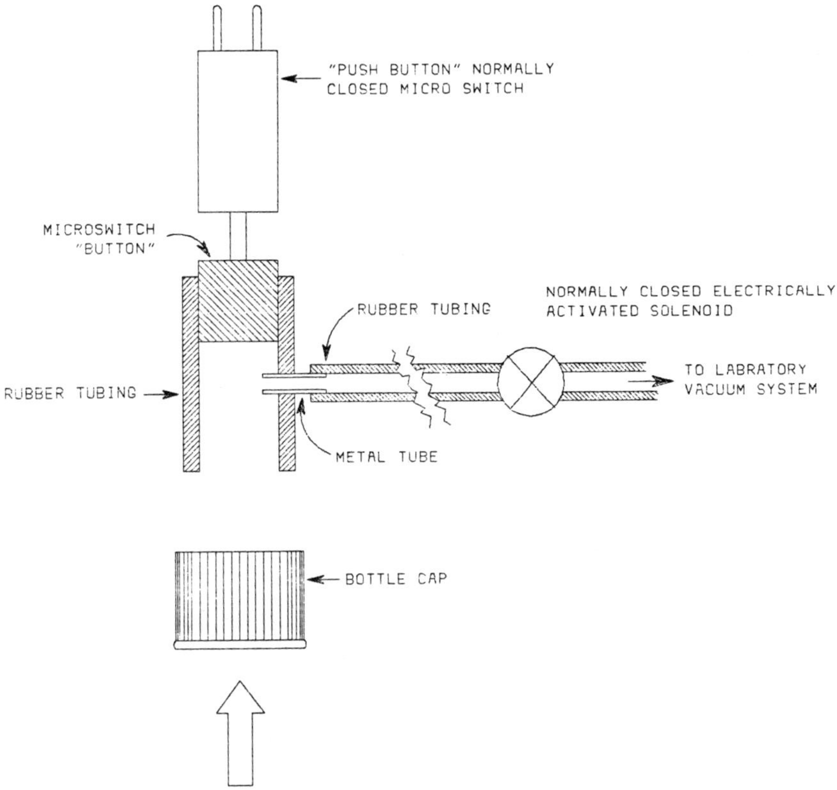

Figure 2. Microswitch/Vacuum Gripper for Caps

it is incorporated into a robot-controlled cap gripping mechanism. Initially, we discarded the original sample container caps, replacing them, at recapping, with new lids. The sheer volume of samples caused us to rethink that policy and the cap gripper was born. It consists of a push-type microswitch, mounted upside down. Its "button" seals one end of a one-half inch long length of rubber tubing. A smaller diameter tube, entering through the large tube's side, is connected via a robot-controlled solenoid to the laboratory's vacuum system.

If a cap is successfully removed, pressing its top against the larger tube's lower (open) end will close the microswitch (an event noted by the power and event controller). Acting upon signals received from the robot's computer, the power and event controller will then activitate a normally closed solenoid converting the tube into a vacuum gripper. The robot releases the cap and leaves it suspended over a drip pan until recapping.

A few points about this system are worth noting: the final location of a cap inside the robot's grippers can vary considerably. If the initial test point returns a negative response, the robot is programmed to move up three millimeters, in two successive steps, repeating the test each time. Of course, the vacuum gripper's outside diameter must be less than the bottle cap's, or the robot's fingers will impact, producing a false confirmation.

One last comment: releasing the cap without breaking the vacuum gripper's hold can be tricky. After switch closure is confirmed, the cap should be lowered to the switch's normally extended position (we are using a spring loaded switch) and the robot's grasp relaxed in a series of steps (example: GRIP=60, GRIP=70, GRIP=80, GRIP=100 and GRIP=200).

Incidentally, the caps used to seal dilutions are obtained from a gravity fed cap dispenser (Figure 3).

That is a fancy name for a vertically mounted clear plastic tube (of a slightly greater inside diameter than the vial cap's outside diameter), stacked full of caps. When the robot removes the bottom cap, the one above falls into its place.

We use a multiple tube dispenser capable of holding 190 four dram vial sized Polyseal caps. The robot maintains a running total of the caps it has used and automatically switches from an empty tube to a full one.

Actually, robotic "bookkeeping" is an integral part of the system's design. The workcell's expendables (solvents, caps and pipette tips) are monitored by the Zymate controller. When requested to perform a task, it first calculates the expendables required and if the supplies are inadequate, refuses to start.

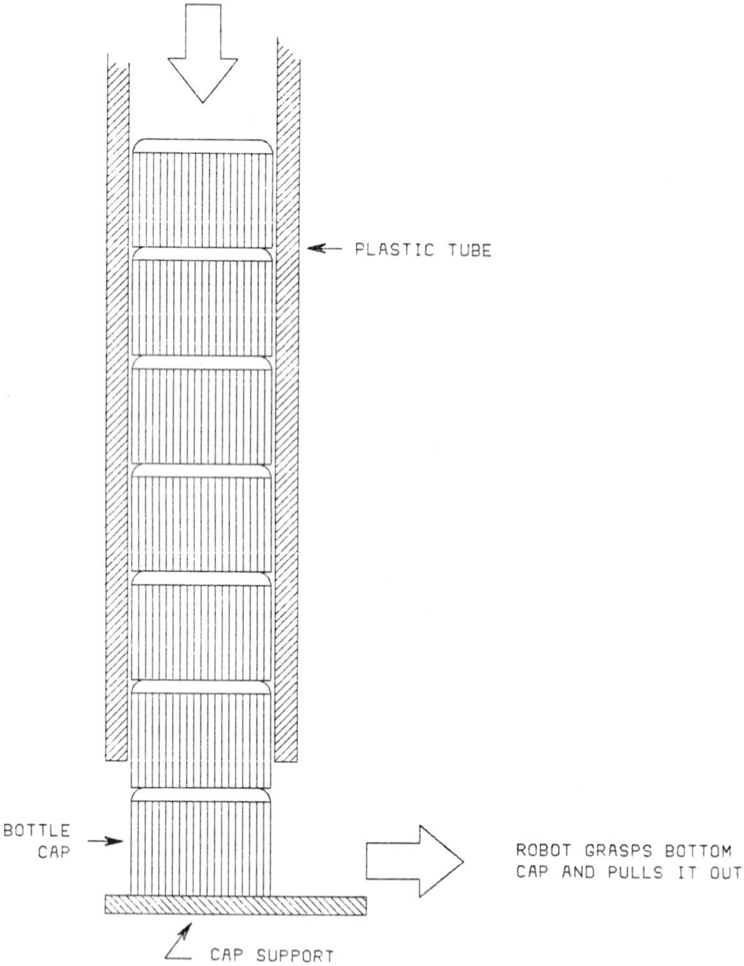

Figure 3. Gravity Fed Cap Dispenser

Three additional workcell features are worth mentioning. First, a safety precaution - the user can, simply by entering DILUTION.FACTOR = "whatever is desired" override the default value for several of the control programs (a value which is automatically reinstated at the program's conclusion). This means as the dilution ratio increases, the amount of sample pipetted must decrease, or the dilution vial could be overfilled when solvent is added. To accomplish this, as the user entered dilution factor increases, a subroutine progressively scales back the final syringe setting, meaning less solution is pipetted.

This procedure is not entirely foolproof, so the robot will (1) not accept a dilution ratio greater than 1:50 and (2) before dispensing solvent, calculate what the total weight would be, given the dilution ratio and sample weight (3) compare this total weight to the maximum safe value for that size of container, and if it is exceeded, dispense only the safe value.

The second feature is a user-designed workstation capable of sealing aqeuous dilutions in test tubes with "press-on" caps (Figure 4). It consists of a gravity fed cap dispenser and two pneumatic actuators. The robot inserts the tube into the workstation, the first actuator positions a cap over it an the second "presses" it on. The whole operation takes about four seconds.

The third feature resulted from sample dumping. Sheer volume swamped conventional racks, so we designed a very unconventional "dumping rack"

(Figure 5).It consists of two levels, an upper and a lower. Samples are packed as densely as possible (average spacing is about 1/8 of an inch).

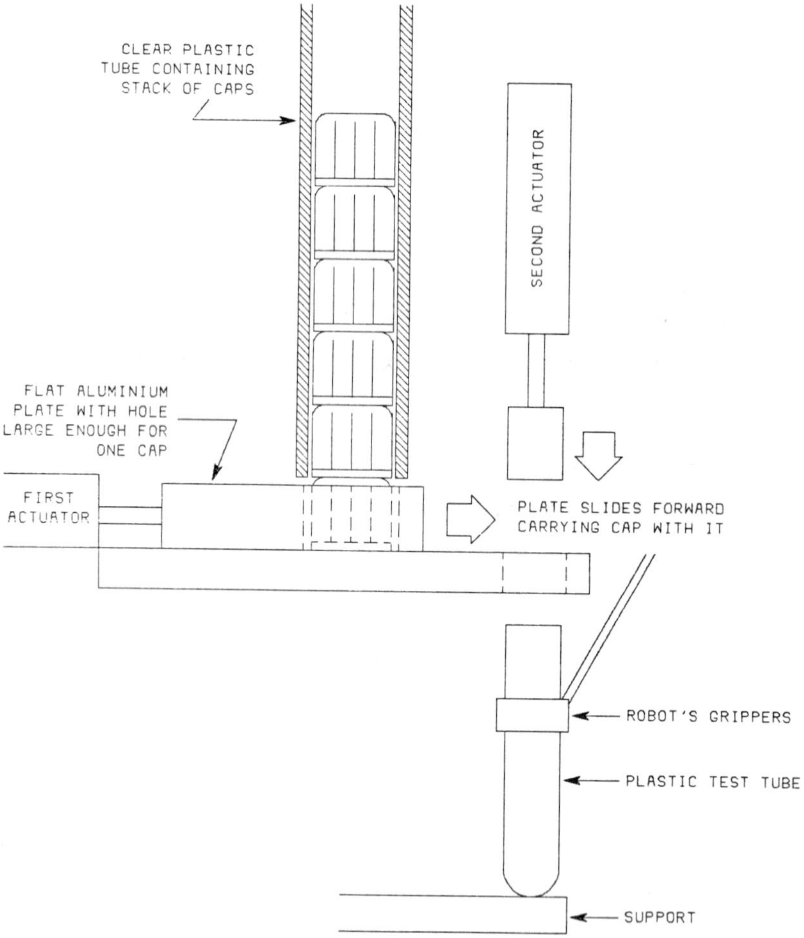

Figure 4. Workstation Designed for Sealing Plastic Test Tubes with "Press On" Caps

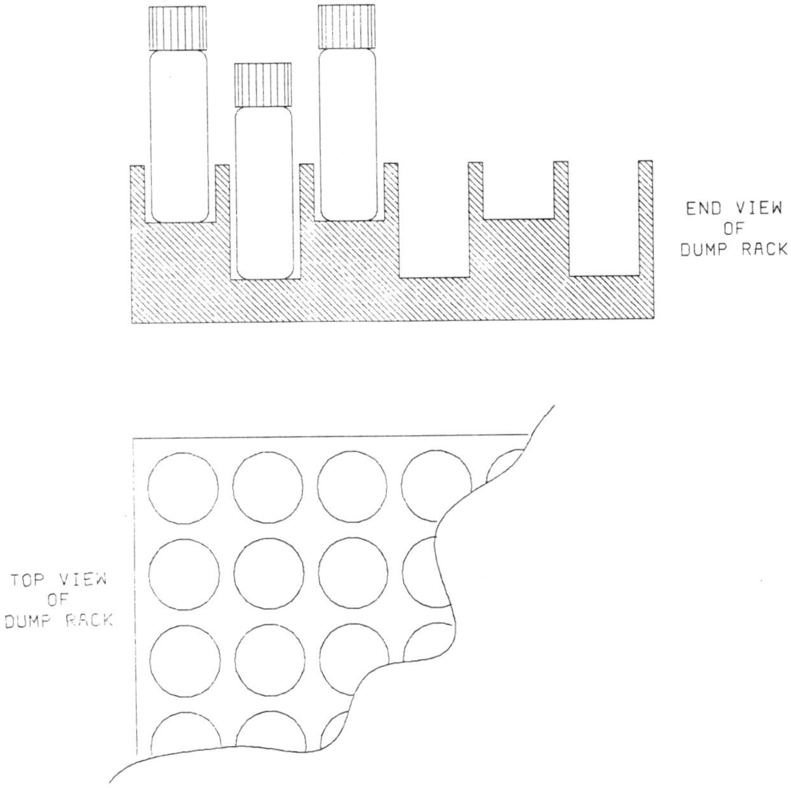

Figure 5. "Dumping" Rack

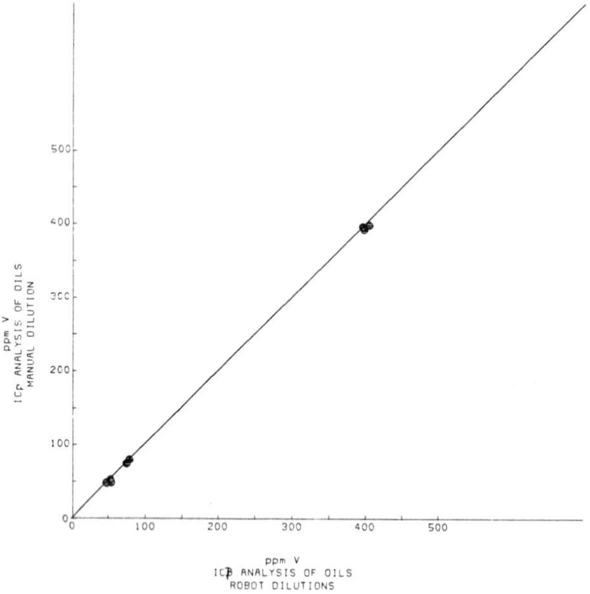

Figure 6a. Inductively Coupled Plasma Analysis of Vanadium in Oils: Robotics Sample Preparation vs Manual Dilutions

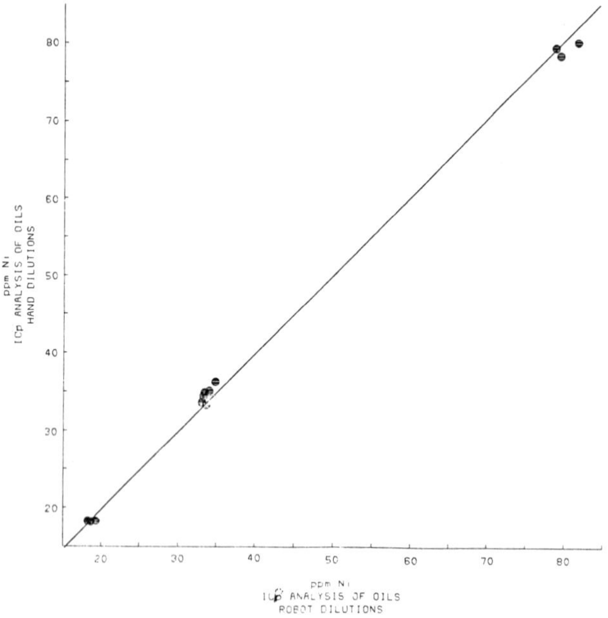

Figure 6b. Inductively Coupled Plasma Analysis of Nickel in Oils: Robotic Samples Preparation vs Manual Dilution

RESULTS AND DISCUSSION

As shown in Figures 6a and 6b, we have found robotic dilutions to compare favorably with those prepared manually. In these figures, data are plotted for analysis of Ni and V in oils by inductively coupled plasma. Sample preparations by the robotic technique are compared to manual sample preparations. The data plotted indicate that the results reported on the sample for either technique is equivalent within the precision of the analysis technique itself.

In both cases, steps must be taken to insure the sample is in a homogeneous state before aliquoting occurs. The robot handles this problem by first shaking the capped sample (via repeated wrist rotation) and then, prior to aliquoting, repeatedly filling and emptying its pipette tip.

The robot does not accomplish either task at a blinding rate of speed. Indeed, one of the most noticeable features of the system is that it is slow. Six minutes are required to dilute a sample, eight minutes to prepare duplicates and about 90 seconds to dump. Obviously, a human can manually perform these operations faster. How much faster? That depends upon the data one wishes to use. In our tests, individual technicians were able to dilute small batches of samples at a rate of as little as 2.5 minutes per sample, with three minutes being the overall average. However, as the batch size increased, so did the time required per sample. Viewed as a continuous operation (which is effectively what we have), the average time to dilute a sample (single dilution), on a day-to-day basis, was typically about five minutes. The robot requires six minutes.

Of course, the robot also requires some human assistance to be precise, an average of 0.6 minutes per sample. This includes inputting data, labeling dilutions and replenishing the system's expendables (solvents, pipette tips and caps). Viewed as a continuous operation, one man-hour invested in supporting the robot will yield a return of about 100 completed dilutions versus 12 if the task is performed manually.
We have experienced significant productivity gains as a result of the robot freeing skilled personnel for more challenging tasks. Any time lost due to the robot's slower pace is more than offset by its capacity to work around-the-clock.

Since the present workload still leaves the robot idle nearly 50% of the time, we are developing additional work routines - automatically "splitting" incoming samples requiring multiple tests and robotically cleaning certain types of glassware being among the most promising. We are also striving to minimize the manpower required to support the robot's operation. Under development are systems and software that will enable it to:

1. Read the incoming samples' bar code labels.
2. Consult our computing system to determine the preparation(s) required.
3. Generate bar code labels containing sample identification and dilution data.
4. Automatically affix those labels directly to the dilution vials.

CONCLUSIONS

Phillips Petroleum Company has developed a robotic workcell capable of processing literally thousands of aqeous and hydrocarbon samples each month. It is expected to save on the order of 1,100 man-hours through automated dilutions, plus an additional 500 man-hours due to automated dumping, during its first full year of operation and pay for itself in as little as ten months. In the future, we shall be striving to both minimize human support (making it more productive), as well as to develop additional work routines. Ultimately, we intend to have the robot performing menial, repetitious laboratory tasks nearly every hour of the day, seven days a week. This is, after all, what robots are all about.

REFERENCES

1. Wharry, S.M., Cross, J.B., Marak, E.J., Cantor, D.M., Amer. Lab., 17 (19), p. 157 (September 1985).

2. Cross, J.B., Marak, E.J., Advances in Laboratory Automation - Robotics 1984, G.L. Hawk and J.R. Strimaitis, eds., Zymark Corp., 1984, p. 181.

3. Cross. J.B., Wilson, L.V., Marak, E.J., and Jones, R.D., "Robotic Automation for X-ray Fluorescence Analysis of Sulfur in Oils", Advances in Laboratory Automation-Robotics 1985, G.L. Hawk and J.R. Strimaitis, Zymark Corporation, 1985.

Zymark, Zymate, and EasyLab, are registered trademarks of Zymark Corp.

AUTOMATION OF THE CALORIMETRY STEP FOR PRODUCTION OF PLUTONIUM-238-OXIDE-FUELED MILLIWATT GENERATORS

D.W. Knobeloch, L.R. Austin, T.W Latimer, D.N. Schneider
Los Alamos National Laboratory
P.O. Box 1663
Los Alamos, New Mexico 87545

ABSTRACT

Milliwatt generator heat sources are routinely produced by Los Alamos National Laboratory in the plutonium facility, TA-55. The production operations are performed manually by technicians working in controlled atmosphere gloveboxes. The passive calorimetry measurement is routine in nature, thus favoring the automated system. A robotic system, based on the Zymark robot, is being developed to perform this step automatically. The robot will be interfaced with a HP-9825 computer that will run the main program fro the calorimeter, provide a data base for the results, and control the Zymate System. Because the robot will be operating in a glovebox, several modifications were made to the robot arm. These include all aluminum parts, remoted electronic control box, an electromagnet incorporated into a dual function hand and dry lubricant bearings. When the system is installed in the glovebox, it will be the first robotic application of this kind at Los Alamos.

INTRODUCTION

Since the discovery of transuranics in 1940, production in large quantities has grown in sophistication. Larger and more complex glovebox lines were constructed to handle the demand for new chemical processing techniques. The need to process the ever-growing amounts of radioactive wastes being generated placed additional burdens on facilities and the people needed to operate them. As more and more studies were made in qualifying and quantifying ionizing radiation, concern about personnel exposures increased. Now that standards are in effect to reduce radiation exposure and demand for special nuclear materials is increasing, new technology is needed to solve both problems. Robotics and automation can have a significant effect on reducing radiation exposure to personnel, while accomodating the nuclear material processing demands.

The fifteenth Annual Report on Radiation Exposures for DOE and DOE Contractor Employees,[1] released in February 1984, summarizes occupational radiation exposures being experienced over the years. The Department of Energy (DOE) outlines basic radiation protection policies that state the "radiation exposures be maintained as low as reasonably achievable". All operations are to be conducted "in a manner to assure that radiation exposures to individuals and population groups are limited to the lowest levels technically and economically feasible."

Out of the ten facility categories reported, the "General Research" group received the largest percentage of collective dose equivalents totaling 1676 person-rem. The "Reactor" category was the second highest category with 1612 person-rem. Los Alamos National Laboratory reported 658 person-rem with 60% received by the Plutonium Handling Facility personnel.

These data indicate the need to establish a technology base to solve exposure problems at Los Alamos. Automation was immediately suggested because it would reduce the number of labor intensive tasks, thereby reducing exposure and improving safety. Additional benefits can be realized in areas such as improved product quality, simplified accountability, improved productivity and improved worker satisfaction. Automation can also allow implementation of new processes to handle more radioactive materials at lower cost.

Milliwatt generator heat sources are routinely produced by Los Alamos National Laboratory in the plutonium facility at TA-55. In these devices, the heat generated by decay of plutonium-238 dioxide is directly converted to electrical energy by means of thermoelectric materials. Radiosotopic thermoelectric generators (RTG's) are finding increased use in applications where reliable, long term, low-wattage power is required and access is difficult. One application for use of the RTG is in the space program for unmanned missions to Jupiter planned for 1986.

The RTG source contains plutonium dioxide with the plutonium being about 84% Pu-238 by weight. In this composition, 99.9% of the alpha activity is from Pu-238 decay, and therefore, it is the isotope responsible for almost all the thermal power of the source. The alpha radiation is stopped by the cladding used to encapsulate the plutonium dioxide, but there are both neutron and gamma radiation fields external to the RTG. The neutron emmisions come from (alpha,neutron) reactions, spontaneous fission and neutron induced fission. In order to reduce the neutron emmision rate from oxygen (alpha,neutron) reactions in the RTG or gloveboxes, the plutonium dioxide can be depleted in oxygen-17 and oxygen-18.

The production operations are performed manually by technicians working in controlled-atmosphere gloveboxes. Time studies have shown that manipulations performed in a glovebox require about ten times the duration of duplicate maneuvers performed on an open workbench. The objective in reducing radiation exposure is elimination of labor intensive tasks required for production of the RTG's.

The simplicity and repetitive nature of the calorimetry measurement made this particular step ideal for an automated system. Described below are the steps to develop an automated procedure and a discussion of the modifications implemented to develop a successful robot system for use in the gloveboxes.

METHODS

Present Method of Operation

Safety considerations necessitate complete containment of the radioactive fuel in the RTG in any credible accident situation. The encapsulation process uses expensive tantalum-based materials and precise welding procedures that require careful detail and precise control. It is necessary then, prior to the encapsulation step, to verify the specific thermal power of the plutonium dioxide. Calculations based on plutonium isotopic analysis are completed to determine the weight of fuel needed for a particular heat output. Individual charges of fuel are weighed, based on these calculations, and loaded into stainless steel transfer containers.

These containers are then moved to another glovebox where each charge is measured by a calorimeter to verify its thermal output. An operator calibrates daily the calorimeter by first obtaining a baseline reading (the sample holder is calibration standard is removed and a transfer container with 238-plutonium dioxide is inserted into the sample holder. The heat output is measured every 100 seconds. When heat output reaches equilibrium (output remains stable for 10 consecutive readings), data are recorded. The sample is removed and the next sample is inserted. The procedure is repeated.

The operations of obtaining a baseline and calibrating the calorimeter normally require two to three hours. Samples are normally completed in 1 1/2 to 4 hours, depending on the calorimeter. The calorimeters are microprocessor-controlled but require frequent technician interfacing to manually transfer to the next step. Because of the variable nature of the determination, it is difficult to effectively use the technicians' time to perform other tasks and maintain effective use of the calorimeter.

Automated Method of Operation

Two calorimeters presently exist in the floor of the glovebox. The automatic calorimeter station will involve positioning the Zymark robot between the calorimeters such that each can be loaded and unloaded by one robot. Additional modules include the Zymate capping station, the power and event controller, and other specially designed equipment that will be discussed later.

A Hewlett-Packard 9825 computer is used to control both calorimeters as well as the Zymate System. The interface between the two computers utilizes the standard Zymate interface card for data input and output. An additional interface card was designed by Los Alamos to facilitate the interface through switch closures and inputs on the power and event controller. This was required due to the random nature of the calorimetry determination.

The Zymate capping station utilizes new aluminum jaws designed to hold the 3" tall sample holder. A 35 degree chamfer at the top of each jaw aides in thread alignment between the sample holder and the measurement assembly.

To assist the robot during the insertion and retrieval of the measuring assembly, two solenoid driven brake devices were designed. The brakes hold the measuring assembly in an intermediate position while the robot re-grips the assembly to complete the maneuver in or out of the calorimeter well. The brakes are activitated by switch closures on the power and event controller, with 24VDC provided by an external power source.

Because the robot will be contained within a glovebox, special care has been taken to develop an easily maintainable system. A special "one-of-a-kind system" has been designed by Zymark Corp. This robotic arm has all the electronic components relocated outside of the arm in a special box. Thus, the electronic circuit boards cannot become contaminated, will not be subjected to radiation exposure and can easily be repaired/replaced without entering the glovebox. In addition, the robot arm will have an aluminum nut housing, dry lubricant bearings, argon gas purge system and special electrical connectors suitable for glovebox conditions.

To eliminate potential problems with hand changing (corrosion of the electrical contacts by the 238-Pu) a dual function hand was designed for this application. An electromagnet is mounted on the bottom of a special aluminum grip hand fabricated by Zymark. Power for the 24VDC electromagnet is obtained through a modification to the wrist circuit board. The magnet is activitated by software commands normally used for the syringe module. A special set of aluminum grippers were designed to accomodate the large diameter of the measuring assembly.

The automated sequence of operation for the system is as follows:

1. An operator will position a tray of plutonium dioxide charges to be measured in the calorimeters. The charges are in transfer containers made from Invar, a magnetic stainless steel.

2. The operator will select the proper numerical code and initiate the cycle via HP-9825 computer.

3. The calorimeter will be run without a sample to establish the base line.

4. The robot will select a standard and place it in the sample holder using the electromagnet.

5. The sample holder will be threaded onto the measuring assembly using the capping station.

6. The measuring assembly will be inserted into the calorimeter to check the calibration.

7. The measuring assembly will be removed and disassembled.

8. The standard will be returned to the storage rack.

9. A sample to be measured will be selected from the feed tray using the electromagnet. Steps 5, 6, and 7 will be repeated until all the samples are measured.

DISCUSSION

Design and fabrication of all the peripheral equipment for the automated system is complete. A robot locating fixture will facilitate glovebox mounting and accessibility when the system is installed. Should the robot arm require maintenance or replacement the locating fixture will allow exact re-positioning of the arm eliminating the need to "re-teach" positions.

The electrical connections between the robot arm inside the glovebox and the remote control box outside the glovebox required special considerations. A service panel has been fabricated using 6-pin, hermetically sealed, stainless steel Cannon connectors. The robot arm will required the use of three 6-pin connectors. This panel was designed to replace an existing panel on the glovebox. Cables used inside the glovebox to link the robot arm with the service panel are being fabricated by Pave Technologies(Van Dalia, OH). The six pairs of insulated solid wire will be bundled inside another standard plastic sheath. This is covered by heat-shrink tubing that has proven to be resistant to corrision in the Pu production line. This will covered by stainless steel flexible mesh for impact resistance, and both ends of the cable will be hermetically sealed to stainless steel connectors. Three of these cables will link the robot arm to the service panel (Figure 1). The hermetic seal is required to prevent the Pu-238 from "creeping" along the wires and destroying the plastic insulation. Three standard cables will be used to link the remote control box with the connectors on the service panel.

Installation of the complete system is planned for the first week in November 1985. The special Zymark robot arm will be delivered to Los Alamos by the first week of October. A complete mock-up of the automated system will be ready by the time the arm is delivered. This includes a glovebox fabricated out of plywood and lucite, special electrical cables modified service panel, two resistance heated calorimeter wells to evaluate any thermal expansion problems, argon gas purge system, Invar transfer containers, and all the electronic computer and control equipment. The system will be tested for several hundred cycles to complete the evaluation. The full scale mock-up should be completed by the end of October. Concurrent with the mock-up, the glovebox will be prepared for the automated system. The service panel will be installed and studs welded to the glovebox floor for bloting and equipment in position.

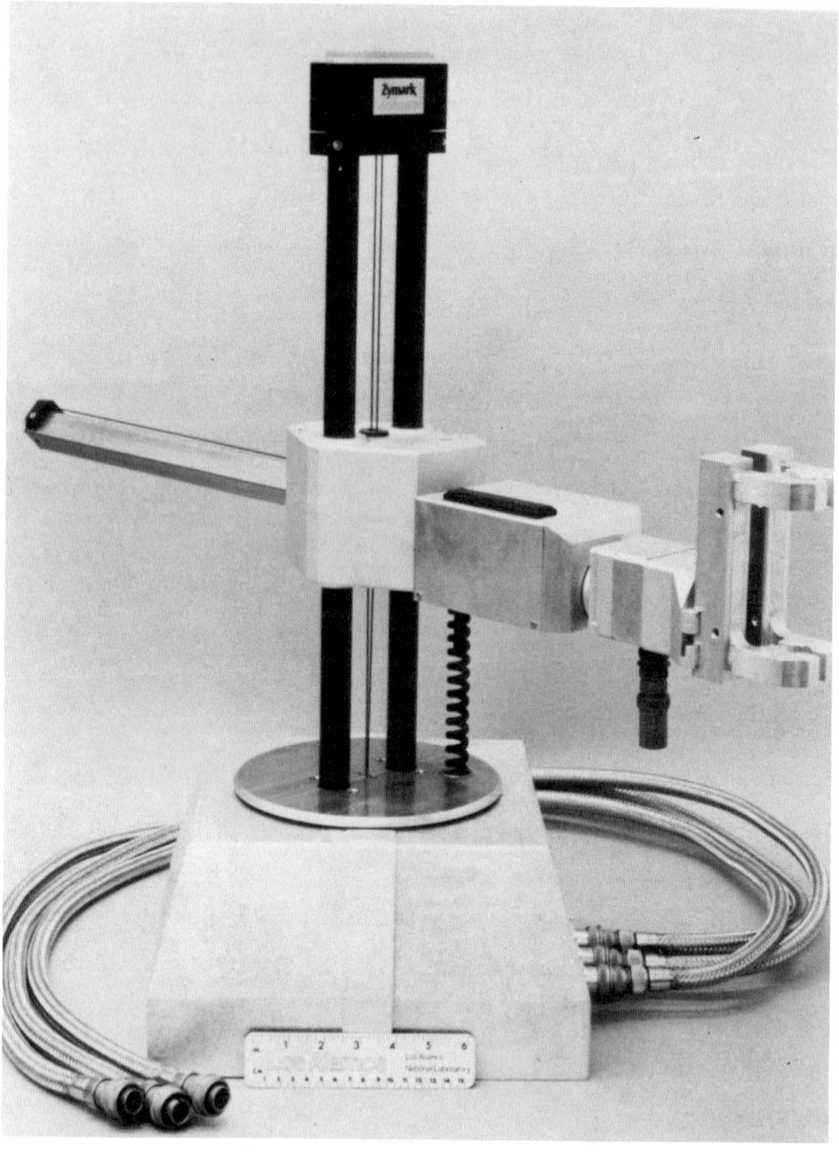

FIGURE 1. Custom Zymate System for Calorimetry.

Introduction of the robot arm and peripheral equipment should be completed in one work day. By the middle of November all software changes should be complete.

CONCLUSION

The current production schedule for RTG fabrication is about 40 charges a month. Historically, it has been a lot higher but a backlog of RTG's has slowed down production. The rate limiting step for the system will be the calorimeter. At the current production rate, the robot will be operated roughly eight hours a day, two days a week. Any increase in the production schedule will not overload the automated system.

The real benefit of the system will be indicated by a decrease in the exposure rates for the operating personnel. The Zymate System operating inside the glovebox can perform tasks faster and more accurately than a human. The reproducible movements of the robot arm will improve safety by reducing mundane labor intensive tasks that historically contribute to the cause of accidents.

ACKNOWLEDGEMENTS

We wish to thank the Zymark Corporation for their continued support of this special project. Special credit is due to Dr. R. Brown and William Buote for their creative input that insured a successful robotic application at Los Alamos.

REFERENCES

1. U.S. Department of Energy, "Fifteenth Annual Report on Radiation Exposures for DOE and DOE Contractor Employees - 1982", pp. 1-14, Appendix B, Department of Energy, Office of Nuclear Safety, Washington, DC.

AUTOMATED HYDROGEN AND NITROGEN ANALYSES USING A ZYMARK ROBOT

L.J. Hilliard, L.G. Alexakos, R.J. Kobrin,
M.P. Granchi, and P. Grey
Mobil Research and Development Corporation
Research Department, Paulsboro Research Laboratory
Research Services Section,
Paulsboro, NJ 08066

ABSTRACT

The first Zymate Laboratory Automation System purchased by Mobil has been placed into routine use to automatically change samples on a continuous wave NMR instrument. This instrument is used to determine the weight percent hydrogen in petroleum samples. This application represents the first routine use of robotics in the Paulsboro laboratories. Use of the robot allows unattended operation of the instrument, which has led to significant improvements in productivity. Unattended operation has also enabled greater flexibility in the data acquisition program, resulting in much better precision for the test.

A second Zymate System was purchased and programmed as a multi-purpose sample preparation system. The first sample preparation method developed was for a test which uses a chemiluminescence technique to determine nitrogen at the ppm level in liquid petroleum samples. Unattended operation of this sample preparation has also improved productivity, and there is excellent agreement of results between robot-prepared samples and manually prepared samples.

PART I: AUTOMATED HYDROGEN ANALYSES

INTRODUCTION

The application of a robot[1] for changing samples on the NMR instrument[2] has been set up in a "Rapid Analysis Laboratory" (RAL). This lab was designed to provide 24-hour turnaround on petroleum samples requiring sulfur, nitrogen, and hydrogen analyses for material balance calculations at our pilot units. The use of the NMR technique for hydrogen determination at Mobil was developed by L.G. Alexakos and J.E. Bowen.

The RAL is attended by only one analyst. The use of automatic sample changers on all three instruments was thus a prerequisite for successful operation of the lab. The sulfur and nitrogen analyzers did have commercially available autosamplers, whereas the NMR-hydrogen analyzer did not. Use of a laboratory robot presented a more cost-effective alternative to an in-house developed autosampler.

In addition, adaptive control of the robot by a PDP 11/44 computer allows its use as an "intelligent" autosampler. Autosamplers are fixed mechanical devices that transfer samples to and from an instrument in the same order every time. The robot arm allows random access of an entire batch of samples and therefore is completely flexible. The computer decides the following:

- How many readings are necessary for satisfactory precision on a standard or sample
- How often to run a standard
- Whether or not individual samples must be rerun

The robot waits for the appropriate signals from the computer regarding which tube to load, and when to load or unload it, based upon the above decisions. This operation has resulted in significant improvements in productivity and much better precision of the test.

EXPERIMENTAL

Robot Equipment

The robotic equipment used for this test was purchased from Zymark Corporation and consists of:

- Robot Arm and Controller
- Printer (used solely in this application for obtaining hard-copies of programs)
- Power and Event Controller (PEC)

Neither the A/D converter nor the AC outlets on the PEC were used in this application. A "teach" pendant was used with the system for programming robot motions. A diagram illustrating the layout of equipment for the method is shown in Figure 1.

Gripper Fingers and Rack Design

During method development, and since the robot has been in routine operation, there have been consistent problems associated with reliability in returning the tubes to the rack after analysis. The tubes being used here are longer and wider (graduated Nessler tubes, 32 mm o.d. and 210 mm long than those commonly used with laboratory robots). Several designs of fingers and racks have been tried. Initially, the manufacturer's fingers were machined from aluminum, were somewhat angular, and approximately 1.2 cm deep. Due to slippage of the

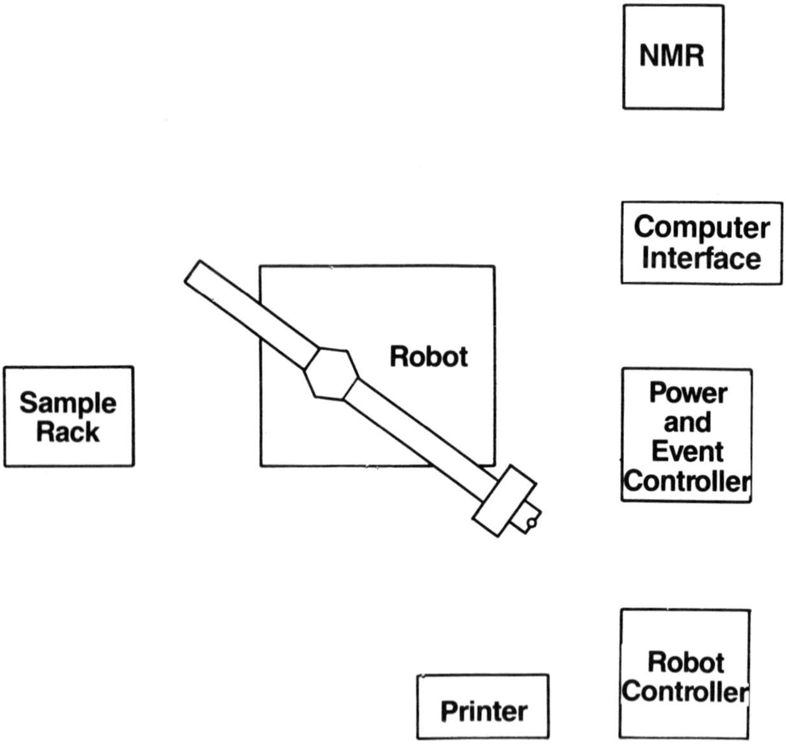

FIGURE 1. Bench layout for NMR sample-changing.

glass tubes in the metal fingers, four rubber pads were added. However, these only allowed four-point contact on the tube. A complete rubber lining was then added. The angling of the rods for the fingers only allowed the robot to grip the tubes at the very top, and there was thus significant swing of the tube as it traveled. A second finger design was machined at Mobil which consisted of fairly straight rods with a 2.1 cm deep piece of spring steel between them. This failed to alleviate the positioning problems, and a third design was machined by Zymark (those in use at this writing). These are 3.1 cm deep cylindrical aluminum fingers allow the tubes to be gripped at a lower height and hold them quite rigid during travel. Because of this, however, the tube must be placed down on the bench and the gripper fingers moved to a higher level on the tube so that it can be properly placed in the NMR sample chamber. The reverse must also be done on the unload cycle.

It was also discovered that the best design of a rack (thus far) seems to have only two layers, one towards the top of the tube with holes about 0.25 inch bigger than the tube o.d., and a second locator plate at the bottom of the rack, with holes that conform to the size of the tube. Both layers must be beveled. This is especially important with these flat-bottomed tubes. This holds the tube rigid enough to allow proper pick-up, but affords the felxibility necessary to return the tube to the rack properly.

Analysis Procedure

Samples are prepared by weighing approximately seven grams of sample into the tubes and diluting to the 30 mL mark with carbon tetrachloride. The sample tubes are placed in a 30-position rack next to the robot. A sealed tube containing cyclohexane diluted with carbon tetrachloride is used as the calibration standard and is always placed in rack position 30. The rack and robot are placed on an elevated bench next to the NMR so that the robot can lift the long NMR tubes into the NMR sample chamber.

Data Acquisition and Calculation of Results

An interfacing device, termed a "MAGIC Box" (Mobil Analytical General-Purpose I/O Console), is used to acquire the data from the NMR and to transfer commands to the Zymate System from a PDP 11/44 computer. A FORTRAN program converts the integrated NMR signal to percent hydrogen. The program also performs statistical analysis on the results to determine how often the calibration standard must be checked, to determine the appropriate number of readings to be taken to achieve a satisfactory precision, and whether certain samples must be rerun due to instrumental drift, etc. Cyclohexane check standards are run with the samples to monitor accuracy and precision.

The run begins with the measurement of the calibration standard; a minimum of three "sets" of readings are obtained on it. A "set" of readings consists of nine to twenty individual NMR readings (one reading may be rejected if necessary), this number being determined when the prescribed precision has been met (standard deviation <= 1.10 in NMR reading units). An average for each set of readings is calculated, as well as sensitivity (calibration) factor for each based upon the known percentage of hydrogen (14.37% for cyclohexane). The standard deviation for the three factors is calculated; if this is below the set limit (.050), the analysis of samples proceeds. If not, additional sets of readings are taken until this criterion has been met. An average sensitivity is also calculated. If greater than seven sets of sensitivity readings are required, only the last seven are used in the calculations.

The measurement of the calibration standard as described above is repeated after every six samples. At this point, the "before" and "after" average sensitivities are compared. If the difference is greater than the set limit (.070), then the intervening samples must be rerun. (One source of this may be large instrumental drift). The first three samples are rerun; if this is successful, the number of samples in a batch between standards is increased by one each time until six are being run. Should the first three re-analyses fail, then the first sample is re-analyzed again (bracketed by standard runs) until it is successfully completed. The number of samples between standards is gradually increased as just described. The data acquisition criteria and NMR parameters are summarized in Appendix 1.

Sample Changing Operation

Because of the need to control the order in which samples are run, it was necessary to transmit tube numbers to be loaded to the robot. The first five inputs on the power and event controller (PEC), have been used to transfer this information in binary, which is then converted to decimal using a small counting program shown in Appendix 2. This COUNT is then used as the rack indexing variable by the robot. The use of the inputs and switches on the PEC for communication between the computer and the robot represented a novel application of this module and has been described in the Zymark Laboratory Auutomation Newsletter.[3]

Input 8 is used to transfer "load" and "unload" signals to the robot. Each time input 8 is checked, the "COUNT.TEST" program is run. A "handshaking" routine between the PDP 11/44 and the robot is initiated at the beginning of the run with signals being sent and confirmed of all its action to the PDP 11/44 computer through switch 8 on the power and event controller. The overall communications program between the two is shown in Appendix 3.

Safety Features

For safety reasons, an uninterruptable power supply (Topaz Powermaker[4]) which will provide six minutes of power was added to the system. Normally, in the event of a power failure, the robot arm drops at its current location; this can damage both the arm and fingers as well as glassware and samples. Input 6 is connected to a status monitor on the power supply. The robot checks this input while waiting for an unload tube signal. If it determines that it is operating on backup power, the arm moves to a safe position and ends the program. The status monitor check was added at this point in the program because it is the rate-limiting step for the program.

Another safety feature built into the program is the use of a small switch mounted on the side of the tube rack and connected to input 7. This switch is used to confirm the presence of a tube in the robot gripper hand, as tubes are being loaded and unloaded. Should the robot fail to detect the presence of a tube, the robo waits for analyst intervention.

Additional safety features are being incorporated into the test, such as an alarm to alert an analyst if necessary during the error detection steps, enclosure of the robot bench in a drain pan to catch spills, and hydrocarbon sensors to warn of a spill.

DISCUSSION

Prior to the use of the robot for this test, the constant presence of a technician was required solely to change samples (after sample preparation was completed). The average analysis time of 8 minutes per sample prohibited the analyst from performing any other tests simultaneously. Commercial sample changers were not available for the instrument and the use of a robot offered the most convenient and cost-effective way to automate the instrument.

In addition, the precision for the test has improved because of the more flexible data acquisition program made possible by the robot. There has been improvement from an observed repeatability of 0.09% to 0.05% hydrogen for a single type of material (jet fuel), and a 13% imporvement in the precision for a variety of samples ranging from gasolines to residuals.

This application, while relatively simple, was deliberately chosen to allow familiarization with a robot system concommitant with a successful, labor-savings application.

FUTURE WORK

Extension of the robot for unattended sample preparation for the NMR test is in progress and is expected to produce further productivity improvements. In addition, the RAL can be modified so that the time that the robot waits to unload an NMR tube can be put to use for sample preparation for another test.

PART II: AUTOMATED NITROGEN ANALYSES

INTRODUCTION

Our second Zymate System is a multi-purpose sample preparation system. The first method developed was the sample preparation for an instrument which uses a chemiluminescence technique[5] to determine nitrogen at the ppm level in liquid petroleum samples. The method invloves diluting a weighed portion of sample with xylene, mixing the sample, and transferring this diluted sample to a crimp-top autosampler vial. This automated sample preparation will yield significant improvements in productivity, and there has been excellent agreement between robot-prepared and manually-prepared samples.

EXPERIMENTAL

Robot Equipment

The robotic equipment purchased from Zymark Corporation for this test consists of:

- Robot arm and Controller
- Printer
- Power and Event Controller (PEC)
- Master Laboratory Station (MLS) with fixed nozzle dispenser
- Capping Station
- Vortexer
- Pipet tip remover
- Syringe hand
- Dual-Purpose hand with small standard capper fingers and nozzle for use with pipet tips and attached via hosing to the MLS

The following equipment was also used:

- Air-powered crimper[6] and associated solenoid valve for control of 60 psi air source by PEC[7]
- Solenoid valve between vacuum source and crimp-cap holding station under control of PEC[8]
- "Load box" consisting of two low-wattage light bulbs between the solenoid valves and the switchable outlets on the PEC
- Balance[9]

A diagram illustrating the layout of equipment for the method is shown in Figure 2.

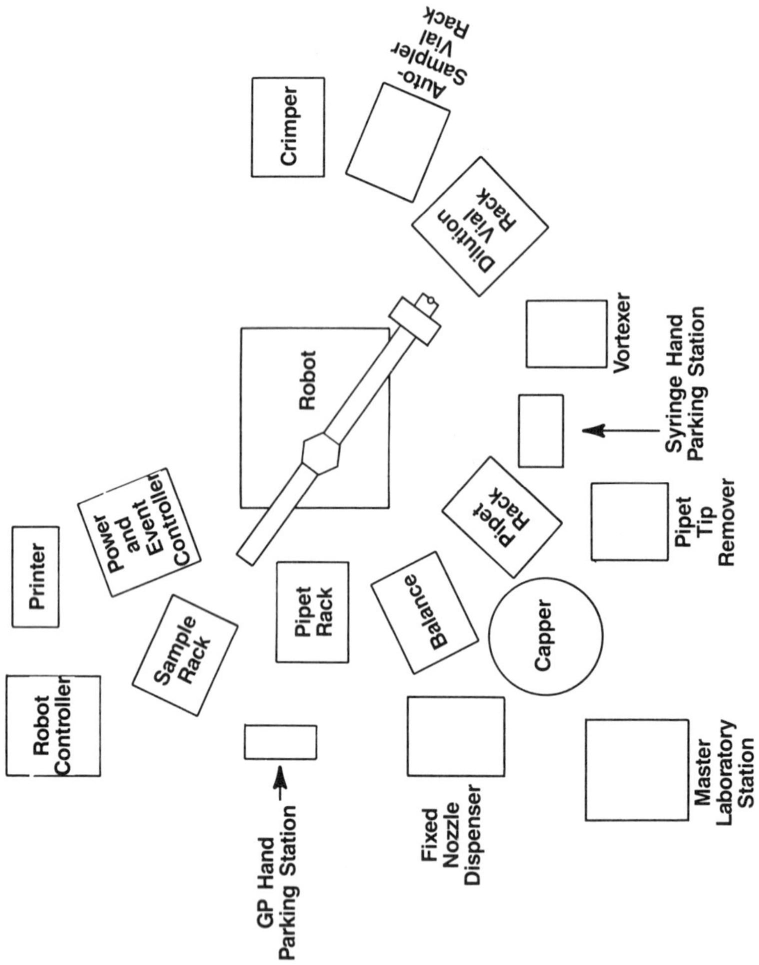

FIGURE 2. Bench layout for sample preparation for nitrogen analyses.

Sample Preparation Method

The robot begins the sample preparation by uncapping a clean dilution vial, placing the cap on a holder, and obtaining a tare weight on the vial. The sample to be analyzed is inverted several times using the wrist action of the robot, then uncapped, and the cap placed upside down on a holder, to prevent cross-contamination of samples.

The robot hand is then rotated 180 degrees, and a 5-mL pipet tip in conjunction with the MLS is used to transfer the appropriate weight of sample (as entered by the analyst at the start of the program, usually from about 0.1 to 0.8 gram) to the dilution vial. The pipet tip is then removed while a weight is recorded for the sample.

The hand is then inverted, the original sample capped and returned to the rack, and the dilution vial carried to the fixed nozzle dispenser where 4 mL of o-xylene are delivered. A total weight is then obtained for the vial and contents, and a dilution factor calculated. Sample weights and dilution factors are printed out for the analyst.

The dilution vial is then capped, inverted several times, and placed in the vortexer for 20 seconds. The vial is removed and uncapped at the end of this time period. The robot then picks up an autosampler vial from the rack, takes it to a station which consists of an inverted rubber septum slightly larger than the crimp cap, that is connected to a vacuum soruce. By actuating a solenoid valve through the power and event controller, the cap is held in place until it is needed. The hand is parked and the syringe hand then attached. This hand is used with a 1-mL pipet tip to transfer an aliquot of the diluted sample to the autosampler vial. The pipet tip is then discarded, the syringe hand parked, and the dual-purpose hand re-attached.

The filled autosampler vial is then picked up, placed on the (screw) cap holding station so that the hand can be inverted and the vial picked up with the wrist 180 degrees. This is done so that the vial can be placed up within the crimper jaws. The cap is retrieved from the suction holding station when the vacuum is turned off via the solenoid. The cap is crimped on by actuating the air solenoid, and the vial is returned to the rack. The dilution vial is also re-capped and returned to its rack. The robot continues the method until all samples have been prepared, parking its hand at the end of the procedure. A final report is printed out for the analyst which also includes date and analyst badge number.

Safety Features

Just as with the previously described method, the use of switches as error detection devices are an important safety feauture. A switch was used in a rather unique manner to confirm the presence (or absence) of crimp-caps on the autosampler vials. If the red cap on these switches is removed, a small white plastic plunger is exposed. This plunger is narrower than the opening on the autosampler vials, and thus can be used to detect the cap presence only if the cap with a rubber septum is there to force the switch to close. If the robot detects the absence of a cap on a vial prior to crimping, it places the vial which will not be used for sample preparation, and places it on the vial to be crimped (there are more autosampler vials racked than are needed in any one batch of samples).

Additional safety features, such as an uninterruptable power supply, an alarm, drain pan to catch spills, hydrocarbon sensors, and a hood to enclose the entire system, are also being incorporated.

DISCUSSION

This application was the first use of robotics for sample preparation at Mobil. While the sample preparation time for the robot is probably twice that of a human as the method is performed now (approximately seven minutes for robotic preparation), significant labor savings can be achieved through 24-hour operation of the robot. Further improvements in productivity will be achieved when the sample preparation is serialized with sample analysis by the chemiluminescence instrument.

To validate the accuracy of robotic sample preparation, approximately thirty samples covering a range of viscosities and nitrogen concentrations were prepared by the robot during method development, and the results were compared to those obtained using manual preparation (Table 1). With one exception, all sample results were well within our prescribed precision limits for the test. The use of the robot is thus improving our precisions as well as the efficiency of our analytical procedures.

TABLE 1. Comparison of Nitrogen Analyses for Robot-Prepared Samples vs. Manually-Prepared Samples.

RESULTS (ppm nitrogen)		
Manual	Robot	Difference
100	100	0
13	16	3
13	14	1
12	15	3
120	120	0
78	80	2
5	6	1
140	140	0
23	26	3
44	45	1
150	140	10
39	40	1
8	13	5
200	200	0
40	41	1
41	33	8
25	28	3
21	24	3
89	88	1
23	21	2
36	34	2
27	24	3
97	87	10
98	89	11
200	190	10
160	150	10
23	23	0
23	22	1
6	7	1

FUTURE WORK

The operation of this robot is being expanded to include two other sample preparation procedures:

- samples for the NMR-hydrogen analysis
- used oil samples for flow injection/ICP analysis

This will be readily accomplished because the robot has been programmed using "generic" subroutines for performing the various unit operations,and will only require the addition of two pieces of hardware to the robot bench, as illustrated in Figure 3.

This flexibility is exemplary of the definition of robots; this definition distinguishes robots from other types of automation used for sample handling: "(A robot is) a reprogrammable and multifunctional manipulator, devised for the transport of materials, parts, tools, or specialized systems, with varied and programmed movements, with the aim of carrying our varied tasks[10]".

ACKNOWLEDGEMENT

The assistance of J.A. Biggerstaff and G.P. Sutton in this work is gratefully acknowledged.

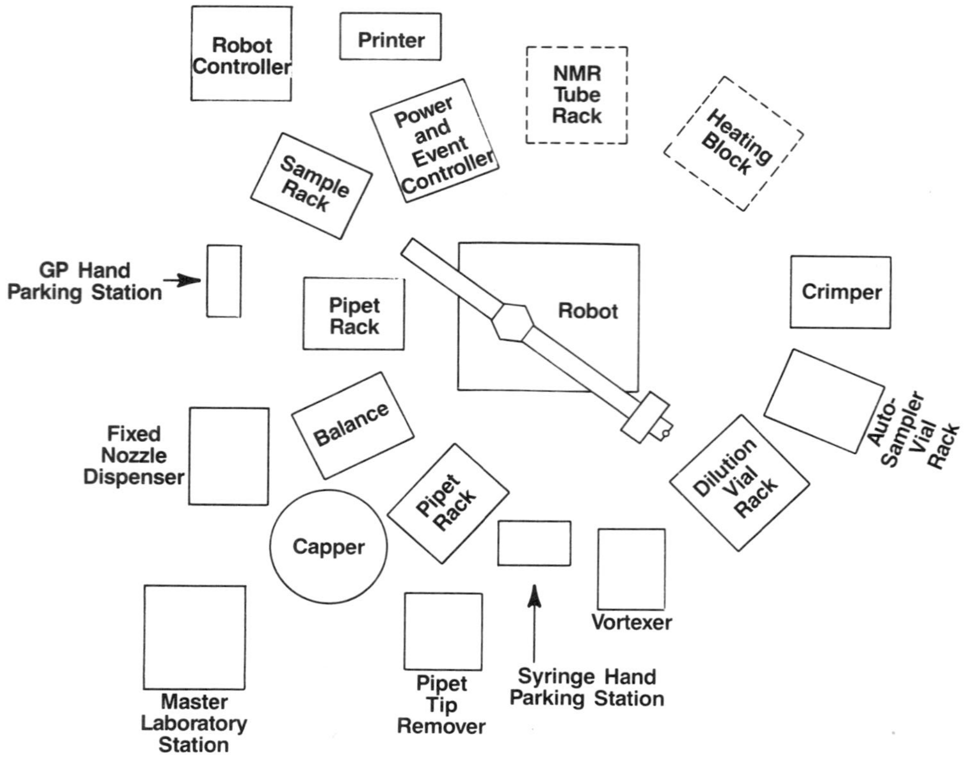

FIGURE 3. Bench layout expanded to accomodate three sample preparation schemes - hydrogen and nitrogen by NMR and ICP analysis.

REFERENCES

1. The Zymark robot and associated modules are manufactured by Zymark Corporation, Zymark Center, Hopkinton, MA, 10748.

2. Oxford Analytical NMR (Newport Analyzer), avaialble from Anlaytical Marketing, One Dundee Park, Andover, MA, 10810.

3. Hilliard, L.J., Zymark Laboratory Automation Newsletter, Vol. 2, no. 2, 1985.

4. Topaz Powermaker Micro UPS, 1000 VA, 60 Hz, Model 84126-01, available from Topaz, 9192 Topaz Way, San Diego, CA, 92123-1165.

5. Antek Model 703C Chemiluminescent Nitrogen System and Model 736-2 liquid autosampler, available from Antek Instruments, Inc., 6005 North Freeway, Houston, TX, 77076. Also equipped with Spectra-Physics Model 4100 Data System with BASIC programming option, available from Spectra-Physics, 333N. First St., San Jose, CA 95134.

6. Wheaton air-powered crimper, attaches 11mm (224211) seals, and stand, P/N 224465, available from Wheaton Instruments, 1301 North Tenth Street, Millville, NJ, 08332.

7. Humphrey 120V Mini Myte 3-way valve, P/N 31E1, 0-100 psi, 6W, available from Winco, Inc., 155 Terwood Rd., Willow Grove, PA, 19090.

8. Skinner solenoid valve, P/N B13 DK9 150, 120V, available from John C. Whiddett, P.O. Box 597, 5151 West Chester Pike, Edgemont, PA 19028.

9. Mettler PE160 balance, available from a Mettler distributor, or contact Mettler Instrument Corporation, Box 71, Highstown, NJ 08520.

10. From the definition of "robot" supplied by the Robot Institute of America, as quoted in An Introduction to Robot Technology, P. Coiffet and M. Chirouze, McGraw-Hill, New York, English Edition, 1983, P. 17.

APPENDIX 1

Summary of NMR Instrumental Parameters
and Data Acquisition Criteria

Integration Time = 32 sec
A.F. Gain = 9.00
Gate Width = 1 Gs
R.F. = 20 A
Minimum number of readings = 9
Maximum number of readings = 20
Maximum standard deviation in average reading = 1.10
Number of samples between standards = 6
Maximum standard deviation for sensitivity = .050
Maximum before/after sensitivity difference = .070

APPENDIX 2

"Count.Test" Program

```
COUNT = 0
IF INPUT.1 = 1 THEN COUNT = COUNT + 1
IF INPUT.2 = 1 THEN COUNT = COUNT + 2
IF INPUT.3 = 1 THEN COUNT = COUNT + 4
IF INPUT.4 = 1 THEN COUNT = COUNT + 8
IF INPUT.5 = 1 THEN COUNT = COUNT + 16
```

APPENDIX 3

"M1252.RAL" Program

```
    TOP180
    CLEAR.1 These are safe starting positions for the arm
    SWITCH.1.ON Signal for no robot problems; the default is off
    SWITCH.7.OFF
1   IF INPUT.8 = 0 THEN 1 Wait for load signal
5   COUNT.TEST
    IF COUNT > 0 THEN 1 Default on inputs is high
    PRINT COUNT
10  SWITCH.8.OFF Tube 0 is loaded
20  IF INPUT.8 = 1 THEN 20 Wait for unload signal
25  COUNT.TEST
    IF COUNT = 31 THEN 60
    IF COUNT > 0 THEN 20
    PRINT COUNT
    SWITCH.8.ON Tube 0 is unloaded
30  IF INPUT.8 = 0 THEN 30 Wait for load signal
    COUNT.TEST
    PRINT COUNT
    IF COUNT = 31 THEN 60
31  LOAD.TUBE Program containing locations for loading tube
    SWITCH.8.OFF Tube is loaded
35  IF INPUT.6 = 0 THEN 70 Check status of UPS monitor
    IF INPUT.8 = THEN 35 Wait for unload signal
    PRINT COUNT
41  UNLOAD.TUBE Program containing locations for unloading tube
    SWITCH.8.ON Tube is unloaded
    GOTO 30 Cycle back and wait for load command
60  SWITCH.8.OFF Tube 31 loaded
    CLEAR.1A
70  TOP180
    END
```

ROBOTIC AUTOMATION FOR X-RAY FLUORESCENCE ANALYSIS OF SULFUR IN OILS

J. B. Cross, L. V. Wilson,
E. J. Marak, and R. D. Jones
Phillips Petroleum Company
Research and Development
Bartlesville, Oklahoma 74004

ABSTRACT

Robotic automation has become an important method for improving the productivity and data quality in modern analytical laboratories. A Zymark Laaboratory Automation System (Robot) is used to automate sulfur analysis of hydrocarbons by x-ray fluorescence. The report describes the workcell designed and built to provide complete automation of the method including sampling, sample preparation, analysis, and report generation. The robot accomplishes the complex tasks involved using workstations commercially available and special workstations built by the authors. A description of the communication link between the x-ray system's computer and the Zymate Controller is included.

INTRODUCTION

The analysis of sulfur in hydrocarbons, especially fuel oils, is an important factor in predicting performance characteristics and potential corrosion problems. Both areas are important in the processing and marketing of petroleum and its products. As indicated in ASTM Procedure D-2622[1], x-ray fluorescence (XRF) is an appropriate technique for this analysis. Modern x-ray equipment has made the application of the technique an accepted and straightforward procedure. However, due to the nature of the technique and the sample preparation involved, there are still ample opportunities for automation. For example, the sample can be analyzed on an as received basis when the sulfur concentration is low, or the sample may be diluted to match the calibration standards if the sulfur concentration is high. Robotic automation, as described in this report, can be used to automate the sample preparation and to interface with the analytical instrument for automated, accurate analytical results.

To measure the sulfur content by x-ray fluorescnece, the diluted or as received sample is placed in an appropriate cell, and the cell covered with a thin (1/4 micron) Mylar film. The sample cell is placed in the x-ray beam and measured. The x-ray intensity is compared to that from a set of calibration standards for quantitation. Errors in the results can arise from either changes in the carbon to hydrogen ratios or from the presence of interfering elements such as chlorine. Particulate or water settling can also affect the results. Often these errors can be eliminated by diluting the sample in a solvent, such as toluene, or more rigorous techniques[2] can be applied if necessary. Sample homogeneity, instrument stability, and differences in the measurement atmosphere from sample to sample (usually a helium flush is used) are other sources of error.

Most modern x-ray instruments are successful in automating the measurement portion of the analysis; however, the sample preparation and data evaluation portions of the technique have remained isolated activities. The robotic workcell described here ties together these activities for a complete analysis package. The procedure includes sampling, sample preparation, measurement, error detection and corrective action, and reporting. The system uses a Zymate Laboratory Automation System to automate the sampling and sample preparation procedures, a computerized x-ray spectrograph that includes software designed to take x-ray measurements, detect errors, report new sample preparation instructions back to the Zymate System, and to generate analytical reports. One report is made to the analyst(s), and a separate report of the final analytical result is made to the customer.

Robotic automation was chosen to tackle the job of measuring sulfur in oils because of its flexibility and reliability in manipulating analytical samples and its capability to communicate with other analytical instruments and computers. The system developed includes techniques to:

1) Identify incoming samples through bar code labels.
2) Aliquot samples from the sample bottle.
3) Weigh, and if necessary, dilute the sample.
4) Load the sample in the x-ray cell, including covering the cell with a thin Mylar window.
5) Shake the sample and vent the sample cell to prevent pressure build-up during x-ray measurement.
6) Load the cell into the x-ray spectrograph.

7) Transmit the sample preparation information to the x-ray computer and receive further sample preparation instructions.

EXPERIMENTAL

Instrumentation

A Philips Universal Vacuum X-Ray Spectrograph (4-positions) was used. The spectrograph is equipped with a flow proportional detector and preamplifer, a graphite crystal, helium flush, a chromium x-ray tube and a x-ray generator. The goniometer position is determined by a computer driven stepping motor. Data are collected with a PDP/8e computer system, and the sample changer is operated by a computer controlled stepping motor.

A Zymate Laboratory Automation System is used to prepare the samples. Commercial work stations from Zymark and other analytical equipment companies, as well as special work stations built in-house by the authors, are needed to assist the robot in the sample preparations. A picture of the workcell is shown in Figure 1.

Accessories purchased from Zymark include: a Z510 master laboratory station for dispensing toluene to dilute samples; a Z840 RS-232 compatible computer interface for communication with the PDP 8/e x-ray system computer; a Z830 power and event controller for operating solenoid valves and detecting input signals from workcell monitors; a Z850 current loop balance interface; and a Z410 capping-uncapping station.

Other commercial equipment includes: a Mettler PE 160 electronic

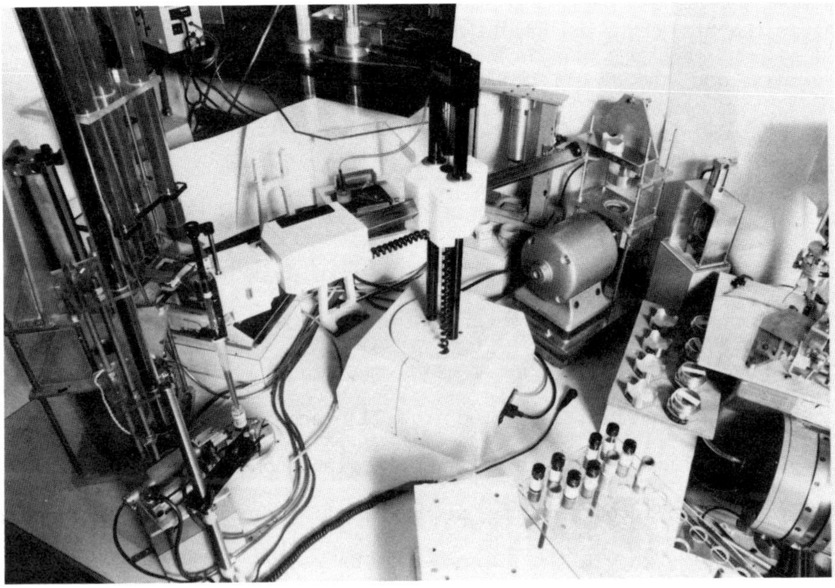

FIGURE 1. XRF sulfur workcell.

balance; and a Corning PC20 hot plate with the top modified to be a rectangular rack suitable for the robot.

Special work stations were custom built to satisfy the method requirements. The robot places a new receiving container (a one and three eights inch diameter cap plug) on the balance, uncaps the sample vial and dispenses an appropriate amount of sample into the cap plug. The sample is diluted or run as is. The cap plug is covered with a thin (1/4 micron thick) Mylar window. The film is stretched over the cap plug using a nylon collar. The collar slides around the outside of the cap plug to form a sample cell. Diluted samples are shaken, then vented to prevent warping the thin window under heat from the x-ray beam.

After the analysis, the sample cell core (cap plug and Mylar window) is removed and the nylon collar returned to its holder. Below are descriptions of the work stations developed to assist the robot in performing these tasks.

Samples that come into Phillips Analysis Branch are recorded in a laboratory management / computer system (AB/LMS).[3] Each sample is given an entry number (AB number). Information from the customer, that identifies the sample, is stored in a file accessible with that number. Samples are labeled with the AB number printed on bar code labels. The bar code label identifies the sample for data reporting either by wand input or by typing in the AB number. The x-ray computer uses a RS-232 interface link with the AB/LMS to retrieve the necessary information to report the analytical results. A bar code wand was mounted onto a pneumatic piston in close proximity to the Zymark capping station for the robot to automatically enter the bar code data. The robot loads the sample into the capping station, positions the wand via the power and event controller, then starts the capping station so that the bar code label can be read. This station is shown in Figure 2.

Sample containers are one and three eights inch diameter by one inch high plastic cap plugs. The cap plugs are picked up from a gravity fed dispenser similar to those described in another report.[4] Caps removed from the sample vials are held by a vacuum grip holder described in a separate report[5] with back-up caps supplied from a gravity fed cap dispenser.

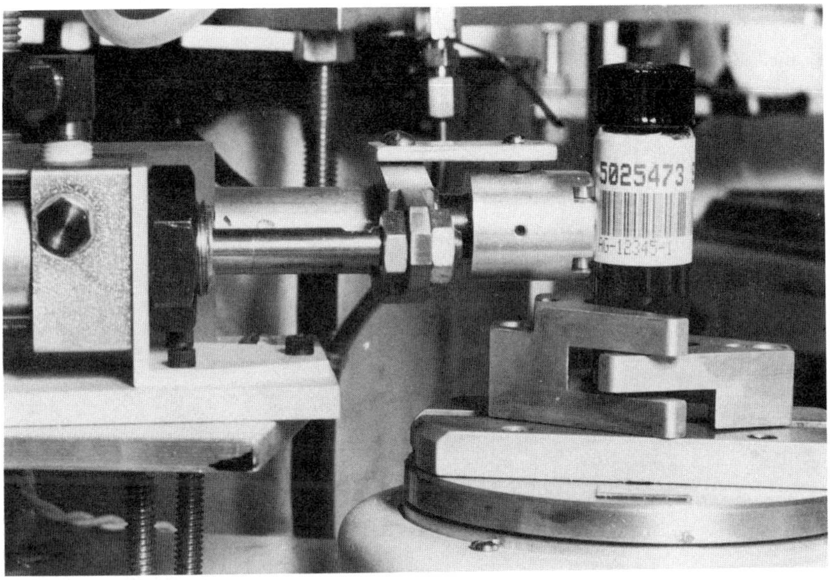

FIGURE 2. Bar code reader.

The cap plug with sample (diluted if necessary) is transferred to a staging area where it is covered with the Mylar film. The robot uses a vacuum gripper to cover the cap plug with a sheet of Mylar film. The gripper consists of a copper tube bent into a three inch circle with one end closed off, and the other connected to a vacuum line through a solenoid valve. The plastic film is held to the tube by vacuum through small holes. The top side is equipped with a handle for the robot to grip. The film is lifted, unrolled about three and one half inches, draped over a nichrome wire that is heated via electical power supplied by the power and event controller, and cut. The resulting sheet of film is placed over the sample and a nylon collar placed on top of the cap plug. The cell is assembled by pressing the collar down around the film and cap plug with a pneumatic piston. The work stations are shown in Figure 3 and Figure 4.

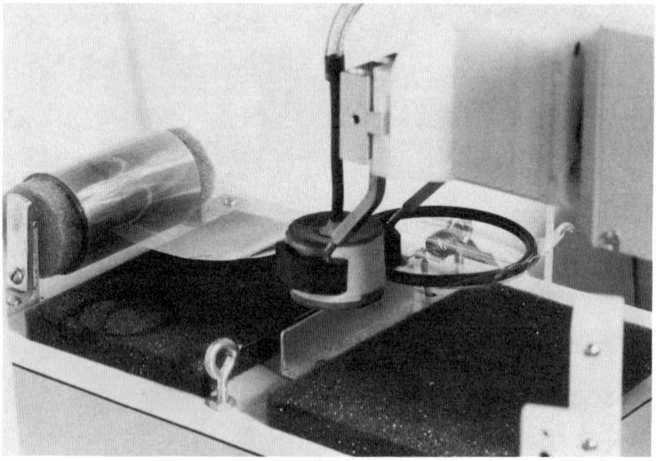

FIGURE 3. Film cutter.

FIGURE 4. Stamper.

The sample is mixed using a modified "wiggle bug" shaker shown in Figure 5. The sample is held into place on the shaker with a pneumatic piston and mixed for thirty seconds at 3000 rpm. Bubbles formed during the mixing are eliminated by turning the sample window side down.

The sample cell is vented by puncturing the bottom of the cap plug (now the top of the sample cell) with the device shown in Figure 6. This operation also serves to ensure that the cap plug is flush with the collar for accurate placement inside the x-ray spectrograph.

FIGURE 5. Shaker.

The robot transfers the sample to a pneumatic robot arm that inserts it into the x-ray spectrograph. The pneumatic robot, shown in Figure 7, consists of a pneumatic piston with a three fingered, pneumatic gripper from Mack Corporation. The spectrograph is sealed off from the atmosphere with a pneumatic controlled shutter. The sample port is monitored using an Opcon 70 Series photoelectric sensor. This confirms successful sample loading and retrieval or shuts the system down.

After the analytical measurements have been made, the robot disposes of the sample cell core (cap plug and Mylar film) using the device shown in Figure 8. This device has a metal plate with a hole beveled to fit the

FIGURE 6. Spike.

cell collar. A pneumatic piston drives the cell core into a dump container below.

The communication link between the x-ray computer system uses Zymark's Z840 computer interface. Five volt signals, sent by the x-ray computer, were used to signal the Zymark controller to configure itself to transmit to or receive information from the x-ray computer. These signals are detected through the A/D input line on the Zymark power and event controller.

FIGURE 7. XRF robot.

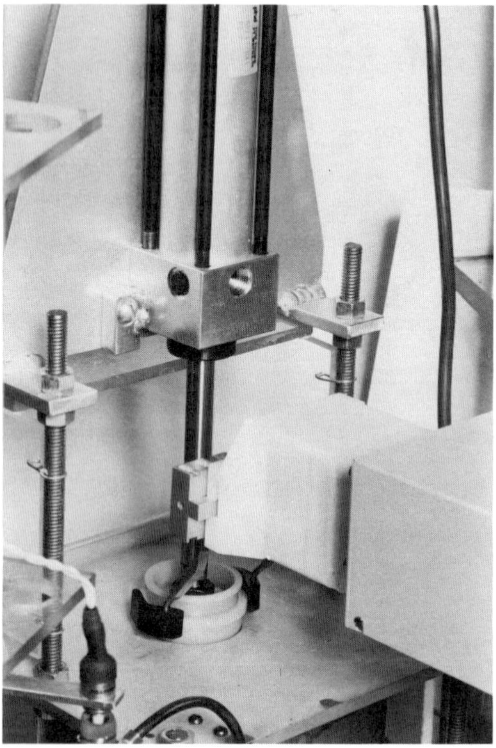

FIGURE 8. Waster.

Communication codes were established to provide usable information packages that both systems can interpret. For example, when the x-ray system is ready for a new sample to be loaded, it transmits a 1.10. Other messages include codes to unload samples, load and unload, calibrate, adjust monitors, etc.. In turn, when the Zymate System loads a sample it transmits a message such as 321.1000, where the 3 is the rack index, the 2 is the type of sample preparation scheme, the 1 indicates the first specimen of a set of duplicates, and the .1000 is the dilution factor resulting from preparing the sample.

Analytical Procedure

Samples are measured as received (if the sulfur content is low enough) or they are diluted with toluene. The sulfur content is quantitated by using standards prepared in either mineral oil, for samples measured as received, or toluene for diluted samples. Secondary standards, made by fusing lithium sulfate in lithium tetraborate to form a glass wafer and analyzed against primary mineral oil and toluene standards, are used for routine calibrations. The general procedure is to use multiple sets of standards to cover different ranges of sulfur concentrations for more accurate results.

Reference monitors (glass wafers) for periodic calibration adjustments were made in the same manner as the secondary standards by fusing lithium sulfate in lithium tetraborate to form a glass wafer. These monitors are used at the start of every run and then re-run at a four hour interval.

X-ray measurements are made on each sample at a peak and a background position. Each peak measurement is corrected for the background contribution by measuring a specimen that represents a blank for the matrix used (mineral oil or toluene). The background correction factor (BCF) is rechecked every time the periodic adjustments are made to the calibration curve using a blank lithium tetraborate wafer. The x-ray intensities at both the peak and background for the glass wafer were measured and compared to that from both mineral oil and toluene. This allows use of the glass wafers in the routine procedure.

Similarily, a "quality assurance" wafer was prepared from lithium sulfate in lithium tetraborate. A concentration was established for this sample versus both mineral oil and toluene standards. It is measured periodically to provide assurance that each curve is still valid.

The instrument calibration is based on a quadratic fit of the x-ray intensity data versus calculated concentrations of the secondary lithium tetraborate wafer standards. The calibration curves are stored on floppy disk and read into the computer memory prior to analysis. Adjustments to the curves are made by measuring the reference monitors.

A report of the analytical results is generated after the measurements are finsished. Samples are batched together, on the basis of customer, for reporting. The reports include the customer name, sample identifying numbers, sulfur content, peak and background intensities, calibration curve used, instrument parameters (measurement angle, x-ray tube, detector, crystal, power settings, etc.) and the data file name.

RESULTS AND DISCUSSION

The automated system is operated via a menu selection on the XRF computer system. Included in the menu are:

Sulfur Analyzer Options

1. Read Sample Rack
2. Analyze Samples

3. Run Calibration
4. Calibrate Versus Primary Standards
5. Run Single Prepared Samples
6. Load Calibrate Adjust Standard
7. Scan A Spectral Region
8. Measure BCF Sample

The x-ray system software then starts operation of the x-ray spectrograph and sends instructions to the robot for its operation through coded messages over the RS-232 communication link.

The instrument is calibrated using the secondary lithium tetraborate wafers. These standards are measured against primary mineral oil and toluene standards. Option 4 calibrates the spectrograph using the lithium tetraborate wafers. Both operations are under automatic control of the robot once the analyst has loaded the standards (already in sample cells) and inputted the necessary information to the XRF computer to set up the spectrograph. This input is entered as answers to a series of questions. Options 3 and 4 are not run on a daily basis. Adjustments to the calibration curve are performed using high and low reference wafers as described by Plesch.[6] Having the instrument calibration performed by the robot makes the procedure less burdensome for the operator. Using the calibrate adjust specimens avoids the time-consuming instrument calibrations for every run. Robot control over the instrument calibration makes it easier for the analyst to updatae the calibration and results in more frequent updates.

One of the keys of the success of the x-ray fluorescence technique for the analysis of sulfur in oils is the recognition by the analyst when adjustments must be made to the sample to provide accurate, reproducible results. Error detection in the data analysis software was essential to automate the method. The software developed looks at several areas where errors can occur and makes decisions concerning the data and determines if the analysis can be improved by using a different preparation of the sample. If the decision is made to try another preparation, a signal is transmitted to the robot controller containing the sample preparation instructions.

Standard matching is important to accurate analysis. Samples that have sulfur concentrations above the upper limit of the calibration curve are diluted with toluene. Also, samples that have a very low sulfur content are best measured against standards that cover only a limited sulfur concentration range. Standard selection is accomplished by using the curve for the proper preparation scheme (mineral oil or toluene dilution) as instructed by the code from the Zymark controller. The results are compared against limits set on the sulfur concentration. If the sulfur concentration exceeds the concentration limit, the Zymate System is instructed to dilute the sample with toluene.

Water or particulate material can settle during the analysis and affect the results. Settling problems are checked by taking x-ray intensity data in a series of ten-second readings. The data are fit using a least squares curve fit to a straight line. The slope of the line is compared to the statistical scatter of the data to see if settling has occurred.

Where settling problems are encountered, the maximum dilution of the sample that keeps the estimated sulfur concentration above 0.1 wt% sulfur is determined and transmitted to the robot. This approach will help overcome minor settling problems but will not handle all such problems. For cases when dilution will not improve the analysis, the sample analysis is rejected with a message printed to the analyst stating why the analysis was rejected. Then the analyst can take the proper steps to measure the sample.

Interference from other elements can also affect the results. The system routinely checks for the presence of chlorine and determines dilutions to handle the effect (if necessary). The system can also be configured to check for elements that might be added to a sample in special analysis requests.

The system has analyzed a variety of check samples and NBS standards to verify that the robot is performing the analysis correctly. Table 1 shows data for these samples. Listed are values determined by the automated system and the NBS certified value. The automated procedure provides results that are in agreement with the accepted values.

TABLE 1. Comparison to NBS Standard Samples.

NBS SRM No.	NBS Certified wt% S	Measured wt% S
SRM 1621	1.05	1.054
SRM 1622	2.14	2.167
SRM 1623	0.268	0.270

The automated sample preparation is somewhat slower than manual preparation. However, by using the robot around-the-clock, its capacity is almost double that of a person in an eight hour shift, plus it has the added benefit of requiring only minimal operator assistance. We estimate productivity gains of about 85% will be achieved using the automated system.

The benefits from automating the procedure have reduced manpower required to determine sulfur in oils and increased capacity of the x-ray system so it can handle more samples per day. The capital cost of installing the robot is less than half that of buying a new x-ray spectrograph.

CONCLUSIONS

This report demonstrates the capability of laboratory robotics to automate a complicated analytical procedure. The system is required to perform complicated tasks, communicate with other analytical instruments, interpret coded messages and adjust to new instructions. The resulting system provided the capability of analyzing the sample from the start to finish with little operator interaction.

Laboratory robotics has become an effective means of automating the analytical laboratory. The robot can be looked upon as a tool to make analysts more productive, to perform complicated tasks or to take the boredom out of an analyst's job. Either applicaton can lead to freeing the analyst for more creative work and provide real dollar savings to the employer.

REFERENCES

1. ASTM D-2622-82, Standard Test Method, "Sulfur in Petroleum Products (X-Ray Spectrographic Method)", Annual Book of ASTM Standards, Part 24, 1983.

2. R. Tertian, Spectrochimica Acta, 24B(8), 447-471 (1969).

3. P. R. Gray and T. V. Iorns, Anal. Chem., 55, 286A-296A (1983).

4. D. C. Shaker, Zymark Newsletter, 2(2), 11 (1985).

5. R. D. Jones and J. B. Cross, "Automating Sample Preparation (and Disposal) With A Robotic Workcell", in Advances in Laboratory Automation - Robotics 1985, J. R. Strimaitis and G. L. Hawk, eds, 1985.

6. R. Plesch, G-I-T Fachzeitschrigt fur das Laboratorium, No. 6, 677-683, (1973).

AUTOMATIC PREPARATION OF FUSED BEADS FOR X-RAY FLUORESCENCE ANALYSIS BY THE COMBINATION OF A PERL'X 2 BEAD MACHINE WITH A ZYMATE LABORATORY AUTOMATION SYSTEM

Jean Petin and Armand Wagner
Laborlux S.A.
L-4004 Esch-sur-Alzette
B.P. 349
Luxembourg

ABSTRACT

A Zymate Laboratory Automation System has been interfaced to a Mettler balance AE 160, a flux dispenser system and a Perl'X bead machine to realize a fully automatic system for the preparation of fused beads for X-Ray fluorescence analysis.

The flux dispenser which can add up to four different fluxes has been specially developed in order to avoid the take-up of humidity by the fluxes.

Crucibles of glassy carbon are used and a special flux has been developed to avoid cross-contamination. The system can be interfaced to an X-Ray Spectrometer and so be the final part of a totally automated system in production control; for instance, in a cement plant. The system works with the same accuracy as a manually operated Perl'X and leads to a drastic reduction in manpower especially for systems operated on a 24 hours basis.

INTRODUCTION

Preparation of oxidic samples in form of fused glass beads for analysis by X-ray fluorescence spectrometry is a widely used sample preparation technique used in a broad range of applications: geology, materials control, process control in major industries such as cement, ceramics, steel and aluminium, for example. The fusion technique is known as being the one capable of highest precision and accuracy of any sample preparation method.[1] The powdered sample is dissolved in a borate glass matrix. After fusion at a temperature of 1000 to 1200 C the melt is cast under controlled conditions to avoid crystallization or cracking.

In many laboratories the bead preparation is still completely manual and the fusion is performed in Platinum-gold crucibles on gas burners and with muffle furnaces. A certain number of automatic and semi-automatic devices are being marketed using generally RF induction heating or gas burners. Perl'X is one of these devices. It has been developed by IRSID, the Research Institute of the French Iron and Steel Industry with the financial help of ECCS (European Community for Coal and Steel).[2]

DESCRIPTION OF PERL'X FUSION BEAD MACHINE

Figure 1 shows the device. The essential features of Perl'X 2 are:

- The platinum-gold or glassy carbon crucible is heated by RF induction. Fusion is performed in about three minutes.
- It is possible to program two different heating intensities in one fusion program
- The crucible is agitated during fusion
- Casting after fusion is automatic
- Six different fusion programs are available

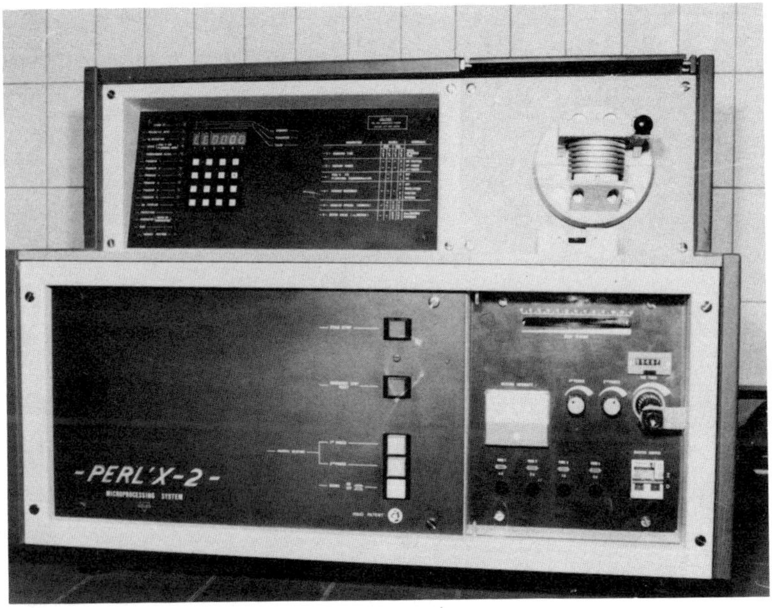

FIGURE 1. Perl'X 2 automatic bead machine for preparing powdered samples for X-ray fluorescence analysis.

In manual operation the operator weighs sample and flux on an analytical balance. After mixing the two components, the mix is introduced manually into the crucible. After the manual start of the bead machine fusion, agitation, casting and cooling are performed automatically according to the conditions programmed. After cooling the casting dish is emptied by hand.

AUTOMATION OF PERL'X 2

A first completely automatic system including a Perl'X 2 bead machine has been developed by Ciments Francais and Philips in France using a programmable automate. This system marketed under the name of MAPP'X is dedicated to process-control in the cement industry. Our approach was aimed at a general solution applicable to different laboratory situations.

So the use of robotics was a must in order to be flexible. Of course, the quality of the beads coming out of the automatic system should be the same as with a manually operated machine. These requirements led to the developemnt of a dedicated flux dispenser avoiding the take-up of humidity by the flux and making it possible to add different fluxes to the same melt. This development was accomplished by Retsch according to our specifications. Figure 2 shows the dispenser. Two completely closed vessels are mounted over two vibratory feeders which are completely closed. The system is equipped with a small pump aspirating air which is dried over silica-gel and flows over the vibratory feeder

FIGURE 2. Flux dispenser.

to avoid the take-up of moisture. The dispenser is interfaced to the robotic system via the power and event controller (PEC). By varying the value of the mains applied, the amplitude of the vibrations are modified, permitting to slow down the feeding speed at the approach of the target weight.

In order to minimize the contamination by dust we use fused fluxes (I.C.P.H. Villerupt France) of controlled grain size. Platinum-Gold crucibles are replaced by composite crucibles[3] where a crucible made out of glassy carbon is positioned inside the Pt-Au crucible. The composition of the flux used has to be optimized in order to be completely non-wetting to glassy carbon[3] and dissolve a broad spectrum of materials. We developed a flux containing 78% of $Li_2B_4O_7$, 17% of B_2O_3 and 5% of La_2O_3 which fullfils these specifications.

EXPERIMENTAL

Figure 3 shows the layout of the system. The following items are connected via the PEC of the Zymate System:

- Perl'X 2
- Retsch flux dispenser
- Vortex
- Cleaning device using compressed air
- Suction device.

The Mettler balance is connected via an interface in the controller.

The samples are placed on two sample racks with a total capacity of 100 samples. Different areas are allocated to different fusion

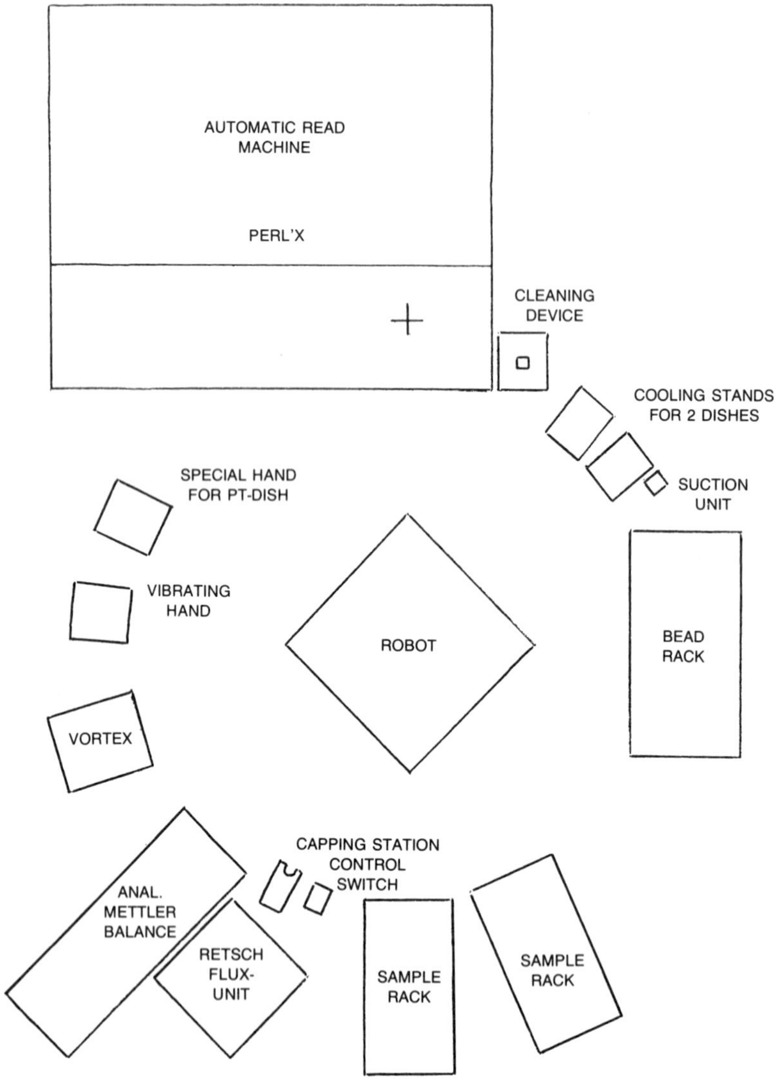

FIGURE 3. Table lay-out for robotic system combined with a Perl'X 2 fusion bead machine.

programs. The sample aliquot is introduced in a test tube which is located on the balance by using a general purpose vibrating hand. The target weight ±10% is reached in 80-120 seconds depending on grain size, humidity and density and is known to 0.1 mg.

Knowing the weight of the sample, the necessary amount of flux is calculated in order to have a constant dilution ratio. This amount is added from the flux dispenser with an accuracy of ±4 mg at 95% confidence level. Flux and sample are mixed in the vortex and transferred to the Perl'X 2.

The working cycle of the Perl'X 2 is started and the transfer tube is cleaned by compressed air. During the fusion cycle the weighing cycle of the next sample is performed. After casting of the first melt, the casting dish is transferred to a cooling stand and a second casting dish is positioned on the Perl'X 2. After casting, the bead is transferred with the help of a suction device actuated by a Venturi connected to the compressed air circuit.

RESULTS AND DISCUSSION

Cycle Time

The time necessary for the fusion cycle is 5.5 minutes whereas the total cycle time is 9 minutes. This compares to a good operator who does the weighing, mixing and transferring operations in 3 to 4 minutes.

Reproducibility

Table 1 shows a comparison of analytical results on beads prepared with

TABLE 1. Comparison of Manual and Automatic Sample Preparation with Perl'X Bead Machine

Iron Ore

	Fe	SiO_2	CaO	Al_2O_3	TiO_2	P	Mn
content %	65.69	2.41	0.18	1.75	0.1	0.053	0.092
manual	0.091	0.041	0.011	0.048	0.007	0.003	0.006
robot	0.131	0.040	0.003	0.103	0.0009	0.0002	0.0007

Slag

	Fe	SiO_2	CaO	P	Mn
content %	11.92	5.55	51.63	17.7	1.91
manual	0.025	0.071	0.13	0.085	0.013
robot	0.040	0.031	0.17	0.029	0.007

Fire-brick

	Fe	SiO_2	Al_2O_3	TiO_2	MgO	P	Mn
content %	2.45	56.5	34.2	1.52	0.78	0.16	0.1
manual	0.053	0.13	0.082	0.014	0.091	0.005	0.02
robot	0.070	0.092	0.121	0.011	0.008	0.003	0.0005

the Perl'X 2 by the Zymate System and by an operator.

The samples used are Certified Reference Materials.

The results on manual operation have been published by Irsid[2] whereas the results on robotic operation have been obtained in our laboratory. The standard deviation has been calculated on the measurement of 10 beads and includes preparation plus measurement error.

The X-ray spectrometer and the operating conditions where not the same for the two experiments. This explains for the varying trend of the results for the different elements measured.

Since it is difficult to get an objective experiment with manual operation which reflects routine conditions, we preferred this comparison even if it is slightly biased. The results show that the robot performs as well as an operator.

Cross-Contamination

Cross-contimination might be a problem if samples with extreme differences in concentration for one element follow each other.

Table 2 shows the result for CaO on a fire brick containing 0.18% of CaO where the samples have been fused alternatively with a slag containing 51.63% of CaO.

TABLE 2. Determination of Cross Contamination of a Fire-Brick by a Slag

Test Number	% CaO found
1	0.18
2	0.18
3	0.31
4	0.31
5	0.33
6	0.19
7	0.19
8	0.22

The bias results from a last drop remaining sometimes at the upper edge of the glassy carbon crucible. This bias represents a maximum cross-contamination of 0.29%. To avoid this error it is necessary to fuse a blank between two series of samples with these extreme differences.

Economic Considerations

The price for a complete system including set-up and engineering is about $70,000. On an eight hour operation this means a pay-back of about 3.5 years. In a production control situation where the system is used over 24 hours the pay back time is only one year.

The manual preparation of beads is an operation which is very time-consuming and not very challenging, and it is difficult for an operator to keep a steady working pace going after hours.

CONCLUSIONS

It has been shown that the Zymate System can be used to automate the weighing and transferring operations necessary to prepare fused beads for XRF analysis with a Perl'X 2 bead machine. The economic interest of the operation is obvious, especially in process-control.

The next step will be the interfacing of the whole system to an XRF spectrometer by using the robotic arm for changing samples and by transferring the necessary information from the controller of the robotic system to the computer of the spectrometer. As a first application, we are working on this problem together with the Philips' Application Laboratory in Almelo (Netherlands) both for their new 1404 and 1606 spectrometers.

ACKNOWLEDGEMENTS

A. Weber has done the electronic interfacing and R. Thines the programming of this application. The people from Retsch did a most valuable work by designing and constructing the flux feeder. We would also like to acknowledge the technical support from Zymark France.

REFERENCES

1. Ron Jenkins in Quantitative X-Ray Spectrometry, Marcel Dekker, New York and Basel, 1981, p. 365.

2. Wittmann A., Willay G. and Seffer R. in Eur 6288 f commission of European Community, Luxembourg 1979.

3. Wittmann A., Willay G., Spectrochemica Acta, Vol. 40 No 1/2, p. 253-265, 1985.

LABORATORY ROBOTICS:
APPLICATIONS IN THE MATERIALS SCIENCE LABORATORY*

Patricia A. Gateff and James C. Abbott
The Procter & Gamble Company
Paper Products Division
6100 Center Hill Road
Cincinnati, OH 45224

ABSTRACT

Originally designed as an aid in preparation of samples for sophisticated chemical analysis, the usefulness of the laboratory robot in the materials science laboratory is rapidly becoming recognized. The manipulation of test portions of material is, in many cases, the key step in a test in the materials science laboratory. Determinations of strength, color, porosity, absorbency, and other less familiar physical properties generally require the placement of a portion of the sample in a suitable test instrument. Paper, polymer films, metal strips, and wire are only a few of the materials whose physical properties must be measured. Completely automated procedures for basis weight, tensile, burst, thickness, and color measurement can now be routinely done using available equipment.

The modern techniques of laboratory robotics, combined with a data acquisition system, can result in a tremendous increase in laboratory productivity as well as improved data quality. Laboratory productivity enhancement of 50% is quite common and, in some cases, may even approach 80%. Here, we provide some general comments regarding the use of laboratory robotics in industrial control laboratories, followed by a discussion of laboratory robotics in the paper laboratory with special emphasis on the materials science laboratory. Specific examples of the use of laboratory robotics in the testing of paper and polymer films, drawn from the authors' own laboratory, will be included. The testing laboratory is fertile ground for the "robolution" of the 1980's!

INTRODUCTION

A better way! A faster way! A more cost effective way! The optimum way to operate a process, perform a test, or make a new product! The beneficial application of technology, however simple or complex, to enable us to improve the way we do things, has been a continuing thread of achievement throughout history. The space shuttle, supersonic passenger travel, the computer, the laser- we could list literally dozens of the advances in technology of the past few decades, and yet record only a small fraction of the really significant applications of technology which have occurred in our own lifetimes.

But periodically, there are advances in technology which are of such significance that they literally revolutionized an industry or a procedure, result in a totally new way of accomplishing specific tasks or produce a totally new product. The steam engine, the light bulb, the automobile, the assembly line, the airplane and the mass spectrometer, each in its own time and way, has been a true revolutionary advance in technology.

We can also identify certain key advances in technology in our laboratories which have had, and continue to have, tremendous impact on the accuracy, cost effectiveness or usefulness of the data we produce. The glass electrode for measuring pH, the microprocessor for instrument control or data recording, the electronic balance for ease in weighing or the use of laser optics are but a few of the many advances in measurement technology which have made major improvements in the productivity, accuracy, or overall quality of laboratory activities. As the twenty-first century science historian examines scientific innovation of the last quarter of the twentieth century, however, the laboratory robot will certainly stand out as one of the very major advances in productivity, accuracy, and reproducibility in laboratory science technology.

Here, we will examine laboratory robotics technology and its application in both chemical and materials science laboartories. We will also examine some "near-robots"- that is, some of the other very useful automated sample handling or analysis equipment which preceded the development of the laboratory robot. Finally, we will try to determine why laboratory robotics is destined to have such a major impact on laboratory science and what projections we can make about the potential impact of this technology on our own laboratories and careers.

THE ROBOT

It was some 60 years ago that the Czech dramatist, Karel Capek, in the play R.U.R. -- Rossum's Universal Robots -- described a device which has only begun to be fully utilized in industry in the past decade -- The

Robot. In Capek's play, one of the characters was a mechanical person. The robot, generally used to describe such a being, is derived form the Czech word "robotit" which can be translated into English as "to drudge."

For the dictionary definition of the word "drudge," we find "...a person who does hard, tiresome, or disagreeable work". We also find several dictionary definitions for robot; however, the most generally useful is "...any machine or mechanical device that operates automatically or by remote control". Thus, a robot is, in general terms, a machine or mechanical device which can automatically, remotely, perform routine, tiresome or even disagreeable tasks.

Certain nonmechanical devices used for controlling other equipment such as thermostats, electric eyes, radio frequency transmitters, microwave motion sensors, etc., might have been included in past discussion of robots. By 1980's standards, however, the robot is generally a mechanical device which has broad capability to physically move things and/or itself on command. The ability to open a garage door or the door at the supermarket, while very useful technology, is not generally considered to be an example of robotics technology!

Much has been written about the impact of robots on our lives. Some of you may be familiar with the place that robots are playing in numerous industries. The term "robolution" has been coined to describe the total impact of the "robot revolution" on our work and personal activities.

But let's look more specifically at the paper industry and the paper analyses or testing laboratory. What kinds of robots are available? What is a laboratory robot? What can it do? How much does it cost? What can I do with it in my laboratory? What will be the probable result if I have a robot in my laboratory? What is the payout?

We will attempt to answer these questions (and others) and describe the increasingly important place that laboratory robots are playing in the modern laboratory. We will include some references to chemical testing; however, that information is increasingly available elsewhere, so the majority of our comments will focus on physical testing. Applications in this area are less widely known but offer tremendous opportunities to optimize laboratory data quality and productivity in paper and related industries.

INDUSTRIAL ROBOTS IN THE PAPER INDUSTRY

For those interested in a general description of robots and their potential use in the pulp and paper industry, the recent paper by Hinson is recommended reading.[1] As he notes, "An average robot 'worker' tends to be an electrically powered, jointed arm robot with servo-controls and fixed axis of motion, additonal memory capacity, and a few software options that sells for approximately $70,000 to $80,000. The least complex robot can sell for as little as $5,000 and some of the more complex units for as much as $200,000."

THE LABORATORY ROBOT

Change in the area of laboratory robotics is occurring at an increasing rate, but two papers published in late 1983[2,3] are still an excellent place to begin one's studies of laboratory robots. This should be followed by attendance at technical meetings where presentations on laboratory robotics are featured or discussions with vendor's of the equipment. Excellent quality robots designed specifically for laboratory work may cost in the range of $10,000 - $25,000, with added peripheral components required for specific procedures. Small, general purpose, industrial robots costing even less are suitable for specific applications.

THE USE OF ROBOTS

As one begins to consider utilization of robots to perform specific tasks, whether in the laboratory or in other manufacturing environs, it is most important that a clear understanding of the particular expectations of the robot be in hand. The field of robotics is an extremely rapidly developing one. As the variety of equipment from which to choose becomes greater, the need for clear understanding of capability and performance characteristics becomes most important. Where failure in a robotics application occurs, it is more frequently the improper match of robot and task rather than a failure of robotics technology per se.

USE OF ROBOTS IN THE LABORATORY

Frequently, new technology in testing/analysis is adopted in the industrial laboratory only after an extended "shakedown" in the research area. In studying laboratory robotics, one thing rapidly becomes apparent- industrial laboratories, rather than those in academia or those dedicated to research studies, are the ones making the major use of laboratory robots. This fact stems from the usefulness of laboratory robots in preparing large numbers of samples for analysis in the chemical laboratory or placing them into test instruments in either the chemical or testing laboratory. Both of these steps are generally personnel intensive and, thus, costly. The major benefit of the incorporation of a laboratory robot, in addition to improved precision afforded by the laboratory robot, is reduction in the personnel effort required for analysis. It is not surprising that robots are finding rapid acceptance in industrial laboratories where procedures which are precise and cost effective are required.

Robots developed specifically for laboratory use are generally traced to the introduction of the Zymate Laboratory Automation System in 1981/82; however, applications of small industrial robots to laboratory tasks before that time were reported. In addition, numerous examples of automated test procedures and instruments such as automatic titrators, diluters, pipeters, automatic chromatographic sample injectors, sample collectors and automated analysers such as the Technicon Autoanalyser (Technicon Instrument, Inc., Tarrytown, NY), to name only a few, were introduced prior to 1980. While not robots in the strictest sense, these devices certainly gave a preview of the efficiencies and cost benefits which are afforded by reducing personnel involvement in routine laboratory tasks.

The Zymate System, shown in Figure 1, is an example of the cylindrical coordinate robot whose general work environment is described in Figure

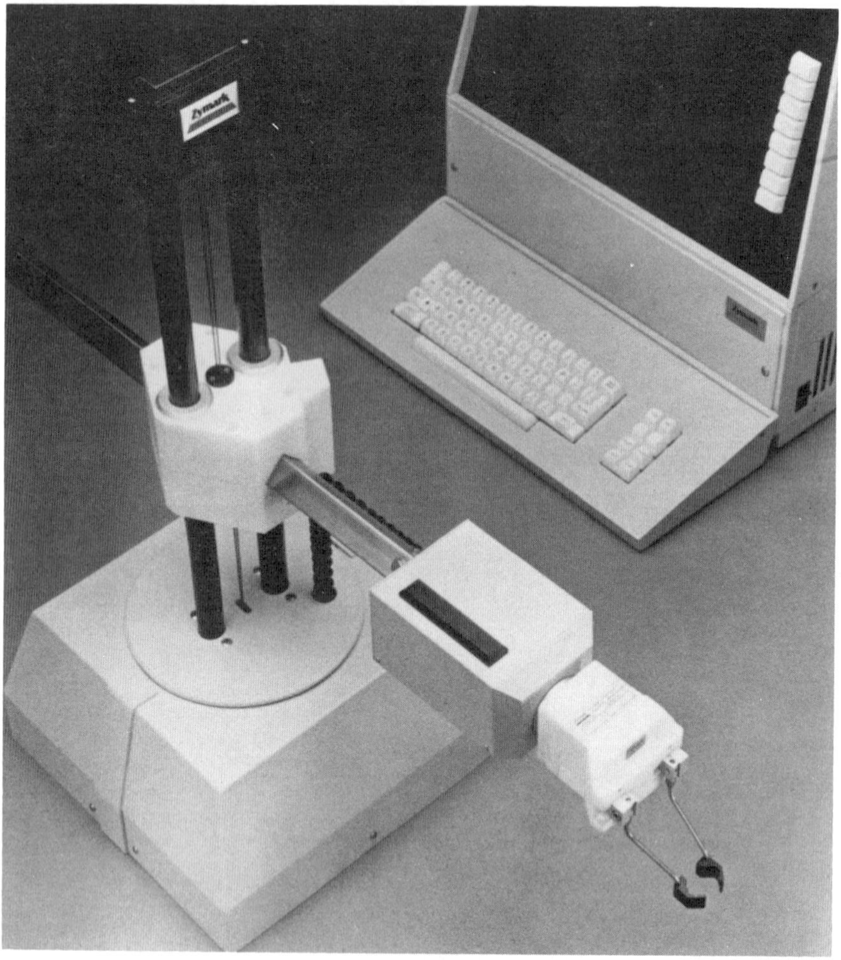

FIGURE 1. Zymate System (courtesy Zymark Corporation).

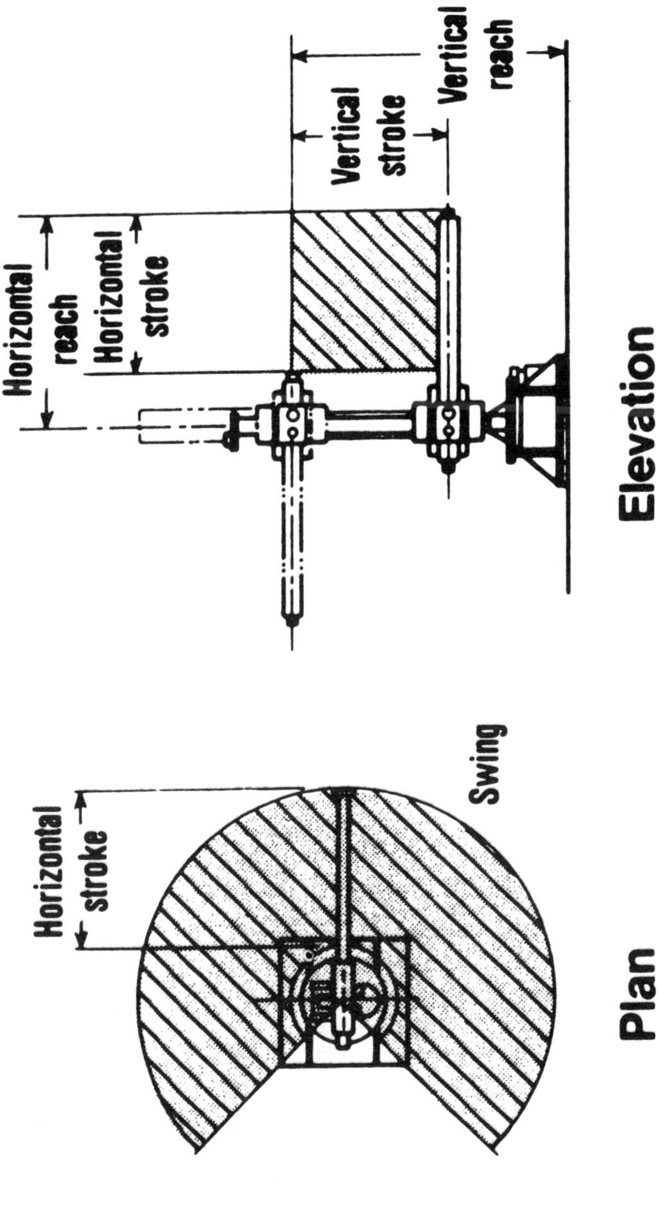

FIGURE 2. Work envelope of the cylindrical coordinate robot as found in Hinson[1].

2, taken from Hinson's paper.[1] The actual effective work envelope of the Zymark robot is, in fact, a 370° donut (10 of overlap), rather than being a cylindrical pie, minus a 90° slice as shown in Figure 2. The Perkin-Elmer laboratory robot, shown in Figure 3, is an example of the jointed arm robot. The work envelope of the jointed arm robot is generally that shown in Figure 4, also taken from Hinson's paper. Another small, jointed arm robot suited for certain laboratory tasks is the Microbot (Microbot, Inc., Mountain View, CA).

WHEN ARE LABORATORY ROBOTS MOST EFFECTIVE?

In considering the use of robots in the laboratory, it is generally found that their effectiveness will be maximized when they are utilized for infrequently performed yet highly complex tasks - at least from the manipulative perspective - or for frequently performed, routine tasks which may be quite simple in nature but must be performed many, many times in a given work period.

In the first case, the robot, once trained, will perform even the most complex task in exactly the same way even after a lengthy period of nonperformance. The human worker on the other hand is much more susceptible to error in performing highly complex tasks infrequently. On the other end of the spectrum, the repetitive performance of very simple tasks by humans is one of the most error-prone operations in the laboratory because of the boredom or "drudgery" factor. The robot does not suffer from boredom.

In the chemical laboratory, the complexity range of tasks varies greatly

FIGURE 3. Perkin-Elmer Laboratory Robot (courtesy Perkin-Elmer Corporation).

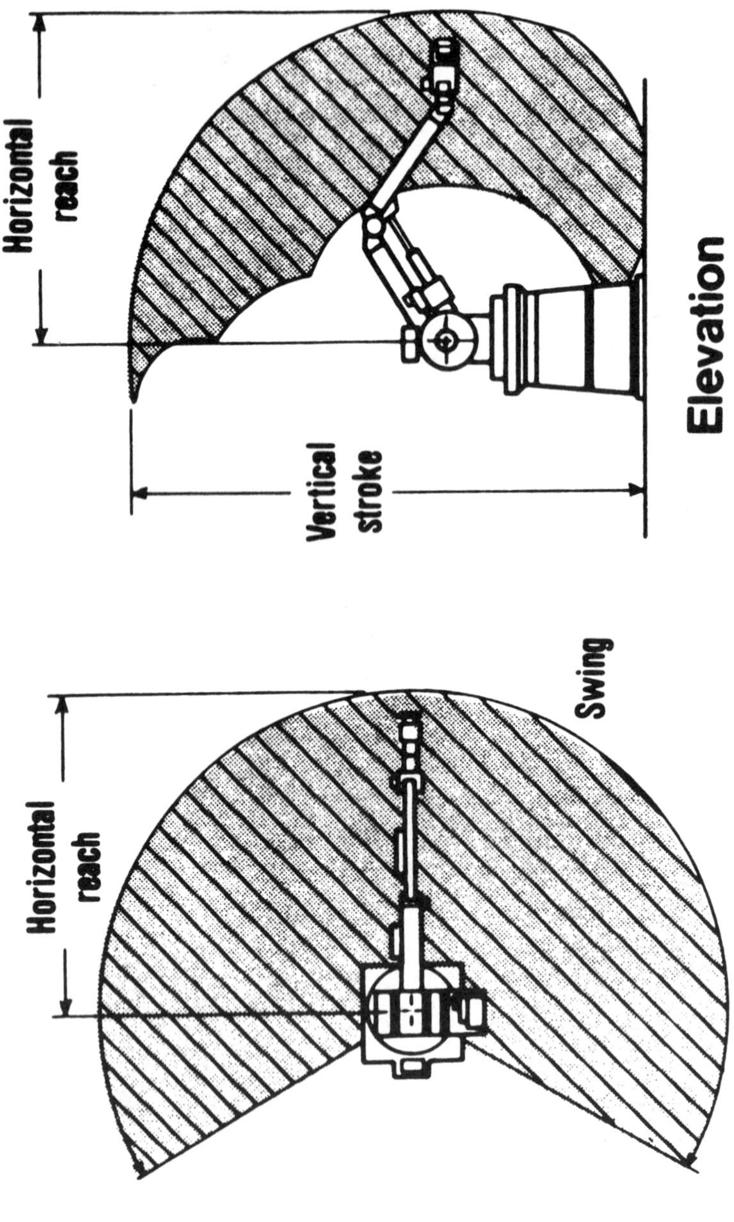

FIGURE 4. Work envelope of the jointed arm robot as found in Hinson[1].

from very simple tasks such as pH measurements on aqueous solutions to the extremely complex reaction/separation/measurement tasks involved in measurement of trace process chemicals or unreacted raw materials in a finished material, a specific compound in an effluent stream, etc.

In the physical testing laboratory, much of the activity is manipulative - placing samples in test instruments, closing instrument grips, and of course making frequent notations of results. While some of the manipulations are quite complex, once taught to the robot, they may be repeated in a very straightforward manner.

The laboratory robot may be effectively utilized for both chemical and physical testing in the research or manufacturing laboartory in the pulp and paper industry.

CHOOSING A LABORATORY ROBOT

The major consideration in choosing a laboratory robot, as I noted earlier, is the requirements of the task to be performed. In the case of chemical testing, the laboratory robots already on the market, as well as those which are just becoming available, are generally well-suited for tasks such as pipetting, weighing, extracting and dissolving. Vendor support for use of laboartory robots in chemical testing is generally very good. Thus, other variables such as cost, availability and compatibility with existing equipment become important in the choice of a robot to be used in chemical testing.

Use of robots for physical testing will depend more heavily on the ease of using the robot to manipulate flexible materials such as paper, polymer films, non-woven fabric pieces, etc. This is generally a new area with regard to robotics, so vendor support can play a significantly larger place in the choice of selection of the robot to be used.

LABORATORY ROBOTS IN THE CHEMICAL LABORATORY

While robots in the chemical laboratory may be a recent developemnt, certainly automated procedures in the chemical, biochemical or clinical laboratory are not new. The flow system AutoAnalyzer of Technicon, whose central component was a constant speed peristaltic pump, was firmly established in the '60's, with modules available for handling the dissolving of solid samples, filtering, heating, extracting, etc. Likewise, automated fraction collectors or sample injectors have been used in chromatography for some time. Added to this are an extensive list of sophisticated automatic titrators, automated moisture analyzers, tensile testers, micrometers, etc., most with micorprocessor control, data system compatibility and direct read-out of the desired answer. What place, then, does the robot have in laboratories which may choose from an already impressive and extensive array of "automated" equipment?

This is similar to the type question asked five or six years ago by a group (Zymark Corporation) interested in "inventing" a new device to simplify the laboratory chemist's work. Simply stated, considering the state of laboratory instrumentation at that time, what were the remaining major barriers to greater effectiveness in chemical or measurement laboratories?

The answer that was developed, after interviewing a number of chemists and thoughtfully considering their input, was that the numerous manipulative steps (sampling, weighing, dissolving, stirring, extracting, centrifuging, etc.) required prior to actually beginning instrumental testing were the major barriers to enhanced laboratory productivity. This point was reiterated recently by Horace G. McDonell, Jr., chairman of the Perkin-Elmer Corporation, who said in an interview,[4] "Further automation of the information coming out of instruments, as desirable as that may be, is being diluted by the lack of productivity aids at the front end of the process - how to get the sample properly prepared, determine the way to analyzed it, and get the analysis done." With some experience with laboratory robots, it will become clear that they complement and make more efficient use of already existing equipment.

A good sampling of the use of laboratory robots in the chemical laboratory will be found in <u>Advances in Laboratory Automation - Robotics 1984</u>[5] where applications ranging from metabolism studies to analysis of industrial chemicals such as glycerine are recorded. In addition, the recent literature includes a number of papers describing applications of laboratory robotics in chemical procedures including sample preparation[6,7] and use in wet chemical analysis, particularly with hazardous materials.[8,9] Figure 5 shows the Zymate System with solution addition, weighing, and mixing peripherals as it might be used in a chemical analysis laboratory. Laboratory robots are also in use in the microbiology, clinical, and pharmaceutical laboratories.[10,11] It must be remembered that the field of laboratory robotics is quite new. Thus,

FIGURE 5. Zymark Laboratory Automation System in the chemical laboratory (courtesy Zymark Corporation).

much of the work appearing in the current literature has not yet appeared in book form.

While in-line monitoring of chemical processes is also rapidly gaining importance in the paper industry, off-line wet chemical and instrumental analytical procedures performed in a laboratory continue to be of importance in a variety of development or control applications. And, in such applications, laboratory robotics offers important opportunities to optimize laboratory contribution to total business productivity and profitability.

LABORATORY ROBOTS IN THE MATERIALS SCIENCE LABORATORY

As in the case of laboratory robots in the chemical laboratory, laboratory robots in the material science or testing laboratory has also been preceded by other forms of automated testing.

With materials being analyzed for physical characteristics in the testing laboratory, the major steps are generally sampling and sample preparation - that is, selecting the sample for analysis, taking an accurately measured test piece for analysis, and then making the actual test measurement using a tensile tester, tear tester, balance, etc. The sample throughput is quite rapid and a high percentage of the procedure is sample manipulation - that is, placing the sample into the test instrument after it has been prepared.

Testing of Continuous Webs

In the case of some non-destructive procedures such as thickness testing, color testing, porosity testing, etc., the cutting of the individual sample test pieces for laboratory testing may be unnecessary. Laboratory, as well as on-line testing, can be done on a continuous strip of sample at some pre-determined frequency -- say every six inches or every three feet. The L&W AutoLine (Scanpro Instruments,), one example of equipment for which performs testing on a continuous sample strip, is shown in Figure 6 where burst, air permeance and surface roughness are being determined on newsprint at the Kaipola Paper Mill in Finland. Similar devices have been manufactured by other companies from time to time as an aid in automating the physical testing of paper or other sheeted materials.

FIGURE 6. Continuous web testing in Finland using the L & W AutoLine System (courtesy Scanpro Instruments).

Completely Automated Testing

Recently, a device called PaperLab was introduced as an aid to the determination of paper properties. This instrument, as designed, features a large number of individual test modules inside a conditioned chamber where test pieces may actually be cut from a sample and evaluated for a variety of properties, both destructive and non-destructive. Interesting in concept, this instrument is quite new, relatively expensive and has yet to be totally incorporated in paper testing. For the laboratory able to justify its expense, it is an interesting concept in automated testing.

Robots in Paper Testing Laboratories

Recently, at least two vendors have offered a combination of tensile testing equipment and laboratory robots for use in the materials science laboratory.

Syntech incorporates an impressive software package run on an IBM PC to control an Instron and a Microbot robot. This system is shown in Figure 7 where the robot is lifting a rigid test specimen for placement in the Instron tensile tester.

Zwick, a German company, likewise offers a combination of laboratory robot and tensile tester for use in the materials science laboratory.

Similar equipment for use in testing rubber and other semi-rigid samples is available in Japan.

Reports on the "Use of Laboratory Robots in Materials Science Laboratories Work" in determining basis weight using a balance and a Rhino robot has already been reported by 3M Company.[12]

More recently, Dow Chemical reported a totally automated tensile testing station using a Zymark robot as the means of introducing the test sample into the tensile tester.[13]

We evaluated robot-assisted or robot-controlled testing in our development laboratory after a review of the technology suggested it had significant potential to improve both data quality and productivity of

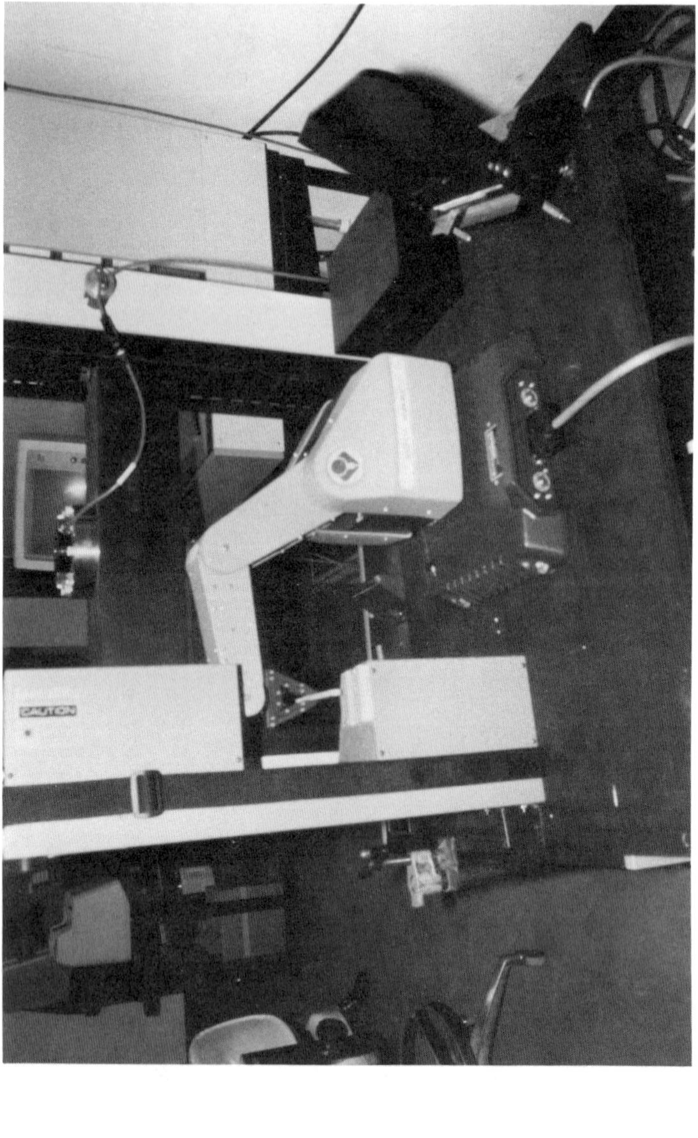

FIGURE 7. A small Microbot robot used in conjunction with in Instron Testing Machine (courtesy Syntech).

our physical testing of paper and related materials. We have described several procedures for the automatd, robot-controlled testing of paper and non-wovens for physical properties[14,15,16] which results from this evaluation. In all of our work, we have used a Zymark robot and peripherals, including a robot controller which controls the robot and other periperals, a power and event controller which is a device used to synchronize operation of laboratory equipment such as stirrers, physical testing instruments, etc., and a master lab station which can be used for liquid handling. We have done much of our own work to develop the specialized racks and grips required to store or handle test pieces of paper.

A data acquisition system for use with laboratory robotics equipment is a must. It makes no sense to minimize personnel involvement in introducing test pieces to a test instrument and performing the test, yet require human intervention to record the test data! For data acquistion we have used both independent units, such as individual Apple microcomputers, or integrated data acquisition systems such as the Thwing-Albert RedLine to acquire and report data from testing instruemnts. Both approaches worked well. The integrated data system has obvious benefits when incorporated with a laboratory-wide information management system.

In addition to the robot and the data system, we have used commonly available testing instruments, with only minimal modification, to make the actual physical measurements. Instruments which we have used include a Mettler Toploading Electronic Balance for the determination of basis weight, a Thwing-Albert Intelect Tensile Tester for determination of tensile strength and elongation, a Thwing-Albert Burst Tester for determination of bursting strength, and a Hunter LabScan Spectrocolorimeter for determination of color.

Figure 8 is a block diagram of the instrumentation used for measuring tensile and elongation, and is representative of the procedures we have used. Operation of the tensile tester is controlled by the robot controller through the power and event control system. Sample disposal is done using a vacuum system - actually a canister ShopVac - also controlled through the power and event control system.

Figure 9 shows the Zymark robot performing a basis weight test on tissue grade paper. Excellent precision and data accuracy have been found when the robot-controlled procedure is compared with the basis weight test done manually. Likewise, Figure 10 shows the Zymark robot, in conjunction with the Thwing-Albert burst tester, performing the bursting procedure which we use. We have also used the Zymark robot in conjunction with specialized equipment which we have developed to perform a test relating to the moisture barrier properties of non-woven fabric. More recently, we have completed our development of procedures for tensile testing, thickness testing, and color measurement which were published in TAPPI Journal in July, 1985. Figure 11 shows the Zymark robot with special fingers placing a tensile strip into special grips in a Thwing-Albert Intelect 500 Tensile Tester for testing.

As noted above, we have developed a number of specialized sample grips

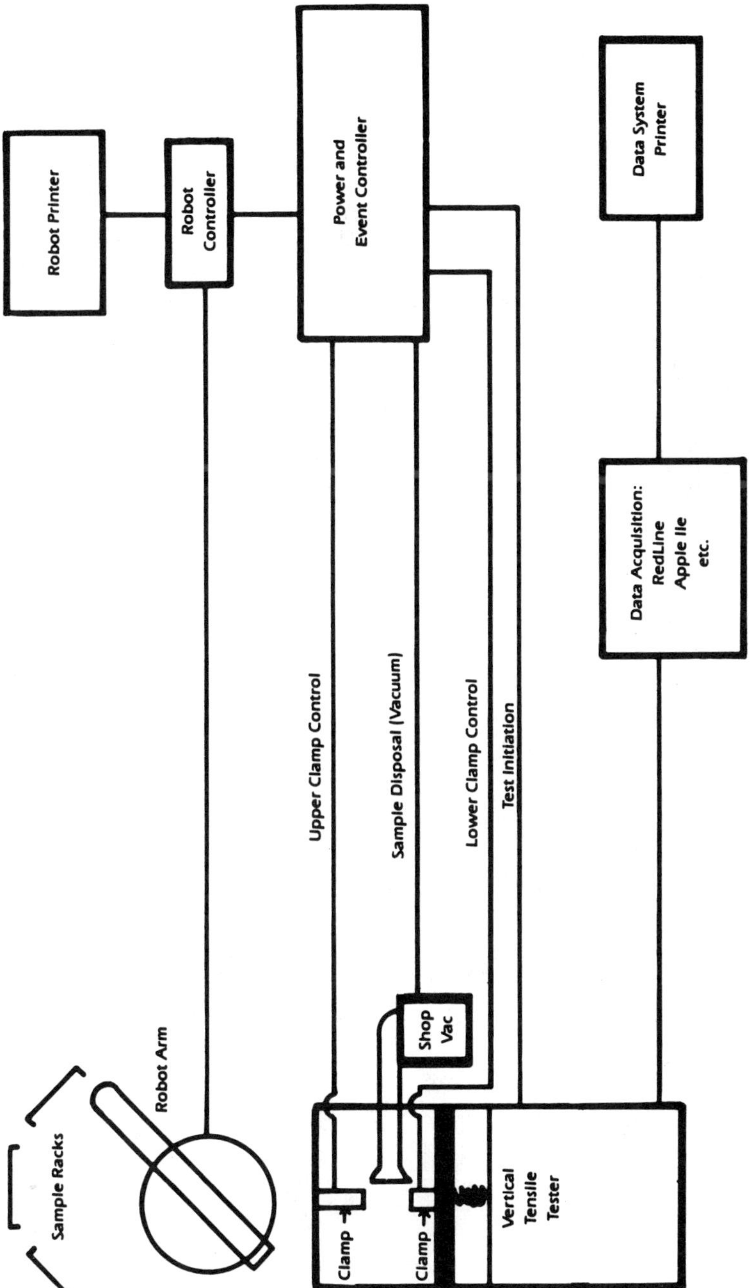

FIGURE 8. Block diagram showing instrumental setup used for making tensile measurements on paper.

FIGURE 9. Zymark Robot placing basis weight sample on a specially constructed balance pan on a Mettler toploading balance.

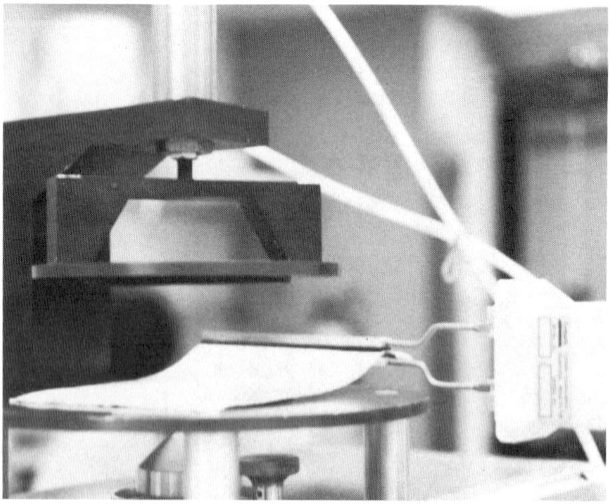

FIGURE 10. Zymark Robot placing sample in Thwing-Albert Burst Tester.

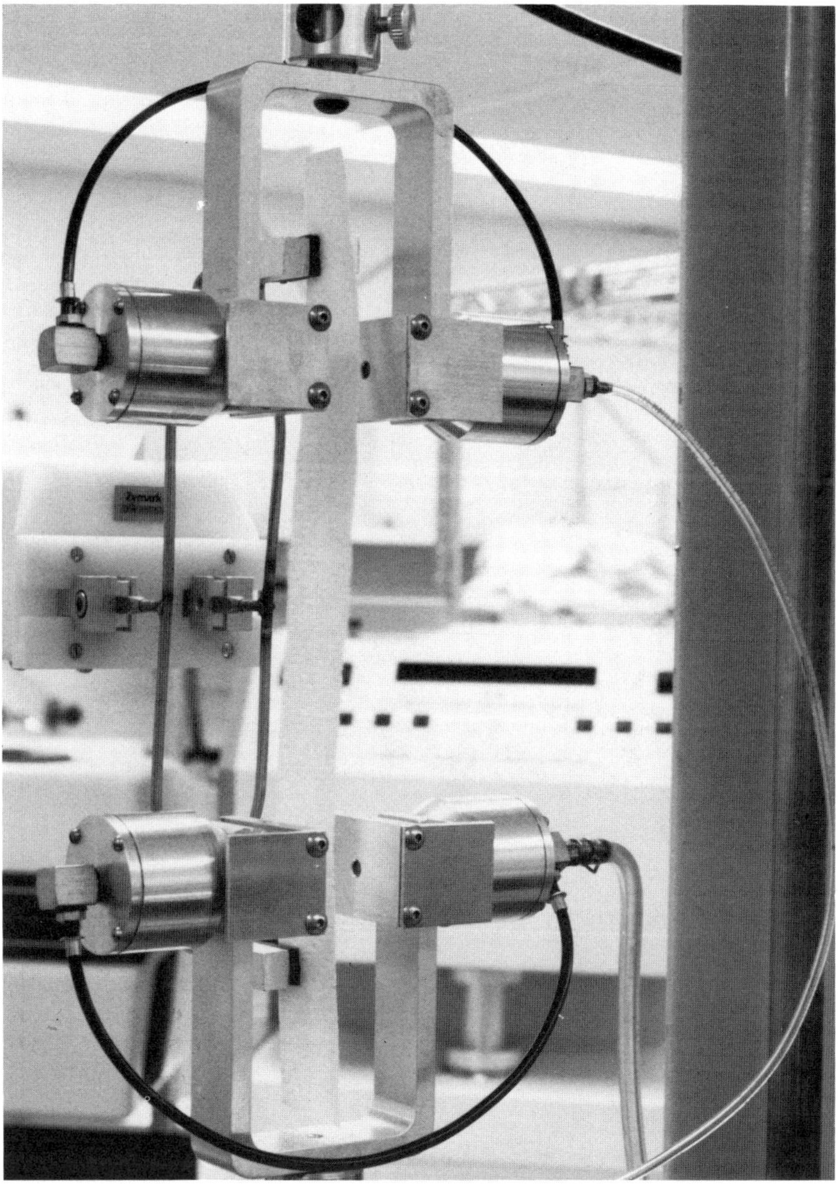

FIGURE 11. Zymark Robot placing tensile strip into grips of Thwing-Albert Intelect 500.

for use in transporting samples. These have worked well with the tissue grade low basis weight materials which are the bulk of our work. Different techniques can be used to handle more rigid samples such as paperboard, including, for example, pick-up of samples using vacuum. We have also placed staples in the corners of paper test pieces and used small magnets to manipulate these test pieces.

The opportunities for use of laboratory robots in the materials science laboratory are really limited only by the creativity of the worker. Undoubtedly a variety of different robots, fingers or grips, and peripheral instruments will be used in materials testing as this technology becomes more widely utilized. The incorporation of laboratory robots, in addition to achieving improved sample throughput, makes it possible to use existing laboratory test instrumentation for making measurements. The flexibility of the laboratory robot makes possible numerous modifications of test procedures in the development laboratory. Thus, the technology is very cost effective and payout is rapid.

PRODUCTIVITY ENHANCEMENT WHEN ROBOTS ARE USED

Productivity enhancement when laboratory robots are used is significant. We have achieved effort reductions of up to 80% in particular analyses when laboratory robots were incorporated into the procedure. Obviously, laboratory robots also enables the one shift laboratory to significantly extend the effective working day by allowing the robot to run unattended after normal working hours. The exact productivity enhancement (payout from the purchase of a robot) must be calculated for the individual laboratory based upon the particular testing being done, frequency, laboratory organization, etc.

THE IMPLICATIONS OF THE USE OF LABORATORY ROBOTS

Laboratory robots give us tremendous opportunities to optimize the effectiveness of our laboratory activities in both the chemical and materials science areas. Combined with data acquistion systems, which are a must when robots are used, laboratory robots have the ability to make many additional hours of effort available to laboratory personnel to spend on new testing tasks or in refining existing ones. Computer skills, improved mechanical skills and the ability to conceptualize and develop testing procedures using new equipment are all opportunities given to the laboratory worker when robots are used. People spend less time on sample manipulation and more time in thinking about the meaning of the data, appropriate tests to be used, and how they may be applied to optimize productivity and profitability.

SUMMARY

The "robolution" of the 1980's will greatly impact all of us. It offers tremendous opportunities for us in our testing and analysis laboratories. Costs are reasonable and payout rapid. What is needed is the ability to see the opportunities that this new technology provides and to respond to them with vigor.

REFERENCES

1. Hinson, TAPPI Journal, 67 (12), 33 (1984).

2. Dessy, R., ed., Anal. Chem., 55, 1100A (1983).

3. Dessy, R., ed., Anal. Chem., 55, 1233A (1983).

4. Hinson, R., TAPPI Journal, 67 (12): 40 (1984).

5. Advances in Laboratory Automation - Robotics 1984, G. L. Hawk and J. R. Strimaitis, eds., Zymark Corporation, Inc., 1984.

6. Hawk, G.L., Little, J. N. and Zenie, F. H., International Laboratory, 12 (7), 48-52 (1982).

7. Owens, G. D. and Eckstein, R. J., Anal. Chem., 54, 2347-51 (1982).

8. Brown, R. K., "Use of Laboratory Robotics for Automated Preparations and Analysis of Hazardous Materials," Proceedings of the 2nd Annual Hazardous Materials Management Conference, Philadelphia, PA, USA, June 5-7, 1984.

9. Burkett, S. D., Dyches, G. M. and Spencer, W. A., "Wet Chemical Analysis with a Laboratory Robotic System," Proceedings from the Pittsburgh Conference and Exposition on Analytical Chemistry and Applied Spectroscopy, Atlantic City, NJ, USA, March 5, 1984.

10. Carmack, L. J., Hight, T. H. and Pyle, W. D., "Applications of a Robot in Aseptic Microbiology Lagoratory Procedures," Proceedings from the American Society for Microbiology 84th Annual Meeting, St. Louis, MO, USA, March 4-9, 1984.

11. Carmack, L. J., Hight, T. H. and Pyle, W. D., "Robotics in Clinical and Pharmaceutical Laboratories," Proceedings from the Association for the Advancement of Medical Instrumentation 19th Annual Meeting, Washington, DC, USA, April 14-18, 1984.

12. Bellus, P., Anal. Chem., 55, 1240A (1983).

13. Scott, R.L., Rieke, J. K., in Advances in Laboratory Automation - Robotics 1984, G. L. Hawk and J. R. Strimaitis, eds., Zymark Corporation, Inc., 1984, p. 151.

14. Abbott, J. C., Jenkins, L. A., McLaughlin, C. A., Presentation, "Laboratory Robotics Applied to the Testing of Non-Woven Fabrics: The Strike-Through and Rewet Test", Presented at LabCon New England/84, October 16-18, 1984, Woburn, Massachusetts.

15. McLaughlin, C. A., Abbott, J. C. and Jenkins, L. A., in Advances in Laboratory Automation - Robotics 1984, G. L. Hawk and J. R. Strimaitis, eds., Zymark Corporation, Inc., 1984, p. 165.

16. Abbott, J. C., Gateff, P. A., Jenkins, L. A. and McLaughlin, C. A., TAPPI Journal, 68 (7), 33 (1985).

Zymark and Zymate are registered trademarks of Zymark Corporation.

FORMULATION AND TESTING FOR A COATING APPLICATION FOR RESEARCH AND DEVELOPMENT

Edward C. Koeninger, Joseph Grano, Jr.
and John F. Heaps
Monsanto Polymer Products Company
730 Worcester Street
Springfield, Massachusetts 01151

ABSTRACT

Research and development of coatings for use in the paint industry requires the preparation and testing of a very large number of samples. The formulation and testing of these coating samples is extremely personnel intensive.

In the application reported, the totally automatic preparation of coating formulations by the Zymate Laboratory Automation System is described using an innovative method of component dispensing. In addition, totally automatic sample handling and testing of film thickness, gloss, and MEK rub by the robotic system is described. Reduction of personnel effort in repetitive tasks has been achieved along with more uniform testing results.

INTRODUCTION

Testing and sample preparation consume a major amount of technical time in an automotive coating laboratory. Much of this work is tedious and monotonous and wasteful of a skilled technician's time.

Sample preparation consists of accurately weighing a number of different crosslinkers, vehicles, catalysts and solvents in various combinations to make paint samples of varying compositions. These samples are then sprayed on test panels and oven cured. The panels are then tested for film thickness using a thickness probe and film gloss using a special glossimeter and then given a methylethylketone (MEK) rub. The MEK rub is done to test for abrasion and chemical resistance of the paint film. The test consists of moving a hard felt marker tip, which has been saturated with MEK, back and forth across a panel until the paint film is broken. This test is particularly tedious to the technician since it may require more than two hundred rubs before the paint film shows any damage.

Although these routines are simple, they are subject to operator variability and/or error. The large amount of data to be manipulated (taking readings and calculating averages and standard deviations) was also subject to operator error.

Previously, the technician would spend a good deal of his/her time weighing material for the paint sample. These would then be applied to the test panels, the panels cured and then the standard tests run of those panels. Only a small portion of the time was used in actually evaluating the results of the test with respect to the composition of the paint formulation.

With the robotic work station, sample preparation requires only loading the system with the proper raw materials (crosslinkers, vehicle, solvent and catalyst) and containers and typing the composition desired into the robot controller. Testing of the panels only requires loading the racks which hold the panels and typing the test instructions into the robot controller. Both sample preparation and panel testing proceed unattended and can be done outside the normal laboratory working hours if desired.

DESCRIPTION OF THE ROBOTIC WORK STATION

The work station layout is illustrated in Figure 1. Key components of the work station are the robot and controller, capping station, power and event controller, master lab station, jar racks, pump bottle racks, scale and pump station, magnetic stirrer, holding station for gloss head, panel racks, panel work station, hand holding stations and printer. In addition, there is the color gloss instrument and the coating thickness gauge.

The sample preparation system is capable of making 20 different samples with a combination of crosslinkers and vehicles. The number of different crosslinkers and vehicles cannot total more than 20. The same catalyst and solvent is normally used for the 20 samples. The system is designed so that all containers are disposable.

In operation, the proper composition for the various samples is typed into the controller together with the location of the crosslinkers and vehicles. The program is initiated and the robot picks the empty

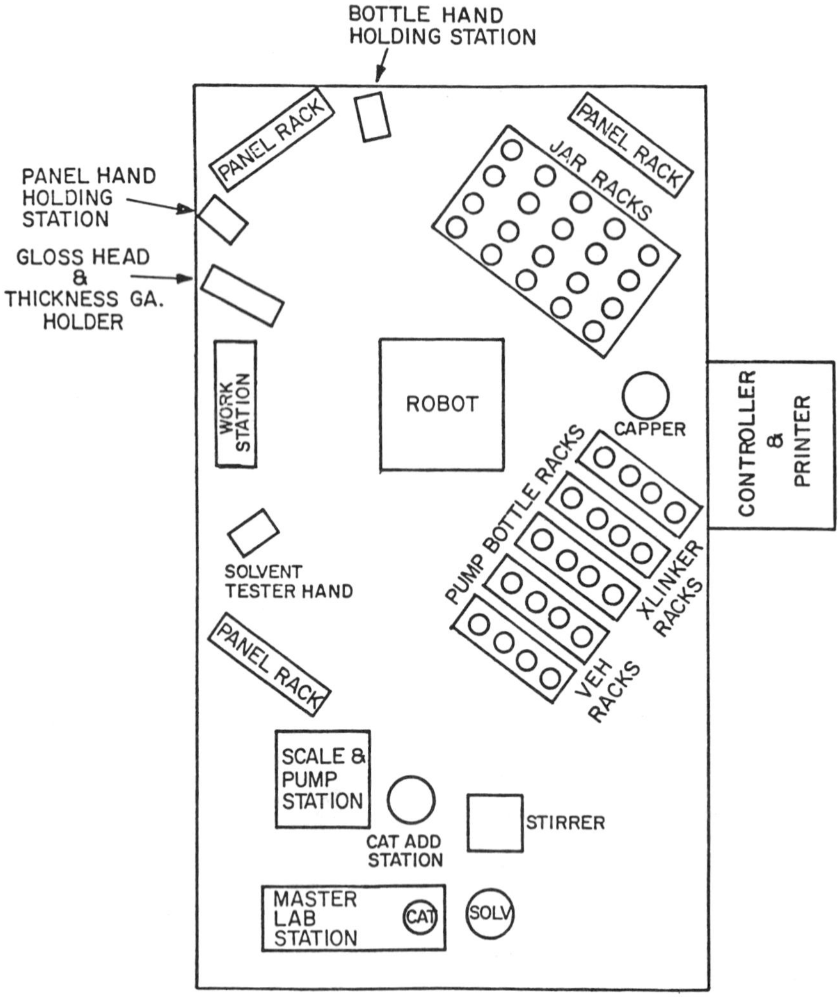

FIGURE 1. Bench layout for formulation and testing of a coating application for the paint industry.

(containing only a magnetic stirring rob) sample jar and removes the screw cap. The jar is then placed on the balance and the tare weight determined. The robot then picks the pump bottle containing the proper crosslinker and moves it to the dispensing station. The pump bottle is a 16 ounce narrow neck bottle fitted with a plastic pump dispenser similar to that used to dispense liquid soap. In the dispensing station, the pump dispenser is operated by an air cylinder through a mechanical arrangement which depresses the plunger of the plastic pump dispenser. It has the capability for a full stroke, which dispenses approximately 1.5 to 1.7 grams of material, or a short stroke, which dispenses about 0.1 gram of material. In operation, the dispenser operates with repetitive full strokes until the actual weight is within 2 grams of the desired value. The dispenser is then switched to short strokes until the desired weight is achieved. The final weight is accurate to within less than 0.15 gram. After the proper weight of crosslinker has been dispensed, the pump bottle is returned to its proper place. The robot then picks the pump bottle containing the proper vehicle and moves it to the dispensing station. The same procedure as described for the crosslinker is followed until the proper weight of the vehicle has been dispensed and the pump bottle returned to its proper place. The robot then moves the sample jar under the catalyst addition station and the proper amounts of catalyst and solvent are added volumetrically form the master lab station. The sample jar is then placed on a magnetic stirrer and the sample agitated for a pre-selected time. After completion of the agitation cycle, the sample jar is returned to the capper. The cap is then screwed on and the sample jar returned to its proper position in the jar rack. The next sample jar is then selected and the procedure is repeated. As each sample is prepared, the actual weights are printed. If a pump bottle runs out of material or a pump fails to work, the sample preparation underway will be aborted (time limit test) and the next preparation started.

The panel test system has the capability for running film thickness, film gloss, or MEK rub tests on a total of 58 test panels. Any single test can be run individually or any combination of the three tests can be run.

In operation, a panel is removed from its position in one of the three panel racks and placed in the panel work station. The combination gloss head and thickness detector is then picked up by the robot and placed on the test panel at six different positions across the surface. Gloss and/or film thickness readings are taken and average values and standard deviation printed out. If MEK rubs are to be run, there are four positions on the panel where the rubs take place. Each one of the four positions can be set for any number of rubs in multiples of ten. Each time ten rubs are done, the felt marker is rewetted with MEK. A preset air pressure establishes the proper force of the felt marker on the panel. The four positions are set for the number of rubs which will be below and above the anticipated breakthrough of paint film on the panel, i.e. 50, 100, 150, 200 rubs. The rub area is then evaluated by a technician to select the proper breakthrough value.

In the sample preparation and panel testing, a total of three separate hands are used by the Zymate System. One hand is used for handling the jars in the sample preparation, another hand is used for handling the panels and the gloss and thickness readings, and the third hand is used for the MEK rub test.

SPECIAL SOFTWARE

The general software package used for robot control in the work station is the standard EasyLab software. However, for the panel testing of film thickness and film gloss and the MEK rub, special software and interfaces were required.

In programming the MEK rub, it was not possible, with the original furnished software, to duplicate the speed of the manual operation. When programmed with the standard software, the felt marker would move back and forth across the panel satisfactorily but would stop and restart at the end of each stroke. Based on discussions with Zymark, this limitation was resolved and only the starting position and the length and the number of strokes needed to be defined. The felt marker then moves back and forth across the panel with no pause at the end of the each stroke. This then came close to duplicating the manual operation and eliminated the problem.

The film thickness guage is manufactured by Elcometer, Inc. It has capability for operating a small printer but is not designed for any computer interface. To get the film thickness data into the Zymate controller, a special interface and software had to be developed. The interface tied directly to the printer output and the software allowed collection of the data. If there was no data signal when the data was to be taken,the robot would then lift the sensor and replace it on the panel in the same location in a second attempt to obtain the data. This would continue until the proper data is obtained or the system times out or aborted. The data is stored for six positions on the panel and then an average and standard deviation are calculated.

The film gloss gauge is manufactured by Byk-Mallinckrodt, Inc. It is fitted with an RS-232 port but still required the development of new software to allow for data pickup and storage in the Zymate controller. The software was similar to that developed for the film thickness measurement. If there was no data signal when the data was to be taken, the robot would then lift the sensor and repalce it in a second attempt to obtain the data and continue as described for the thickness test. The data is stored for the six positions on the panel then an average and standard deviation calculated as described above.

The physical configuration of the system is such that the thickness and gloss readings are in essence taken simultaneously. Therefore, the robot need move the sensor to the six positions only once.

DISCUSSION

This project was developed because of the increasing amount of testing to be done. It became necessary to improve laboratory productivity and accuracy by increasing the automation of our operation. Each panel is handled many times - ususlly once for each test. The sample preparation and panel testing now require approximately 80% of technician time. Our automated robotic work station is still quite new and full-time operation only just been initiated. There has not been sufficient operating time to fully evaluate the productivity increase. However, there are some conclusions that can be drawn. The new system can run unattended during the dark hours as well as during the day. Under normal conditions a technician could do the thickness, gloss, and MEK rub tests on approximately 10 panels per hour or average approximately 60 panels per 8-hour day including the calculations. This work is tedious, involved exposure to MEK, and was error prone in the data handling. The robot can do 100 panels/24-hour day for the same testing. In essence, the robot can almost double the panel testing productivity which eliminates a tedious job, decreases the exposure to MEK and decreases the calculation errors.

Thus, the investment of approximately $30,000 for the robot has the capability to increase productivity and accuracy of our previous manual operation and also to free technician time for other productive work.

AREAS FOR FUTURE IMPROVEMENTS

While we are very satisfied with the initial performance of this robotic work station, there are areas where the overall performance could be improved. Operation of the robotics hardware has been good. Handling of curved test panels without error needs to be improved. Improved roundness of sample jars is required. Reliability of jar handling is good, if the jars are round. The interface for the dry film thickness instrument (Elcometer) is tied in through the printer supplied with the instrument. Reliability of this printer has been poor and has decreased the overall system reliability. Purchase of a more reliable printer, or elimination of the printer interface entirely, is now being investigated with both Zymark and Elcometer. Elimination of this problem area would substantially improve the reliability of the dry film test.

At the present time, all data are printed as hard copy as the information is generated. Data transfer to and from our central computer systems remains to be completed. No standard software for purchase seems available at this time.

CONCLUSIONS

The robotic test station described in this paper, although not in long-term operation, has the capability for substantial improvement in test accuracy and productivity of our coatings application laboratory. It can increase the number of samples prepared and the tests performed while reducing the amount of time our technician staff is required to use doing these tasks. The robotic system has sufficient reliability for essentially unattended off-shift operation. It also has the flexibility to be adapted to revised tasks as our testing requirements change.

ACKNOWLEDGEMENTS

We would like to acknowledge the technical support, developmental effort and helpful suggestions provided by the staff of the Zymark Corporation in bringing this project to completion.

ROBOTIC AUTOMATION IN ORGANIC SYNTHESIS

Gary W. Kramer and Philip L. Fuchs
Department of Chemistry
Purdue University
West Lafayette, IN 47907

ABSTRACT

Technology is now available for automating several aspects of synthetic organic chemistry. We are attempting to create an automated system for reaction development, analysis, and optimization. Our system couples a central computer control network with a Zymark laboratory robot, automated variabale temperature reactors, and automated chromatographs. Our goal is to create an instrument which will allow a researcher to initiate a series of automatically executed experiments.

INTRODUCTION

Selecting the precise conditions under which an organic reaction is run is often determined by the chemist's intuition. Yet the reaction environment is critical to the success of the experiment. Finding the optimal set of reaction parameters is usually limited by finite resources, and the effort expended toward this end is normally proportional to the amount of material required.

In production scale synthesis, knowing just the optimal set of conditions is often not enough. Inadvertent variances in the procedure may require "mid-course corrections." To make these types of adjustments, the chemist or chemical engineer must know the effects of the important variables on the overall reaction rate and yield. When the number of reaction variables is large (the usual case), the determination of optimal conditions or reaction profiles can be a prodigous task. Procedures are replicated with only small variants to determine the effects of the variables. Fortunately, there are techniques in experimental design which can be used to identify the more important variables and reduce the number of trials. Even so, the experimental burden remains heavy.

To reduce the experimental effort required of the practicing organic chemist, we are attempting to create an automated system for reaction development, analysis, and optimization. In other areas of chemistry, automation of routine processes is common. This is not so in organic synthesis where experimental procedures are more diverse. If automation is to be viable for the organic laboratory, it must be flexible enough to allow facile reconfigurations.

DISCUSSION

Phase I - A Demonstration System

Our prototype for automated synthesis consisted of a Zymate System, automated room temperature reactors, and a semi-automated liquid chromatograph as the analyzer (Figure 1). The system was directed solely by the Zymate controller. Reactions studied with this system

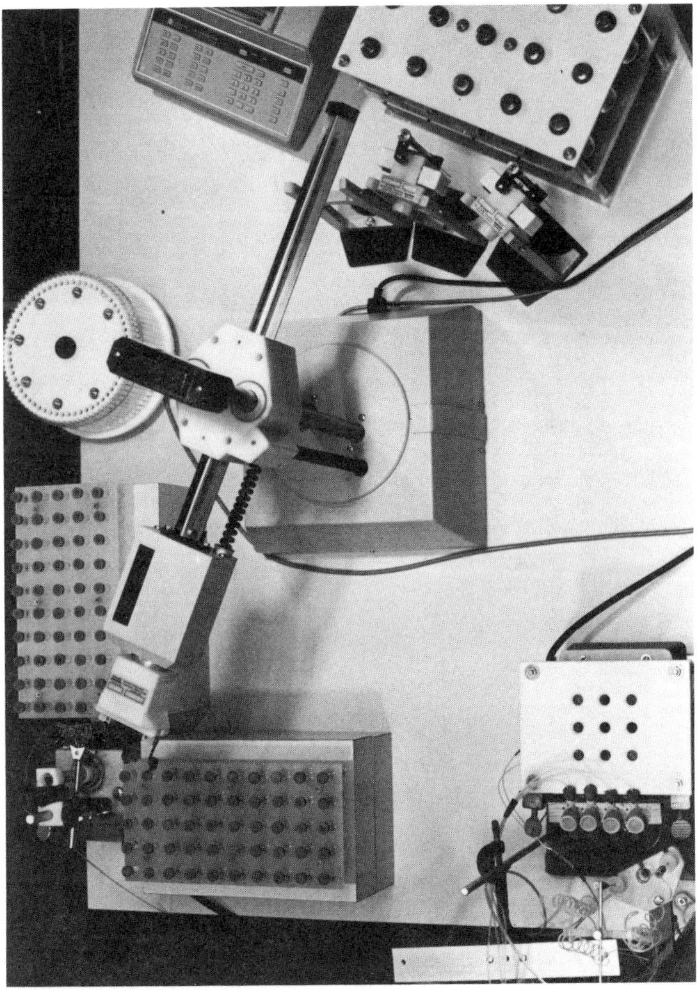

FIGURE 1. Purdue automated organic synthesis system - phase I.

were carried out according to a predefined agenda. The system was open loop: the results from the experiments were printed out for subsequent interpretation by the operator. Many real-world problems were ignored. Any reaction could be studied as long as it involved only liquids or solutions, ran at room temperature, had a half-life greater than one hour, involved reactants and products which were UV-active, produced no solids or intractable tars. Despite these constraints, this system has provided data which led to the improvement of a real synthesis.[1,2]

Of the many restrictions on our Phase I system, limitations of the control system appeared to be the most severe. We exhausted the capabilities of two power and event controllers and the analytical instrument interface and ran seriously short of memory space in the Zymate controller even after installing the expansion module. We were trying to push the Zymate System beyond its design. Although we had anticipated this from the outset, the need for expanded control capability forced us toward the next generation system.

Phase II - Designing for Change

Open-ended projects, such as ours, make design changes inevitable. Building in the flexibility to accommodate these alterations becomes a major design challenge. Modularity and portability in both hardware and software are essential. The control system architecture (Figure 2) for our current and subsequent systems reflects these design goals. The structure of the control system, which looks very much like an organization chart for a small business, allows for orderly system growth. The executive processor contains the user interface, application program,

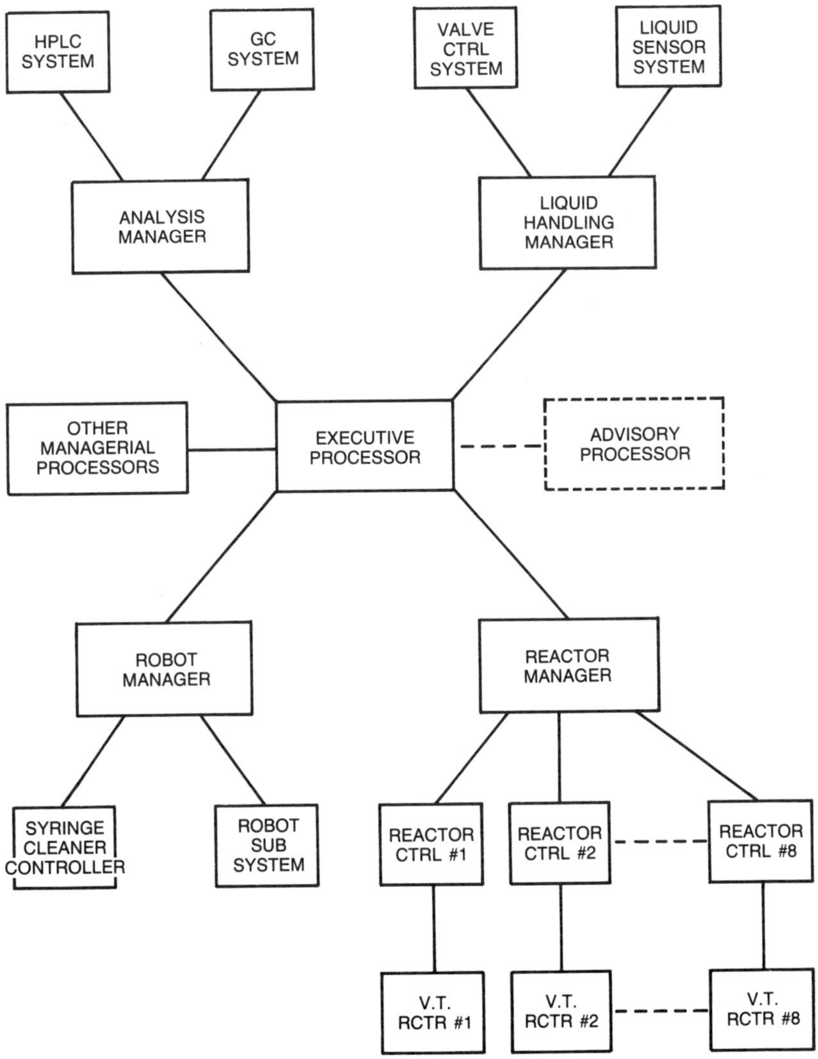

FIGURE 2. Control system architecture.

and main multi-tasking control routines. It interacts with the rest of the system through 8-bit managers.

These managerial processors are the keys to the control system. By serving as buffers, translators, controllers, and isolators, they allow realtime control to be relegated from the executive processor. This allows true concurrency, largely freeing the system from timing constraints. The inherent modularity permits replacement, upgrading, or redesigning of subsystems with only minor penalties. Isolation of the executive from the realtime environment of the managers, permits the application code to be written in a portable, high-level language where tasks are handled as logical concepts. Mapping logical constructs to physical actions is a job for the managers.

Figure 2 shows an advisory processor in a dotted box. This position in the architecture is reserved for future addition of artificial intelligence (AI) to our system. Since this technology usually requires special hardware and software environments, a separate computer seems appropriate. Initially this AI machine will function as a consultant to the executive in an expert system capacity; however, in time, the user interface and application programming functions may be transferred to this processor.

The Zymate controller appears to have been down-graded in the hierarchy of Figure 2. This reflects a shift in our philosophy of using the robot for everything to a principle of employing the robot for automating only what would be otherwise difficult. This change results from our realization that the Zymate controller was seriously overworked in our previous scheme. We still find that robot time is a very valuable resource and often a factor limiting system capability. To make more robot time available, we envision moving other tasks currently being handled by the Zymate controller to dedicated systems. For example, we have designed and are currently developing a separate subsystem to wash the syringes in the syringe hands while parked in their holders, detached from the robot arm. Since we carry out many manipulations with the syringe hands (reagent addition, aliquot removal, sample preparation, etc.), we anticipate a substantial improvement in system throughput.[3]

External Robot Control - The Taming of a Zymate

The Zymate System is well suited in many respects for its function as a system peripheral. Unfortunately for us, its software was written under the assumption that it sould be a controller, not a peripheral. The Z840 computer interface provides a hardware I/O channel, but it was designed to permit data transfers to and from the Zymate System, not to allow external control. Using this interface, data transfers must be initiated by the Zymate System and are limited to numerical data only. Creating a custom system interface might improve the situation, but at present we do not have the resources to pursue this approach. Instead we have developed a less satisfactory, but expedient method. All robot operations are written as macro functions, each being a complete subunit performing one or more tasks. For example, the macro to add a reagent to a reactor involves attaching a clean syringe hand, filling the syringe with the proper reagent, delivering the reagent to the appropriate reactor, and parking the syringe hand (assuming that the automatic syringe cleaner subsystem is in place). These macro calls and their parameters are transferred from the external computer to the Zyamte System as arrays of floating point numbers.

Since the Zymate System retains local control, an EasyLab procedure which first requests, then decodes, and finally executes the data from the external processor is run as the main process in the Zymate System. The external system generates robot tasks and places them in a queue. When the Zymate System requests a new task, the external processor ships out the task at the head of the queue. The robot system receives, decodes, and carries out the task using the parameters supplied with the request. On completion, the Zymate System returns a status message and requests another task. The details of this procedure are described elsewhere.[4]

Variable Temperature Reactors - Description of a Prototype

Individual, variable temperature control for each reactor is a necessity for our system. The variable temperature (VT) reactor must be capable of rapid, computer-controlled cool-down and warm-up over the traditional "organic chemistry temperature range" from a dry ice/acetone bath to refluxing dimethylformamide (DMF), (-80 C to +150 C) with a resolution of ± 2 C . Since a minimum of sixteen reactors is anticipated, each VT unit must be of reasonable cost.

The scheme employed in our prototype design was inspired by the method used with variable temperature nuclear magnetic resonance and EPR probes. As shown in Figure 3, nitrogen gas is the heat exchange fluid and, for sub-ambient work, is initially cooled by passage through a coil immersed in a liquid nitrogen bath. The cold gas is brought to the reactor area through a vacuum-jacketed transfer line. Surrounding the reaction vessel is a Dewar containing a heater coil and a platinum RTD temperature sensor (Figure 4).

By applying the correct amount of power to the heater, the gas flowing over it is warmed to the proper temperature. The nitrogen flows by the feedback RTD sensor and impinges in the reactor vessel bringing it to the desired temperature. In the case of super-ambient operation, a diversion valve allows the nitrogen gas to bypass the cooling coil.

The reactor itself is made from a glass O-ring seal joint. Its total volume is about 10 mL but its usable capacity is 3 to 5 mL. Since the variable temperature Dewar precludes conventional magnetic stirring and also since having an eductor tube which goes completely to the bottom of the reactor is desirable, an alternate means of stirring is necessary. An annular magnetic stirring bar was constructed which spins about the eductor tube like a wheel on an axle. The required rotating magnetic field is generated using the field coils from a gutted stepper motor driven by conventional circuitry. The top of the reactor, constructed from Teflon-coated stainless steel, contains a septum-capped port, tubing inlets and outlets, and a small water-cooled condenser coil.

Each reactor must have its own temperature controller. This unit will be directed by a slave microprocessor with an 8- or 10-bit analog to

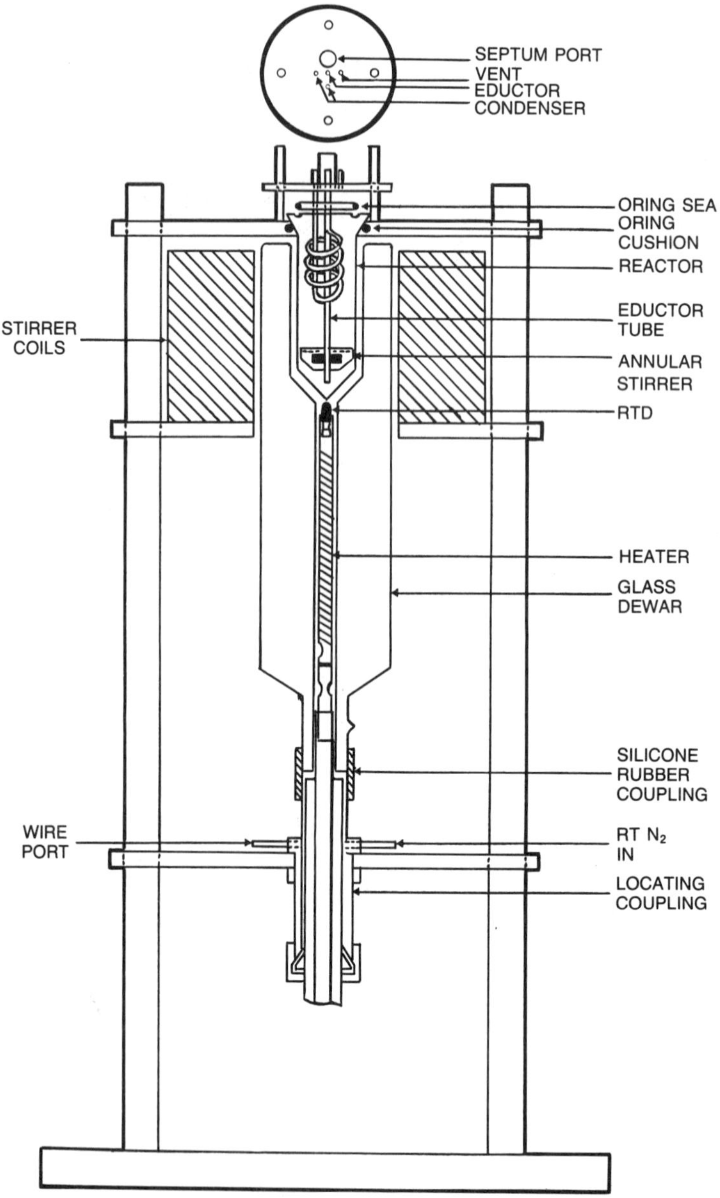

FIGURE 3. Variable temperature reactor control scheme.

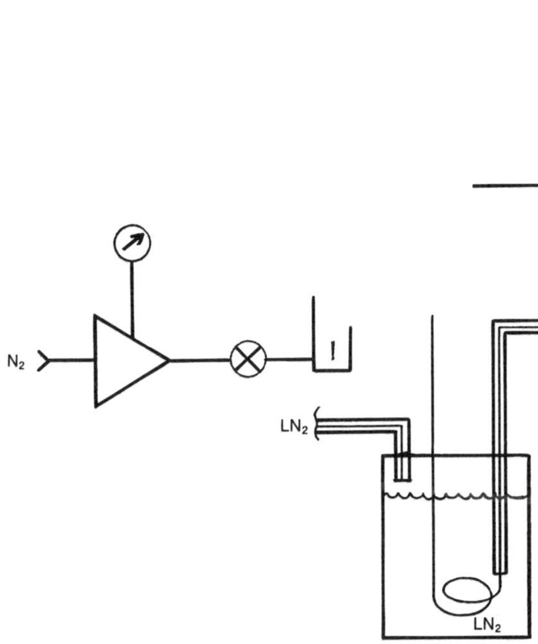

FIGURE 4. Prototype variable temperature reactor.

digital converter to read the temperature, a digital to analog converter to drive a monitoring device, and a 12-bit rate multiplier to activate a solid state relay which meters current to the heater. The controller will have other outputs which operate the chiller bypass valve, provide pulses to the stepper motor driver, drive a front panel temperature dispaly, etc. Each of these intelligent temperature controllers appears as an I/O device to the reactor manager computer. A temperature controller will receive temperature set points from the managerial CPU, but will carry out a PID (proportional, integral, derivative) temperature control algorithm on its own.

The reactor manager computer handles communications with the executive processor, receiving temperature set points, heating/cooling rates, and stirrer commands. The manager can ramp the temperature of any reactor by downloading set points to the proper temperature controller at the appropriate rate. The manager must also maintain the liquid nitrogen level in the chillers, monitor the nitrogen gas source pressure, and report to the executive processor the temperature of any reactor, the estimated time to reach any new set point, and any fatal error conditions.

A prototype of the VT reactor (Figure 4) has been built and appears to function quite well. There are, however, several unresolved issues with this current design. Chief among these is the inconvenience of dealing with liquid nitrogen as a refrigerant. We are currently pursuing an alternate scheme which uses electrical refrigeration units to cool the gas.

SUMMARY

Our efforts to automate organic synthesis are coming closer to fruition. Although we have presented here only an overview of the system hardware design, parallel efforts are occurring in the development of the system software, both at the application and user interface stage and at the control level. Many hurdles remain, the finish line is not yet in sight, but at least we now have a map of the course.

ACKNOWLEDGEMENTS

We would like to make special note of the efforts of Mike Trueblood, Roger Frisbee, and Doug Lantrip in making this project a physical reality. Financial support by Hoffmann-LaRoche Inc., The Dow Chemical Co., Eli Lilly and Co., and the National Science Foundation (CHE-8406115) is gratefully acknowledged.

REFERENCES

1. Frisbee, A. R., Fuchs, P. L., and Kramer, G. W., in Advances in Laboratory Automation - Robotics 1984, Hawk, G. L. and Strimaitis, J. R., eds, Zymark Corporation, Hopkinton, MA, 1984, p. 47.

2. Frisbee, A. R., Nantz, M. H., Kramer, G. W., and Fuchs, P. L., J. Am. Chem. Soc., 106, 7143 (1984).

3. Kramer, G. W. and Fuchs, P. L., Byte, in press.

4. Frisbee, A. R., Lantrip, D. A., Fuchs, P. L., and Kramer, G. W., Zymark Laboratory Automation Newsletter, Vol. 2, No. 3, 1984, p. 3.

SYNTHETIC DNA: APPLICATION OF ROBOTICS TO THE PURIFICATION OF OLIGONUCLEOTIDES

Simon S. Jones, John E. Brown,
Darlene A. Vanstone, David Stone
and Eugene L. Brown
Genetics Institute, Inc.
87 Cambridge Park Drive
Cambridge, MA 02140

ABSTRACT

Three procedures for the purification of synthetic oligodeoxyribonucleotides have been automated using a commerically available robotics system. These include a rapid purification protocol on a disposable reverse phase C-18 column, isolation of oligonucleotides from polyacrylamide gel slices and the preparation of crude samples of oligonucleotides for purification by polyacrylamide gel electrophoresis. The system has been fully automated, including evaporative centrifugation, to process several sets of samples (up to six samples per set) by repeating a single procedure or by concurrently executing any two procedures.

INTRODUCTION

Until recently the synthesis and purification of oligodeoxyribonucleotides (DNA) was a time-consuming process.[1] This fact inevitably limited the potential usefulness of synthetic DNA in biochemistry and molecular biology.

Due to a number of important improvements in the chemistry of DNA synthesis this situation has dramatically altered in the last few years. The development of solid supported DNA synthesis,[2,3] improvements in the phosphotriester approach[2-5] and the advent of nucleoside phosphoramidites in the phosphite approach[6] permits the rapid synthesis of a large number of oligonucleotides[7] in relatively high yield. In turn these advances have allowed for the development of numerous commercial DNA synthesizers.[8]

Paradoxically these developments have made the purification and isolation of oligonucleotides the rate-limiting steps. Indeed this problem has been exacerbated by the quantity of oligonucleotides which can be rapidly produced by a DNA synthesizer. Up to this time there has been no parallel imporvements in oligonucleotide purification in terms of speed and/or the ability to handle several different compounds at the same time.

We describe in this report the use of a robotics system to automate two purification and one isolation procedure in the preparation of synthetic oligonucleotides for biological use. These procedures are:

i) The purification of short ($\leq$ 20 bases), single sequence oligonucleotides, by reverse phase chromatographic separation of the 5'-O-dimethoxytrityl (DMTr) protected product from the 'failed', 5'-hydroxyl sequences[9]. This procedure is called lipophilic selection.

ii) The isolation of oligonucleotides from polyacrylamide gel slices by electroelution followed by desalting on a reverse phase C-18 column.

iii) The preparation of crude oligonucleotides for general purification by polyacrylamide gel electrophoresis by liquid-liquid extraction and filtration.

EXPERIMENTAL

Equipment

The robotics unit consisted of a Zymate Laboratory Automation System (Zymark Corporation, Hopkinton, MA). The integral parts of the system (Figure 1) consists of a controller including EasyLab application software, floppy disk and printer, a robot arm, power and event controller, three master lab stations each containing three syringes for solvent delivery and a cannula wash station. The custom built items consisted of two vortex stations, two liquid dispensors, one aspirator, three work stations, two liquid distribution hands, one dual purpose hand containing a pair of gripping fingers and a cannula for delivering and transferring liquids as well as assorted racks. The evaporative centrifuge was a Speed-Vac, Model SVC100H adapted by Savant (Farmingdale, NY) for full integration into an automated robotics system.

The electroelution equipment and DEAE-anion exchange resin were from Epigene, Inc. (Baltimore, MD). The plastic tubes for the evaporative centrifuge were designed and produced by Plastic Design, Inc. (Middltown, CT). Spice disposable C-18 columns were purchased from Analtech. Acrodisc (0.45 um) filters were from Gelman Sciences.

Methods

All oligodeoxyribonucleotides were synthesized on an Applied Biosystems DNA Synthesizer, model 380A.

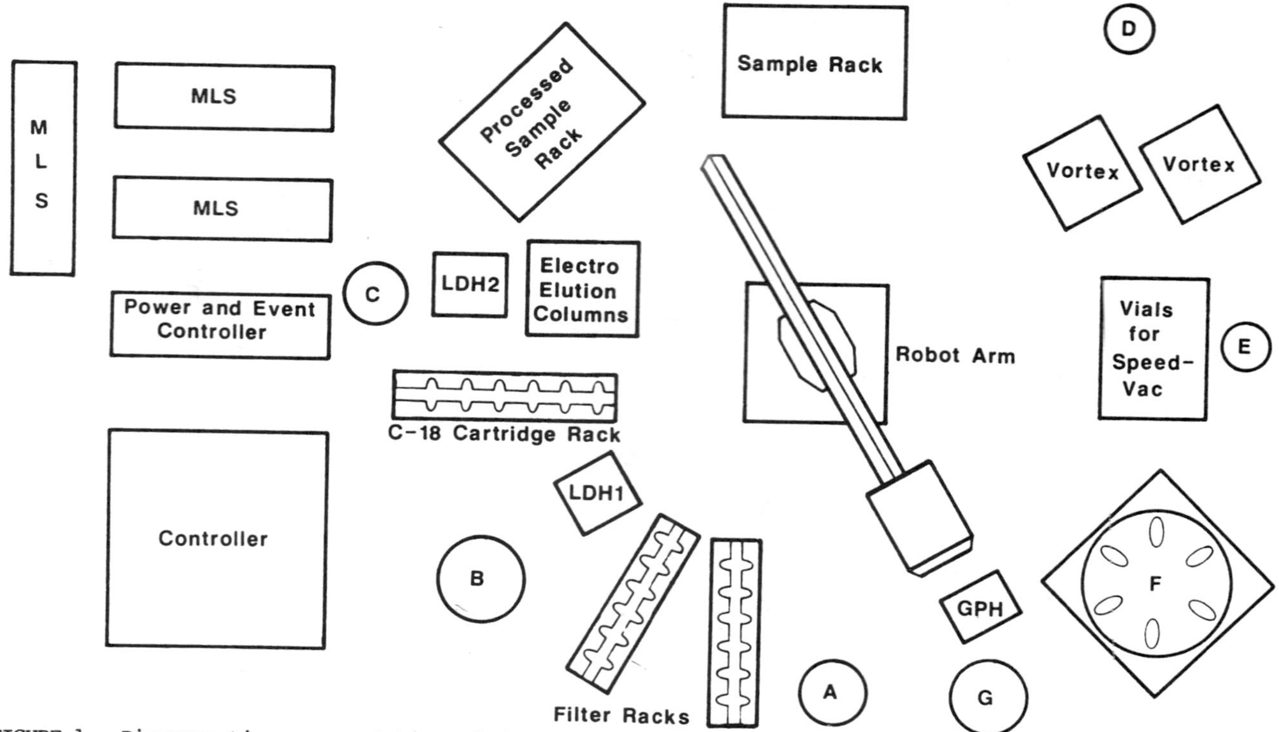

FIGURE 1. Diagrammatic representation of the robotics system. MLS: master lab station, GPH: general purpose and hand, LDH: liquid distribution hand. A - C: workstations for filtration, C-18 columns and electroelution columns, respectively. D: a dispenser for 25 mM TEAB pH 8 and ethyl acetate, also an aspirator. E: a dispenser for acetic acid: water (4:1 v/v) and ethanol: water (1:1 v/v). F: evaporative centrifuge. G: cannula wash station.

Lipophilic selection of 5'-O-DMTr-oligonucleotides was carried out as described[9] using disposable C-18 columns connected to 5 mL syringe barrels. For manual experiments the system was pressurized by a syringe plunger, while argon, via a liquid distribution hand, was used in the automated system. The product fraction was evaporated to dryness, detritylated,[6] evaporated twice from 50% aqueous ethanol, redissolved in water (1 mL) and filtered.

Polyacrylamide gel slices were obtained by purifying an equal number of crude A_{260} units for a given oligonucleotide for each isolation procedure (Table 2). The major band, detected by u.v. 'shadowing', was excised (ca. 1.5 x 25 x 3 mm^3), and the oligonucleotide was manually electroeluted from the gel slice onto DEAE-resin according to the protocol from Epigene. The column containing the resin was presented to the robot which washed the column with 50 mM NaCl, 20 mM Tris-HCl, 1 mM EDTA pH 7.6 (2 X 3 mL) and eluted the DNA with 1 M NaCl, 20 mM Tris-HCl, 1 mM EDTA pH 7.6 (0.4 mL). Subsequently, the diluted (10-fold) salt solution was desalted on a disposable C-18 column as described[9]. Manually the equivalent procedures were followed except the DNA was isolated either by desalting (as above) or by ethanol precipitation (Table 2).

The preparation of crude samples of deprotected oligonucleotides for polyacrylamide gel electrophoresis consisted of the following standard manipulations: dissolution of the sample in 25 mM TEAB pH 8 (1 mL), ethyl acetate extraction (3 x 1 mL), evaporative centrifugation, redissolving the sample in water (1 mL) and filtration.

RESULTS

In most oligonucleotide synthesis methods the crude product is protected at the 5'-terminum with the DMTr protecting group, while the 'failed' or truncated sequences contain terminal 5'-hydroxyl functions, following partial deprotection.[6] It has been demonstrated[9] that the 'failed' sequences can be separated from the DMTr containing product, due to the greater lipophilic character of the latter, on a disposable reverse phase C-18 column.

The procedure is suitable for the rapid purification of single sequence oligonucleotides (⩽ 20 nucleotides) to be used as sequencing primers and probes - e.g., in plasmid constructions (C. Shoemaker, personal communication). It must be noted that this procedure is not as efficient for purifying longer oligonucleotides since the difference in lipophilicity of the product and 'failed' sequences decreases with increasing chain length.

This procedure was readily adaptable to an automated system and the comparison of the various acetonitrile (CH_3CN) elution fractions from the lipophilic selection of 5'-O-DMTr sequences from failed sequences on C-18 columns is given in Table 1 (robot example in triplicate). Similar results were obtained either manually or by the automated system, with a silghtly lower relative recovery of the product fraction (30% CH_3CN fraction) in the latter. A qualitative assessment of the oligonucleotides in the 10% and 30% CH_3CN fractions, after removal of the 5'-protecting group[6] and subsequent labelling with [γ-^{32}P]ATP in the presence of T4 polynucleotide kinase,[10] is shown in Figure 2. Compared

TABLE 1. Liphophilic Selection of 5'-O-DMTr Oligonucleotides from the Failed Sequences.

	Manual (A_{260} units)	Robotic (A_{260} units)		
Applied Sample *	65.0	49.8	48.7	52.7
Non-bound material	1.1	1.2	0.1	0.1
25 mM TEAB+ pH 8 (4 mL) wash	0.6	0.4	0.5	0.5
10% CH_3CN/25 mM TEAB+ pH 8 (4 mL) fraction	15.8	9.7	11.3	10.2
30% CH_3CN/100 mM TEAB+ pH 8 (3 mL) fraction	39.5	25.4	24.6	29.0

* 50% of the crude material, 17-mer d(GATCCCCATGTAATTTT), from a 1 umol scale synthesis.

\+ Triethylammonium bicarbonate (TEAB)

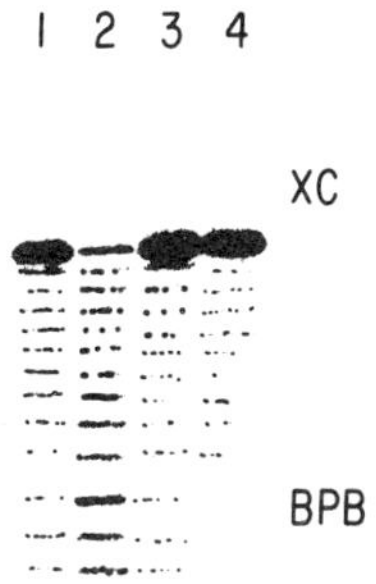

FIGURE 2. A comparison of a [5'-^{32}P] labelled oligonucleotide 17-mer d(GATCCCCATGTAATTTT), prior to and after purification on a reverse phase C-18 column or by polyacrylamide gel electrophoresis. Lanes 1, crude sample before purification; 2 and 3, 10% CH_3CN/25 mM TEAB pH 8 and 30% CH_3CN/100 mM TEAB pH 8 fractions from the C-18 column, respectively; 4, material purified by polyacrylamide gel electrophoresis prior to labelling. The fractions were analysed on a 20% polyacrylavide gel run under denaturing conditions. BPB: bromophenol blue; XC: xylene cyanol.

with the crude product (lane 1) there has been an enrichment for the product in the 30% CH_3CN fraction (lane 3) with a substantial portion of the failed sequences being eluted in the 10% CH_3CN fraction (lane 2). For comparison, lane 4 contains material previously purified by polyacrylamide gel electrophoresis. These results are similar to a manually processed sample (data not shown).

There are various methods for isolating DNA from acrylamide gels.[11] We have had extensive experience with an electroelution procedure which consists of the following steps. Firstly, the oligonucleotide is electroeluted out of a gel slice onto an anion exchange resin column, such as diethylaminoethyl (DEAE) resin. Next the resin is washed with a low salt buffer and then the DNA is eluted with a high salt buffer (see Methods). Finally, the DNA can be precipitated[12] or desalted on a disposable reverse phase C-18 column.[9] In the latter case the DNA adheres to the support, while the salt is washed through and the DNA is then eluted with a volatile buffer containing an organic solvent (see Methods).

The automated version of the electroelution/desalting procedure was compared to the manual equivalent and in parallel, to isolating oligonucleotides by precipitation. As can be seen in Table 2, the amount of recovered DNA for three oligonucleotides of differing lengths is comparable in most cases (robot examples in duplicate). Also the material isolated by the automated system, in comparison with ethanol precipitated material, was labelled to the same extent with $[\gamma-^{32}P]ATP$ in the presence of T4 polynucleotide kinase.[10]

TABLE 2. Isolation and Desalting of Oligonucleotides from Polyacrylamide Gel Slices.

A	B	C	D	E	
15	4[6]	0.80	1.05	0.76	0.79
27	9[12]	1.33	1.45	1.02	1.15
42	25[13]	1.60	1.38	1.39	1.83

A = Length of oligonucleotide (nucleotides)

B = Crude material purified per experiment (A_{260} units) [% of total crude material^{+}]

C = Material recovered: precipitation (A_{260} units)

D = Material recovered: manual desalting (A_{260} units)

E = Material recovered: robotic desalting (A_{260} units)

* To 1M NaCl, 20 mM Tris-HCl, 1mM EDTA pH 7.6 (0.4 mL) three volumes of absolute ethanol were added. Incubated at -78°C for at least 30 min. Precipitate was collected at 10K r.p.m. for 15 min. at 4°C and washed once with aqueous ethanol (1:4 v/v)

$^{+}$ Material from a 1 umol scale synthesis.

TABLE 3. Preparation of Crude Oligonucleotides for Purification by Polyacrylamide Gel Electrophoresis

Length of oligonucleotide (nucleotides)	% recovery * manual	robotic
17	98	80
21	93	75
26	86	61
29	90	75
33	92	64
45	77	61

$$\text{* \% recovery} = \frac{A_{260}\text{ units after extraction}}{A_{260}\text{ units crude DNA}} \times 100$$

Before purifying samples of crude deprotected oligonucleotides, it is advantageous to extract non-nucleotidic and organic material, as well as to remove particulates. This is carried out by a series of standard manipulations (see Methods) which are readily amenable to automation. For purificaiton by polyacrylamide gel electrophoresis a representative set of crude oligonucleotides of varying lengths were divided into two parts, one part for the robotics system and one for the manual equivalent. The amount of material recovered by both procedures was compared (Table 3). In most cases the recovery from the robotics system, although satisfactory, was ca. 20% less than in the manual experiments. This difference probably reflects the varying rates of separation of the aqueous phase from the organic phase in different samples, which can be adjusted for in the manual version by the experimenter. It must also be noted that there was no cross contamination of samples as indicated by polyacrylamide gel electrophoresis (data not shown).

DISCUSSION

The applicaiton of robotics to these three procedures illustrates two important aspects of automation in the laboratory. Firstly, is the concept of "robotic equivalence" which is the translation of manual actions and the corresponding equipment into an automated system that performs essentially the same functions as the manual version. Secondly, the system must be reliable and compare favorably with the manual equivalent in terms of quality and quantity of samples processed in a given period.

Figures 1 and 3 illustrate two facets of 'robotic equivalence' with a diagrammatic representation of the automated system and an example of one of the programs, respectively. The robotics system can be divided into two areas, passive and interactive, with regards the robot arm. The former contains the controller which operates the whole unit through the program, the master lab stations for solvent delivery and the power and event controller which directs electrical input and output. The interactive set consists of several groups of equipment:

i) The robot arm which has three types of motion - vertical, reach and rotary.

ii) Two types of hands, general purpose and two liquid distributors. The latter can deliver solvents and/or pressurize column systems for purification or filtration.

iii) Work stations at which vials and columns or filtration units are positioned while the procedures are executed.

iv) Liquid dispensors.

v) Racks for holding equipment and vials.

vi) The vortexes and the evaporative centrifuge.

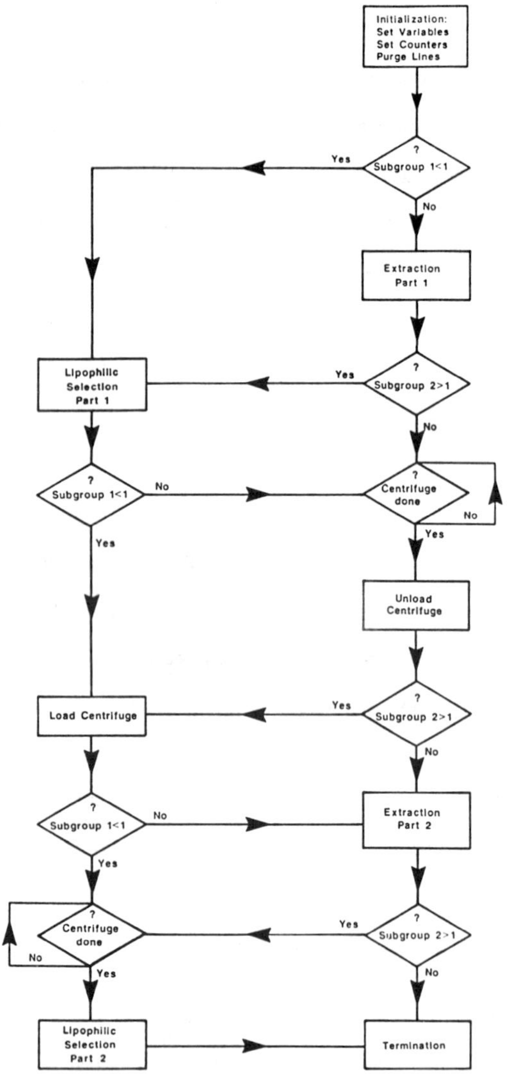

FIGURE 3a. A simplified flow diagram for two purification procedures. Subgroups 1 and 2 are the input variables to define the number of samples (maximum six samples per process) to be processed by extraction and/or lipophilic selection, respectively. Extraction Part 1, is ethyl acetate extraction and evaporation of the aqueous phase; Part 2, is filtration. Lipophilic selection Part 1, is reverse phase purification on a C-18 column; Part 2, is detritylation, aqueous ehtanol wash and filtration.

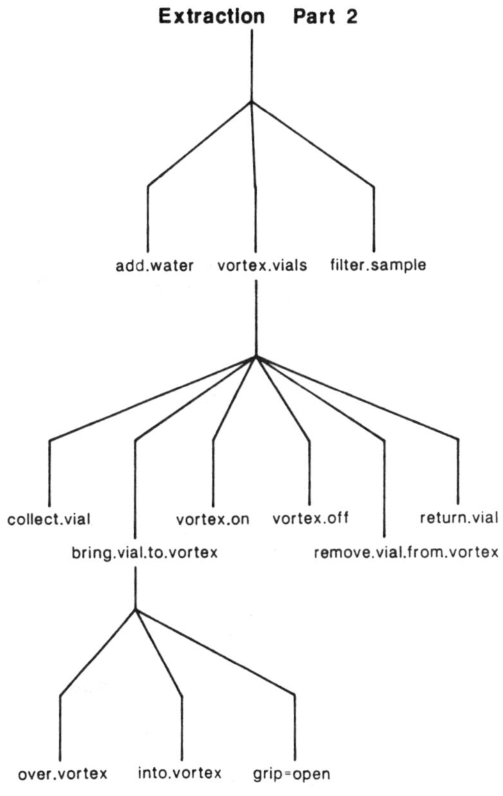

FIGURE 3b. The internal program structure of Extraction Part 2. One branch, namely vortex.vials, is traced down to the lowest programmed level, where lines are equivalent to spatial positions.

One of the novel and essential features of the system is the fully automated evaporative centrifuge operated by the controller. Firstly, the centrifuge is automatically started once it is has been loaded by the robot arm. Secondly, the run is terminated after the samples have evaporated by sensing a rapid drop in the vapor pressure in the centrifuge chamber. Thirdly, the chamber is vented, the lid is raised and the rotor is stopped.

Lastly, the rotor is electrically pulsed to the original starting position. This indicates to the controller through an electrical switch that samples are ready to be removed.

Evaporation is one of the crucial steps in all three procedures, as well as being the most time-consuming. This fact lead to a 'top level' program which was structured so that any one batch of samples can be processed while another set, prepared by the same or a different procedure, is being evaporated to dryness (Figure 3a). It also played an important role in defining the batch size, since the time required to process a second set of samples should be approximately equal to the evaporation time of the first set. Empirically it was found that six samples per batch met this requirement.

The 'top level' program, designed by the operator, is basically a statement of the procedure or procedures to be accomplished. By a series of questions (diamond boxes, Figure 3a) related to the number of samples for each process the robot is directed to different parts of the program to be executed. In turn each program line (rectangular boxes,

Figure 3a) contains a series of sub-programs (e.g. Extraction Part 2 in Figure 3a and b), at varying depths or levels (inverted tree or pyramidal structure) aimed at performing specific tasks. By this downward programming in levels of detail a 'bottom' level is reached where program lines are equivalent to a position for the robot arm or an action by a module. This is shown in Figure 3b where the program lines, over.vortex, into.vortex and grip = open, denote the spatial positions and actions to place a vial, held by the general purpose hand attached to the robot arm, into a vortex and release it by opening the fingers. Spatial positions are equated with 'bottom' level program lines by positioning the arm, with the appropriate hand, at the desired location and then storing these coordinates in the memory under the relevant program line. Similarly actions by modules, e.g. engaging an electrical switch, are equated with the relevant line e.g. vortex.on.

The system has been in operation for about six months and has reliably executed all three processes. From the results seen in Tables 1 and 2 and that gathered by processing over three hundred oligonucleotides, the reproducibility of the automated system was equivalent to the manual version of the procedures. Recovery of samples approaches or is equal to that of the manual experiments. Some losses in recovery may reflect slight variations in the depth of certain vials resulting in incomplete removal of the sample during the transfer of solutions.

Although the system is not necessarily as fast as a manual operator it is considerably more efficient in terms of almost continual operation with minimal attention. As designed, it can process two sets of samples by any two procedures in approximately six hours. Lastly, it removes the tedium of isolating and purifying a large number of samples by repetitive and standard manipulations.

In summary the robotics system described in this report provides a new and efficient method for the simultaneous purification and/or isolation of a large number of oligodeoxyribonucleotides.

ACKNOWLEDGEMENTS

We thank Dr. Robert Kamen for his support of this project.

REFERENCES

1. Khorana, H. G., Science, 203, 614 (1979).

2. Wallace, R. B. and Itakura, K., in Nucleic Acid Research: Future Developments, Mizobuchi, K., Wantanabe, I., and Watson, J. D., eds., Academic Press, New York, 1983, p. 227.

3. Sproat, B. S. and Gait, M. J., in Oligonucleotide Synthesis, Gait, M. J., ed., IRL Press, Oxford, 1984, p. 83.

4. Ohtsuka, R., Ikehara, M., and Soll, D., Nucl. Acids. Res., 10, 6553 (1982).

5. Narang, S. A., Tetrahedron., 39, 3 (1983).

6. Caruthers, M. H., in Chemical and Enzymatic Synthesis of Gene Fragments, Gassen, H. G. and Lang, A., eds., Verlag Chemie, Basel, 1982, p. 71.

7. Frank, R., Heikens, W., Heisterberg-Moutsis, G. and Blocker, H., Nucl. Acids. Res., 11, 4365 (1983).

8. Smith, J. A., American Biotechnology Laboratory, Dec. 1983, p. 15.

9. Lo, K.-M., Jones, S. S., Hackett, N. R. and Khorana, H. G., Proc. Natl. Acad. Sci. USA., 81, 2285 (1984).

10. Brown, E. L., Belagaje, R., Ryan, J. J. and Khorana, H. G., in Methods in Enzymology, Vol. 68, Wu, R., ed., Academic Press, New York, 1979, p. 109.

11. Smith, H. O., in Methods in Enzymology, Vol. 65, Grossman, L. and Modave, K., eds., Academie Press, New York, 1980, p. 371.

12. Rossi, J. J., Ross, W., Egan, J., Lipman, D. J. and Landy, A., J. Mol. Biol., 128, 21 (1979).

Zymark and Zymate are registered trademarks of Zymark Corporation.
Spice is a trademark of Analtech.

ROBOTIC SAMPLE PREPARATION FOR AUTOMATED BATCH-ORIENTED ANALYSIS IN THE CLINICAL CHEMISTRY LABORATORY

William J. Castellani, C.E. Pippenger and R.S. Galen
Cleveland Clinic Foundation
Department of Biochemistry
Cleveland, Ohio 44106

ABSTRACT

Two sample preparation methods for cardiac isoenzymes were implemented on a Zymate Laboratory Automation System with the intent that the final preparations would be analyzed on a Cobas-Bio batch-oriented centrifugal analyzer (Roche Instruments, Nutley, NJ). The two methods, packaged in kit form as Isomune-LD and Isomune-CK (Roche Diagnostics, Nutley, NJ), were a one-tube immunoprecipitative procedure for cardiac-derived lactate dehydrogenase isoenzyme 1 and a two-tube method combining immunoprecipitation and immunoinhibition to determine cardiac-derived creatine kinase isoenzyme MB. The robot was programmed to perform both preparations on up to 24 patient samples in a single run.

The immunoprecipitation procedure was identical for both methods, involving the sequential addition of two antibody preparations, each followed by a minimum five-minute room temperature incubation, and a final five-minute centrifugation. The limiting factor for the robot batch size was the centrifuge, which had six rotor positions; manual performance of the procedure was done as a single batch up to the maximum of 24 patient samples. The robot processed a set of one to three patient samples as a single batch through the centrifuge, the immunoprecipitative tubes of both procedures being centrifuged together. The time to completion ranged from 22 to 31 minutes for the robot versus 23 to 24 minutes manually. As the run size increased, a stepwise increase in robot run time was noted at the initiation of a new batch, with a more gradual increase as the batch was filled out with additional samples. The maximum run of 24 patient samples took 169 minutes on the robot versus 40 minutes manually. This represents a gradual shift from a procedure-limited run time at small sample numbers to a robot-limited run time at large sample numbers comprising full batches, where the robot arm is in constant motion with no idle time; this establishes the minimum run time for batch processing of large sample numbers using such a robot.

INTRODUCTION

Automated analyzers provide the clinical chemistry laboratory with an efficient, cost effective, and labor-saving means of performing a vast array of analyses, including the most often ordered and most critical determinations. These machines perform multiple analyses in batch mode, with great accuracy and precision and little technologist interaction. Although there remain a number of tests that must be performed in a single-channel, sequential manner, the role of robots to perform as front ends to the highly specialized batch oriented analyzers. This interaction raises several considerations. Firstly, the analyzer introduces its own imprecision that must be factored out when evaluating the capability of the robotics system to perform the sample preparation. Secondly, sample preparation run time may be a significant factor since the nature of a batch analyzer forces every sample to be ready for analysis before any are run at all (especially if a physician is anxiously awaiting the result from one of the samples). Finally, these analyzers depend upon the adaptability of the human technologist for their proper setup and operation; the robotic front end must minimize the role of the human interface to be cost effective (and acceptable to the technologist).

Two common analytes in the clinical chemistry laboratory are the enzymes lactate dehydrogenase (LD) and creatine kinase (CK) as measured in patient serum. These enzymes are useful to indicate damage to the heart muscle.[1] A subfraction of each, the electrophoretic fast moving fraction of LD (LD-1) and the intermediate migrating fraction of CK (CK-MB), are more specific for myocardial damage.[2] Several automated analyzers are available that can determine total LD and CK activity in a given sample; however, the determination of a specific isoenzyme subfraction activity depends upon prior separation of the isoenzyme fractions. Immunologic methods are available in kit form to pre-treat patient serum to remove all but LD-1 and CK-MB activity prior to analysis by automated methods. The robotic implementation of this sample preparation method on a Zymate Laboratory Automation System using precision, performance time, and amount of technologist interaction.

MATERIALS AND METHODS

The Isomune-LD and Isomune-CK kit methods for LD-1 and CK-MB (Roche Diagnostic Systems, Nutley, NJ) were used for sample preparation.[3,4] The serum is incubated for a minimum of five minutes at room temperature with a goat-derived antibody to an LD subunit present in all fractions except LD-1 (Figure 1).

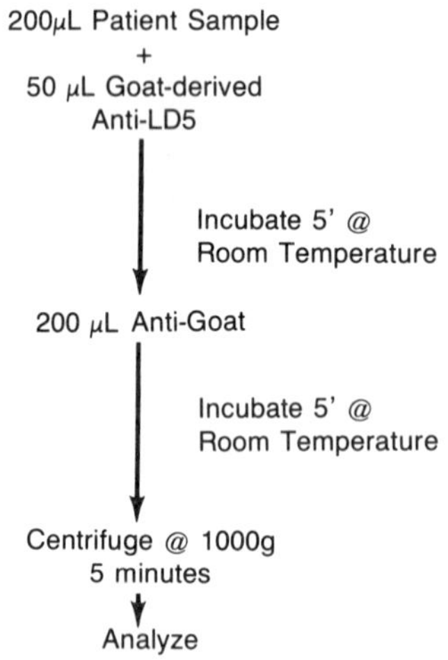

FIGURE 1. Isomune - LD procedure.

Latex bound anti-goat antibody is then added, and the mixture is again incubated at room temperature for a minimum of five minutes. After centrifugation, the supernatant contains only the LD-1 fraction. The Isomune-CK method is a two-tube technique in which the serum in one tube is incubated with an access of goat-derived antibody to one of the two isoenzyme subunits of CK (CK M) and residual activity is determined. The second tube is treated in a similar manner as the Isomune-LD protocol; the difference in activity between the two tubes is then equal to half the total CK-MB activity (Figure 2).

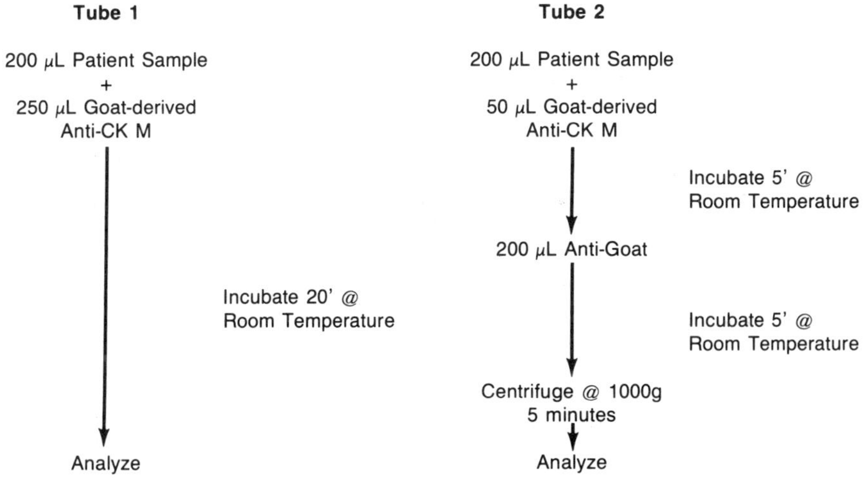

FIGURE 2. Isomune - CK procedure.

A Cobas-Bio centrifugal analyzer[5] (Roche Analytical Instruments, Nutley, NJ) was used to determine residual CK or LD activity in each sample. The sample holder for this instrument was limited to 24 positions, and the instrument is capable of performing only one assay method at a time. The instrument is preprogrammed to perform the analysis for both sample preparation methods.

A Zymate System was programmed to accept from 1 to 24 preclotted patient samples in standard vacuum blood drawing tubes (Figure 3). A six inch

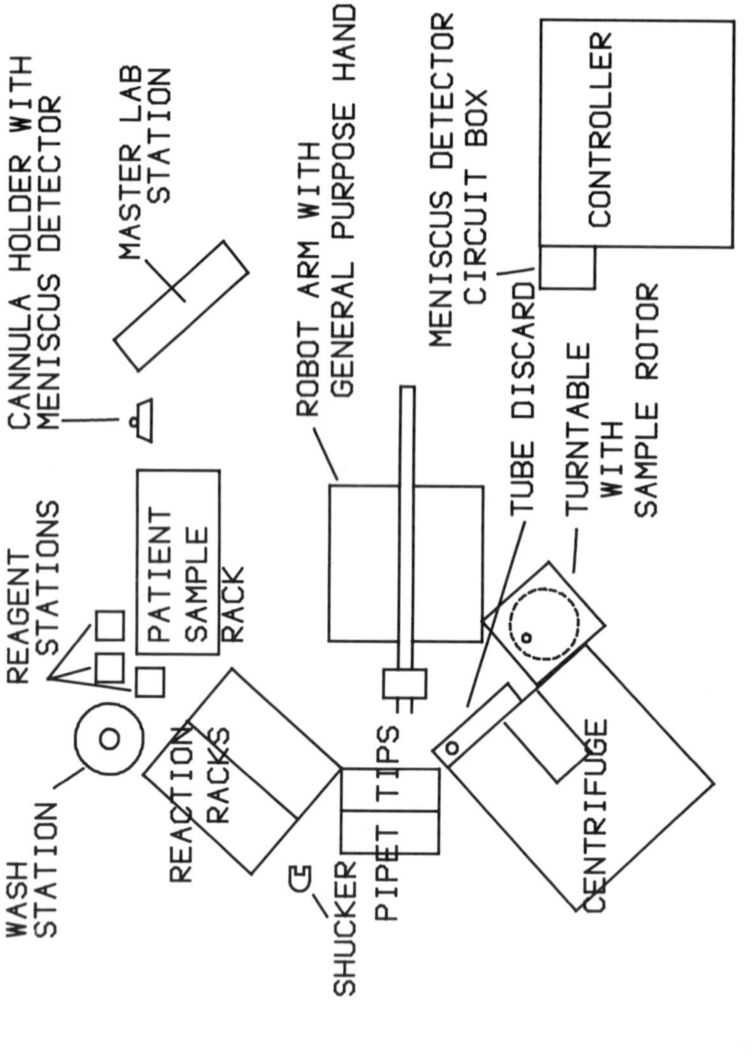

FIGURE 3. Bench configuration for the Isomune-LD and Isomune-CK sample preparation procedures.

cannula was modified to act as a conductive meniscus detector. Each patient sample was aliquoted into three tubes for performance of both sample preparation methods. The robot then performed the sample preparation methods on the appropriate aliquots, completing the Isomune-LD procedure on all samples before initiating the Isomune-CK procedure. The patient samples were divided into batches containing up to six tubes, the maximum number of positions in the centrifuge rotor. Aliquots of the prepared samples were placed into individual cups in a Cobas sample rotor mounted on a modified Zymate System turntable, ready to be capped and place on the analyzer by the technologist.

Robotic precision was determined for the Isomune-LD assay using pooled patient serum samples spiked to four different levels of LD-1 activity. Paired samples from each level were run twice a day for six days following a shortened version of an ANOVA-based protocol for evaluation of imprecision;[6] deterioration of the aging enzyme prohibited performance of the full twenty day protocol. A similar protocol was used to determine Isomune-CK precision, but rapid deterioration of enzyme activity in pooled samples necessitated use of lyophilized control serum (Cardiotrol CK, Roche Diagnostic Systems, Nutley, NJ) at two levels, undiluted and 1:2 dilution. The precision statistics were compared to manual performance of the same methods using the same samples run in parallel with the robot as an attempt to compensate for the imprecision introduced by the Cobas-Bio analyzer. One hundred postoperative open heart patient samples were also run in parallel by both the robot and manual method to determine test correlation. Run time for various sample sizes were determined for both manual and robotic methods from bench setup to bench cleanup, inclusive, for both techniques.

RESULTS

Precision of the robotic implementation closely paralleled that of manual preparation. Table 1 shows the analysis of variance breakdown for within-run, between-run, and between-day precision for the two methods at representative levels of enzyme activity for both LD-1 and CK-MB. Means and total standard deviations for the four LD-1 and two CK-MB standards are shown in Table 2 for both methods, and the variances are compared using the F-test. Precision was comparable at all activity levels except for the low CK-MB control, where the robotic system standard deviation was approximately half that of the manual method. This difference stems from widely disparate results on two separate manual runs, in contrast to the more consistent results obtained on the robot.

Figure 4 shows the scattergram for the one hundred patient samples for both LD-1 and CK-MB activities; manual and robotic results are highly correlated, with correlation coefficients of 0.9741 and 0.9663, respectively.

Time to completion for various run sizes for both manual and robotic implementations are illustrated in Figure 5, together with lines of

TABLE 1. ANOVA Comparison of Representative Levels of LD-1 and CK-MB Activity Between Robot and Manual Methods.

	LD-1		CK-MB	
	Robot	Manual	Robot	Manual
Within-Run S.D.	4.16	3.97	2.25	2.53
Between-Run S.D.	2.62	1.33	1.92	6.08
Between-Day S.D.	0	0.14	0	0
Total S.D.	4.84	4.19	2.96	6.59
Mean	57.8	51.9	71.6	69.0

TABLE 2. Comparison of Means and Standard Deviations for all Levels of LD-1 and CK-MB Activity Between Robot and Manual Methods.

	Mean		S.D.		F-Statistic
	Robot	Manual	Robot	Manual	(23, 23 df)
LD-1 #1	130.7	126.4	5.81	5.24	1.21 ($p > .05$)
LD-1 #2	57.8	51.9	4.84	4.18	1.34 ($p > .05$)
LD-1 #3	49.9	43.7	4.35	5.78	1.76 ($p > .05$)
LD-1 #4	36.3	31.1	5.17	4.59	1.27 ($p > .05$)
CK-MB #1	214.7	210.0	6.04	6.70	1.23 ($p > .05$)
CK-MB #2	75.0	72.9	1.53	2.83	3.42 ($p < .05$)

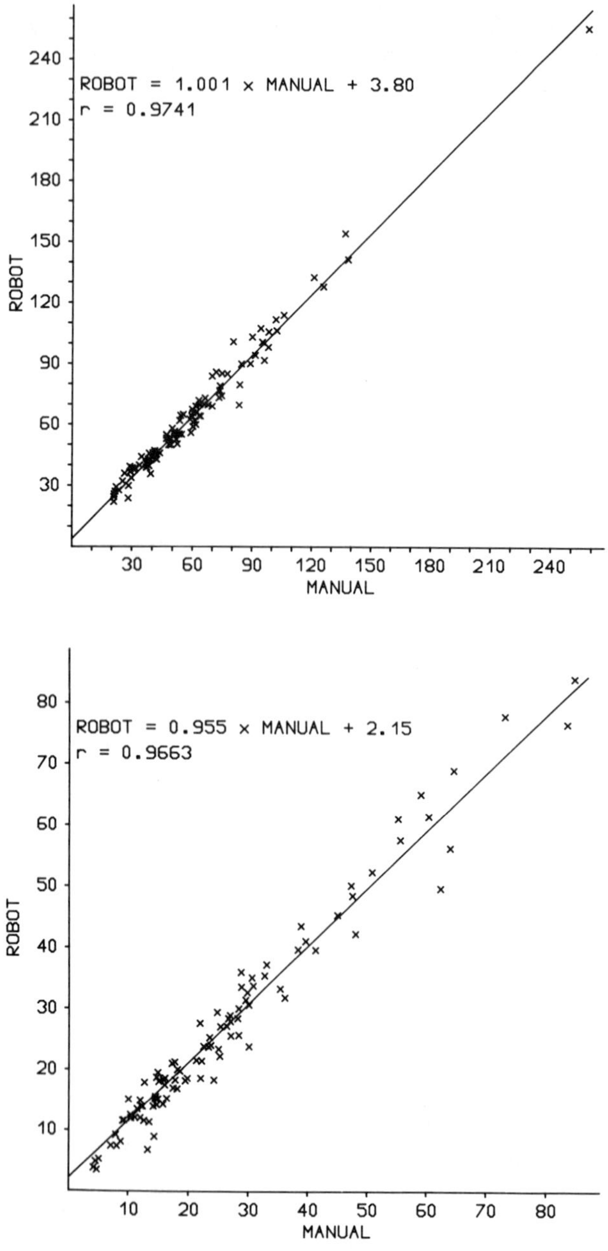

Figure 4. Scattergrams and lines of regression for one hundred postoperative patient samples performed manually and robotically using the Isomune-LD (above) and Isomune-CK methods (below).

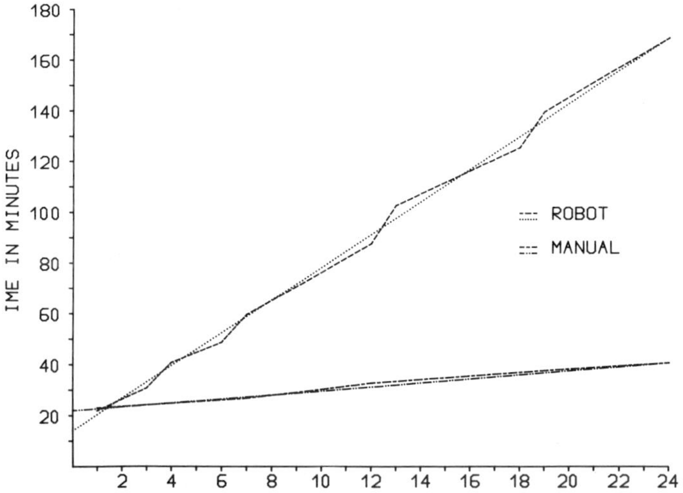

Figure 5. Comparison of Times to Completion for Manual and Robotic Implementations of the Isomune Sample Preparation Methods.

regression for both methods. For small sample sizes, the robot run time compares well with manual performance. As sample numbers increase, and are separated into multiple batches by the robot, run time deviates markedly from manual performance, with maximum times for 24 samples of 169 minutes for the robot versus 41 minutes manually. However, preparation of this sample size demands the full attention of the technologist for the entire 41 minutes if done manually. As shown in Table 3, only 9 minutes of human interaction is necessary in the robotic implementation at four different times during the run.

TABLE 3. Breakdown of technologist's time for maximum robot run of 24 patient samples.

Activity	Time
Robotic workstation setup	6 minutes 30 seconds
Switch sample rotors 1 and 2	30 seconds
Switch sample rotors 2 and 3	30 seconds
Remove sample rotors 3 and cleanup	1 minute 30 seconds
Total Time:	9 minutes

DISCUSSION

The immunologic sample preparations for LD-1 and CK-MB were successfully implemented on the Zymate System, demonstrating precision and accuracy comparable to results obtained by manual performance of the same procedures with a marked decrease in technologist involvement. However, this was obtained with a significant increase in run time for large sample sizes. The use of an automated centrifical analyzer for final enzyme activity determinations made batch implementation of the procedure the most reasonable approach.

The nature of the analyzer forced certain constraints on the robotic system and the procedure. Automated analyzers are effective and rapid because they constrain sample and reagent loading so as to minimize instrument manipulation of either. This is achieved by using the flexibility of the human interface - the technologist - who adapts his or her actions to fit the need of the machine. For a fully-robotic system, the robot front end must also be adapted to the requirements of the analyzer. This is usually not possible without prohibitive effort, so some degree of human interaction is necessary. The design of the robotic system must minimize this interaction to be effective.

The analyzer also affects the workflow through the robotic front-end. For this system, the most efficient implementation dictated that the two sample preparation methods be run sequentially, with the DL-1 procedure performed first, followed by the CK-MB procedure. Therefore, no LD-1 sample could be analyzed until all such samples were ready and no CK-MB sample could be completed until all the LD-1 samples were finished and ready to be analyzed. This delayed reporting of any patient results until the end of the robot run, which approached three hours at maximum run size. Automated analyzers are available that process samples in a discrete manner; these instruments can be programmed to treat each sample in a batch individually, performing any or all of a group of tests on each. With this type of analyzer, it would be possible to program the robot to process batches in a patient-oriented rather than a test-oriented manner, grouping all samples from any one patient together. All results needed to produce a patient's report would be available at the same time; in this manner, patient samples processed early in the robot run could be analyzed and completed while the robot finished the later patient samples. The individual laboratory must decide what machine it can adapt to the robot system, which will in turn influence the flow of patient results.

Both volume and turnaround time will determine whether robots will have a role in the performance of any given test in the clinical laboratory. Those tests that are ordered in high volume with a demand for rapid reporting of the results usually are adapted to automated analyzers capable of processing a sample directly and quickly to a finished result; little improvement can be made on these systems. Other tests are ordered in high volume, but their nature either does not demand or prevents rapid turnaround; urinalysis is rarely diagnostic in an emergency and microbioligic cultures won't grow any faster regardless of their importance. Although specialized automated devices have appeared for some of these tests, most of these procedures remain as manual tasks, and may be adaptable to robotic systems. The greatest impact may be with those tests that are ordered in low volume. Some of these tests may be critical only in special cases, but there is a demand for rapid turnaround; in these instances, one or a few samples must be processed rapidly. A robotic system may be capable of performing such methods as rapidly as a person. The other low-volume tests which do not require rapid reporting are typically batched and run at intervals; in this case, the time it takes a robot to complete these tests does not significantly affect the utility of the results. In both instances of low-volume tests, the use of a robotic system to process the samples may, in fact, be the difference between a laboratory performing the test itself or sending it out.

ACKNOWLEDGEMENT

We would like to gratefully acknowledge the support given to us by Hoffman-LaRoche, Inc. and Zymark Corp. In addition, Frederick Van

Lente, David Chou, and Lenox Abbott provided useful advice and assistance in devising the evaluation protocol and various attachments for the robot. Ann McHugh ably served as the "gold standard" for part of the robot evaluation.

REFERENCES

1. Zimmerman, H.J. and Henry. J.B., "Clinical Enzymology", Clinical Diagnosis and Management, 17th ed; J.B. Henry, ed. W.B. Saunders, 1984.

2. Wicks, R.W., Usategui-Gomez, M., Miller, M., "Immunochmeical Determination of CK-MB Isoenzyme in Human Serum: II. An Enzymatic Approach", Clinical Chemistry 28, 54, 1982.

3. Galen, R.S., "The Enzyme Diagnosis of Myocardial Infarction", Human Pathology, 6, 141, 1975.

4. Usategui-Gomez, M., Wicks, R.W., Warshaw, M., "Immunochemical Determination of the Heart Isoenzyme of Lactate Dehydroganase (LDH_1) in Human Serum", Clinical Chemistry 25, 729, 1979

5. Maclin. E. and Young, D.S., "Automation in the Clinical Laboratory", Textbook of Clinical Chemistry, N.W. Tietz, ed; W.B. Saunders, 1986.

6. Bauer, S., and Kennedy, J.W., "Applied Statistics for the Clinical Laboratory: IV. Total Imprecision", Journal of Clinical Laboratory Automation, 2, 129, 1982.

COMPARISON OF AUTOMATED AND MANUAL EXTRACTION OF DRUGS FROM BIOLOGICAL FLUIDS AT TRACE LEVELS

Steven F. Kramer, Monte J. Levitt
and Mary M. Passarello
Biodecision Laboratories
Pittsburgh, PA

ABSTRACT

In order to measure drugs at trace levels by chromatography, it is usually necessary to first extract them from biological matrix. These extractions, which remove contaminating compounds, are frequently very tedious to perform. We have used the Zymate Laboratory Automation System to perform extractions, resulting in a tolazamide assay procedure that has equivalent reproducibility and accuracy compared to manually performed extractions.

INTRODUCTION

Our company performs bioavailability tests under contract for pharmaceutical manufacturers. These tests are performed by administering small doses of medication to 18-30 normal, healthy volunteers. Blood and/or urine samples are collected at various times after dosing during the entire period that the drug remains in the body. Measurements of drug concentrations in the collected samples after different brands of the drug are administered allow the bioavailability of the versions to be compared statistically. These tests reveal whether different versions of the drug (such as a branded product and a generic copy) behave the same physiologically.

A single bioavailability study can generate, typically, 400 to over 1,000 samples, each of which must be assayed for the same drug. Since only a single dose of medication is usually administered, trace level analyses are required, capable of measuring concentrations as low as 1 ng/mL (1 ppb).

We first became interested in the Zymate System because of its potential to perform extractions more reproducibly than human analysts can perform them. If this potential could be realized, it would permit us to increase our ability to distinguish the behavior of similar but different versions of a given drug product.

We have previously reported on the use of the Zymate System to perform extractions in assays of theophylline and tolazamide[1]. These extractions required excessive time since system samples were processed in a batch mode - one operation was performed on each of the specimens in sequence.

This paper describes how the Zymate System has been used to perform extractions in a serial mode, where all the operations are performed on a given specimen before the next specimen is processed.

The hardware and software changes necessary for these extractions will be described. Finally, data will be presented to compare virtually identical manual and automated extractions of the drug tolazamide.

ADAPTATION

In adapting manual extraction procedures to the Zymate System, the first problem we faced was a safety concern. Most of our manual extraction procedures involve liquid-liquid extractions requiring large volumes of flammable solvents (Figure 1). This does not present a major problem since our analysts perform these operations in fume hoods. Unfortunately, we did not have a fume hood that we could devote to the Zymate System. Furthermore, we were not certain that it would be safe to operate the non-explosion proof Zymate System with flammable solvents, even if a fume hood had been available.

FIGURE 1. View of bioavailability analytical laboratory.

Our solution was to convert our liquid-liquid extraction procedures to liquid-solid extraction procedures. This technique involves the use of disposable columns of either 1 mL or 3 mL capacity (Figure 2). These columns are pre-packed with one of a number of different materials with varying adsorption characteristics. The specimen is added to an appropriate column chosen so that the drug of interest is bound to the packing material. Contaminating components in the sample are removed by washing the column with various solvents. Finally, a relatively non-flammable solvent is added to the column to elute the drug of interest. The eluate is then analyzed by a subsequent chromatographic procedure.

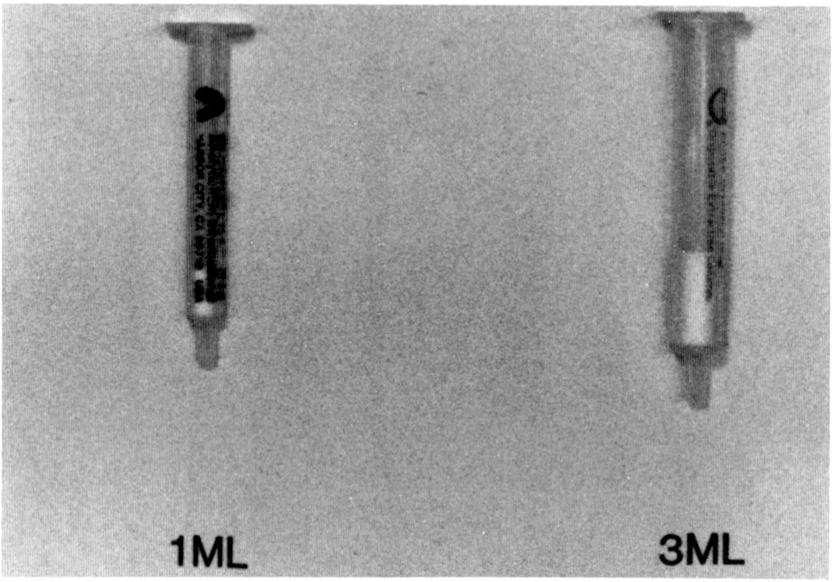

FIGURE 2. Liquid-solid extraction columns.

These columns are packed so tightly that gravitational force is not sufficient to permit liquids to flow easily through them. Using centrifugal force to cause liquids to flow through the columns was not a satisfactory solution. We tried having an analyst place the columns in a centrifuge, but this made an otherwise automated procedure a partially manual procedure. We elected not to purchase a centrifugal work station for the Zymate System because we did not want to invest substantial additional money in a technique that still had to prove its usefulness.

Instead of centrifugation, we used air at 6 psi delivered through the liquid dispensing hand to force liquids through the columns. Delivery of air from a tank of compressed gas was controlled by the master lab station. Af first, this procedure resulted in unacceptably high variability in results. We observed that different volumes of liquid were left in different columns. We eventually realized that the problem resulted from the fact that the liquid dispensing hand was not making a consistently tight seal with the top of each column. As a consequence, a variable amount of the delivered air was escaping into the atmosphere. This variability resulted in an assay with a completely unacceptable within-run coefficient of variation of approximately 40% (Table 1).

TABLE 1. Within-run Coefficient of Variation of Bioavailability Assay Using Zyamte to Perform Extractions.

CONCENTRATION (MCG/ML)	0.30	7.50
MEAN	0.38	4.66
C.V.	34.0%	38.9%
	(N=8)	(N=10)
	$\bar{x}$ = 36.4%	

We were only able to get reproducible elution after we purchased a 3 mL liquid-solid extraction column work station (Figure 4). This station contains a U-shaped slot for holding a column and a sealing nozzle that can be forced down by application of approximately 30 psi of air pressure to form a tight seal with the top of the column. The nozzle contains an air line with a 3-way valve and two stainless steel tubes for introduction of liquids. With the nozzle affixed to the column, the valve is open to the atmosphere, to prevent excessive pressurization of the column while liquids are being added. With the valve closed, air can be introduced into the column to force a liquid through the packing material. The nozzle can be lifted free of the column by air pressure in order to add samples or to remove the column when elution is completed.

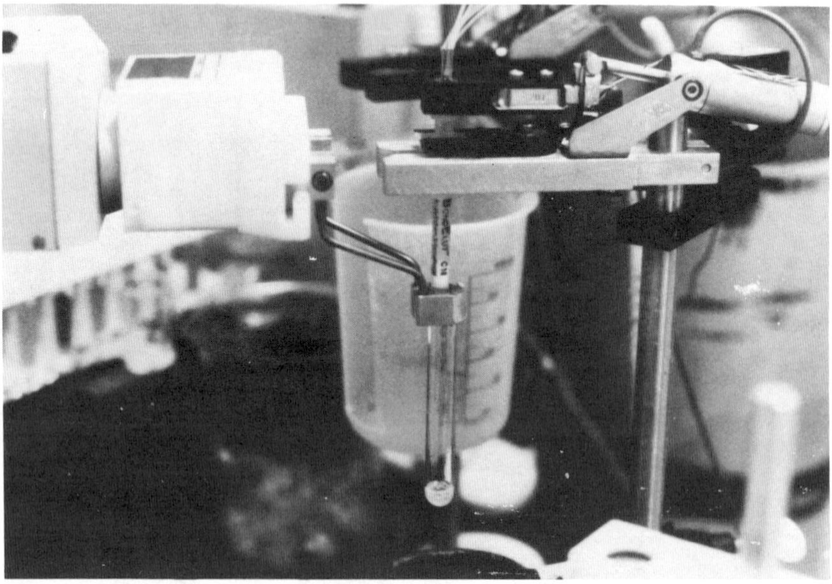

FIGURE 3. Liquid-solid extraction column work station.

With the 3 mL liquid-column work station, we were able to obtain acceptable elution reproducibility from the liquid-solid extraction columns.

Some of our extraction procedures require a 1 mL liquid-solid extraction column, which presented an even greater engineering challenge. The smaller capacity columns have a diameter of only 5 mm, compared to the 9 mm diameter of the larger columns. Fortunately, the Zymark engineers were able to meet the challenge, and they developed for us a 1 mL liquid-solid extraction column work station. This unit works in exactly the same way as the 3 mL one, but is even better designed. It contains 4 stainless steel tubes for adding liquids to the columns.

The smaller diameter extraction columns made positioning of the syringe hand crucial. To assist in this process, we use the syringe hand to push the extraction column to the back of the U-shaped slot that holds it. This action places every extraction column in exactly the same position. It is then possible to have the pipet tip on the syringe hand add reagents without touching the inner wall of the extraction column.

METHOD DEVELOPMENT

With the acquisition of the 1 mL liquid-solid extraction column work station, it required only about two days to develop an automated extraction procedure for tolazamide (Figure 4). This development went rapidly because we already had a manual extraction procedure that utilized liquid-solid extraction columns. (To convert a manual liquid-liquid extraction procedure to an automated liquid-solid

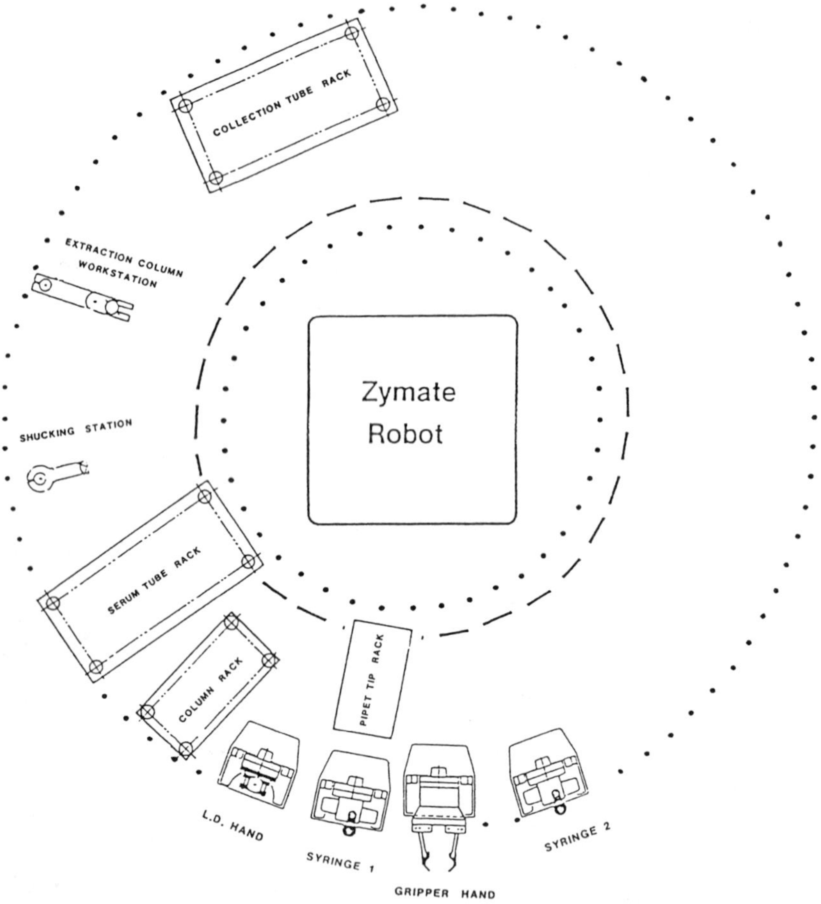

FIGURE 4. Tolazamide extraction layout.

extraction procedure can require up to two months of work to select the proper column and the right solvent composition and volume.)

At first, using automated extraction produced a tolazamide assay procedure with too much variability to be acceptable. Our within-run coefficient of variation was approximately 10%. The poor reproducibility was found to be the result of variability in adding the internal standard, tolbutamide. The internal standard was added by pipetting with the syringe hand fitted with a 1 mL disposable pipet tip and a 1 mL gas-tight syringe. The syringe drew up 0.1 mL of this 50-50 mixture of methanol and water, but some of the liquid dribbled out of the pipet tip before it could be added to the extraction column. This problem was solved by drawing up 0.02 mL of air into the pipet tip after the solution was drawn up. With this modification, the within-run coefficient of variation was reduced by half, to approximately 4% (Table 2).

TABLE 2. Within-run Coefficient of Variation of Tolazamide Assay Using Zyamte System to Perform Extractions.

CONCENTRATION (MCG/ML)	2.00	8.00	20.0
MEAN	1.77	7.50	18.64
C.V.	3.74%	4.74%	4.19%
	(N=8)	(N=8)	(N=5)
	$\bar{x}$ = 4.22		

Another problem resulted from the fact that the specimen tubes contained different volumes of serum. To get an adequate volume of sample from tubes with low volumes, it was necessary to program the Zymate System to bring the pipet tip far into the tubes. However, when this procedure was applied to tubes with large volumes, serum was displaced from the tubes and spilled over the sides. To solve this problem we wrote two sampling sub-routines; the low-volume sub-routine instructs the syringe hand to draw the sample from the bottom of the specimen tube, while the other sub-routine draws the sample from the middle of the specimen tube. An analyst groups the low-volume and high-volue tubes in separate regions of the specimen rack. Each time a batch of tubes is processed, it is necessary to specify the positions of the two types of tubes. This allows the Zymate System to select from the two possible sampling heights, as appropriate.

One other problem did not manifest itself until a fairly large number of tolazamide assays were performed. We found that our HPLC analytical columns were developing high operating pressures (necessitating replacement) much sooner than expected. Upon close examination, we found that the eluates from the liquid-solid extraction columns contained fine, white particulate matter. To overcome this problem, we had to add an additional centrifugation step prior to chromatography. By centrifuging at 15,600 x g it was possible to collect the particulate matter as a pellet, thus avoiding its introduction into our HPLC analytical column. Unfortunately, this additional centrifugation step requires transfer of the eluate to a plastic centrifuge tube, then transfer of the centrifugate to a WISP vial. Both of these steps require manual intervention, and extend the time required to process the samples.

APPLICATION

After overcoming all the described problems, we were finally able to use the Zymate System to extract serum samples collected from a tolazamide bioavailability study. The Zymate System performs seven of the nine steps required before tolazamide can be measured by HPLC:

1. Position 1-mL, C-18, solid-liquid extraction column
2. Wash with 1.0 mL of methanol
3. Wash with 1.0 mL of 1% phosphate buffer at pH 6.5
4. Add 0.1 mL (10mch) of tolbutamide
5. Add 1.0 mL of specimen
6. Wash with 0.5 mL of 1% phosphate buffer at pH 6.5
7. Elute with 0.5 mL of acetonitrile/water (75/25, v/v)
8. Centrifuge eluate for 2 minutes at 15,000 rpm
9. Inject portion onto HPLC column

COMPARISON

Our manual and automated extraction procedures for tolazamide are identical except for one aspect. In the manual procedure, centrifugation is used to force liquids though the extraction columns, whereas the automated procedure uses air pressure. The essential similarity of the two procedures allows a direct comparison of automated and manual extraction techniques.

The automated procedure has not confirmed our expectation of greater precision. The run-to-run coefficient of variation for the entire assay over 12 days averaged 8.13% for the automated procedure, compared to 6.62% for the manual procedure (Table 3).

TABLE 3. Between-run Coefficient of Variation of Tolazamide Assays with Extractions Performed Manually or by the Zymate System.

CONCENTRATION (MCG/ML)	2.00	8.00	20.0
		MANUAL (N = 24)	
MEAN	1.96	8.35	20.31
C.V.	13.23%	9.45%	11.43%
AVERAGE C.V.		11.37%	
		ZYMATE (N = 24)	
MEAN	1.95	8.78	20.58
C.V.	11.65%	7.59%	9.77%
AVERAGE C.V.		9.67%	

In regard to accuracy, both techniques are essentially identical: 101.6% for the automated one, and 100.0% for the manual one.

We process 47 specimens at one time, whether the extractions are performed manually in a batch mode, or by the Zymate System in a serial mode. The comparative processing times per sample are 4.5 minutes or the manual extractions, and 7.7 minutes for the Zymate System.

CONCLUSION

We were surprised - and a little disappointed - when we evaluated our tolazamide assay results and discovered that the Zymate System did not produce better reproducibility than our analysts could produce. We suspect that using liquid-solid extraction columns may not be a fair test of the Zymate System. It is very likely that the greatest source of variation are the columns themselves, in which case the Zymate System could not be expected to do better than an analyst. We plan to use the Zymate System in a liquid-liquid extraction procedure, where we anticipate that it will produce better reproducibility than an analyst will achieve.

We were not surprised that the automated procedure took longer, since the Zymate System is known to process samples at a slower rate than humans for most assays. In our application of extracting drugs form biological fluids, the Zymate System took approximately 70% longer than an analyst to process a batch of 47 samples. The time required to get the first analytical result would be much shorter for the Zymate System if we had an automated HPLC injection station.

Even without an HPLC injection station, we are satisfied with the rate at which the Zymate System processes samples. We are able to extract and chromatograph samples from two bioavailability subjects in a single 9.5 hour day (Table 4). Of this total time, 3 hours of an analyst's time are required for the two batches of samples. Each batch requires approximately 0.5 hour of setup time, 0.5 hour to centrifuge the eluates produced by the Zymate System, and 0.5 hour to transfer the centrifugates into HPLC sampling vials.

Even though we are using the Zymate System currently in a partially automated procedure, we are satisfied with the pay-back period we have calculated for the equipment (Table 5). Allowing for the fact that

TABLE 4. Time Required to Perform Tolazamide Extractions.

	ZYMATE	MANUAL
SET-UP		0.5
EXTRACTION	4.5	
CENTRIFUGATION		0.5
HPLC PREPARATION		0.5

TABLE 5. Cost of Drug Extraction Performed by the Zymate System.

MANUAL HOURS PER DAY WITHOUT ZYMATE			8.0
MANUAL HOURS PER DAY WITH ZYMATE			3.0
HOURS SAVED PER DAY WITH ZYMATE			5.0
DIRECT LABOR COSTS SAVED PER DAY			$50.00
APPROXIMATE EQUIPMENT COST			$25,000.00
DATS TO PAY-BACK	$\frac{\$25,000.00}{\$50.00}$	=	500.00

human intervention is required for three out of nine hours per day, we calculate that the Zymate System saves us $40 each day that we measure tolazamide. Assuming that the replacement cost of our equipment is currently in the neighborhood of $25,000, we calculate that the pay-back period is only two years.

We expect to utilize the Zymate System to perform liquid-solid extractions for use in other drug assays. The development of different extraction procedures for different drugs will be simplified and accelerated by a program we wrote for the Zymate System. This program allows a pair of eluting solvents to be added in varying proportions to an extraction column. By examining the eluates the next day, it is possible to select the best proportions to use for either washing contaminants off the columns, or for eluting drugs of interest.

As a result of our experience with the Zymate System in extracting drugs from biological samples, we fully anticipate further use of robotics in bioavailability testing. This increased use of automation will certainly prove to be cost-effective, and may lead to lower charges for bioavailability testing.

REFERENCE

1. Myers, D. J., Szuminsky, N., and Levitt, M. J., in Advances in Laboratory Automation Robotics 1984, Hawk, G. L. and Strimaitis, J. R., (eds.), Hopkinton, MA, 1984, p. 71.

CENTRALIZED SAMPLE PREPARATION USING A LABORATORY ROBOT*

John E. Brennan, Matthew L. Severns and Linda M. Kline
American Red Cross
Biomedical Research and Development
9312 Old Georgetown Road
Bethesda, Maryland 20814

ABSTRACT

A centralized sample preparation station was developed to transfer samples of plasma from test tubes to multi-well plates for the testing of donor blood. This operation, currently performed manually, is the initial step in testing blood for hepatitis B surface antigen (HBsAg) and for antibodies to Human T-cell lymphotrophic virus type III (HTLV-III). The station was built using a Tecan Sampler 505 (Tecan U.S., Ltd., Chapel Hill, N.C.) and incorporated a laser scanner to read bar coded labels on tubes and plates to ensure positive sample identification. The station can concurrently prepare samples for both tests at a rate of approximately 135 samples/hour. Because the station can pipette more precisely than a technologist, test results should be more consistent and accurate. By automating data management, labor savings are realized while reducing the risk of misidentifying a sample's test results. Additional cost savings are anticipated since the station eliminates the need for disposable pipette tips.

* Contribution No. 659 from the American Red Cross

INTRODUCTION

There are approximately 12 million units of blood collected each year in the United States. Each unit that is collected must be tested for, among other items, hepatitis B surface antigen, syphilis, antibodies to Human T-cell lymphotrophic virus type III (HTLV-III) and irregular antibodies before it or any product made from it can be released for transfusion.

All testing is done on blood collected in 13 x 100 millimeter sample tubes at the time of donation. In most cases, the tubes are centrifuged and a sample of plasma is pipetted into a multi-well plate where the test is performed. The multi-well plates come in a variety of sizes and configuations depending on the type of test and the manufacturer. For 1985-86, the American Red Cross elected to use test kits manufactured by Abbott Laboratories (Chicago, IL) for HBsAg and HTLV-III testing. Both test procedures are done using plates that are available in 60 (5 x 12) and 20 (5 x 4) well configurations. The first step in both tests is the pipetting of a plasma sample from each donor tube into a well on the test plate. The HTLV-III test requires that the sample be diluted by 441:1 before it is added to the test plate. This dilution is difficult to achieve precisely over a large number of samples and is a labor intensive task. Although at least one system has been developed for performing the HBsAg assay,[1] no system suitable for routine clinical performance of these tests has been developed.

Affixed to all sample tubes and associated whole blood units or blood products is a bar coded label containing a number which uniquely identifies that unit. This is the means by which donor information and blood test results are associated with the appropriate blood or blood products produced. Automated equipment has been shown to produce advantages in managing this data.[2] In a non-automated environment, the data management is highly dependent on the laboratory staff. This fact increases the chances of an error occurring. A more reliable and desirable means to identify samples is to read the label using a bar code reading device such as a wand or laser scanner.

Analysis of workload recording studies on HBsAg testing has shown that sample preparation and data management account for more than half of the labor required to perform the test. Significant advantages can be realized by automating these portions of the procedures using a laboratory robot.[3,4,5] Previously, the sample preparation for ABO/Rh testing of blood was automated in this manner.[6]

DESIGN CONSIDERATIONS

A decision was made to automate only the HBsAg and HTLV-III procedures for several reasons. First, HBsAg and HTLV-III tests are the most time-consuming and are the rate-limiting steps in processing. In addition, syphilis testing is currently performed using a plastic coated card rather than a test plate; this would be difficult to automate using the same fluid handling mechanism as for plates. Antibody screening is usually done in a 96 well test plate as opposed to the 60 or 20 well plate for the HBsAg and HTLV-III tests. In order to maintain batch processing that is typical in a large testing facility, the number of samples prepared by the sampler during a single run should be constant. An alternative would be to only process the number of samples equal to that which can be accommodated by the smallest test. This, however, would waste space in the 96 well plates, resulting in an increase in disposables cost.

In general, manipulation of the test plates, which is neither labor intensive nor involves data management, is done much more rapidly by a technologist than by an automated instrument. For this reason, it was decided to only automate the pipetting of samples from donor tubes into the test plates. Several inexpensive dedicated instruments already exist for doing other phases of testing such as reagent addition and sample washing. Having a technologist move the test plate between these specialized fluid handling instruments would provide the most efficient use of human labor while optimizing system throughput.

It was determined that the system should process at least 120 samples/hour. This is based on the time that it would take a skilled technologist to perform the tasks manually. In addition, HTLV-III tests, once started, must proceed without significant delay. Problems have been reported when diluted plasma samples sit for more than 1 1/2 hours before subsequent steps in the test are performed[7]. With a rate of 120 samples/hour, the sampler could process a batch of three 60 well plates in sufficient time to avoid such problems. In this case, a specialized liquid handling system was indicated that would yield higher throughput than a "general purpose" laboratory robot.

If the use of disposable pipette tips could be eliminated, additional cost savings might be realized. This meant that a station had to be developed that would wash a non-disposable probe sufficiently to eliminate contamination between samples (which would eliminate the possibility of performing reliable retesting should a sample test positive) and between wells in the test plate (which may result in false test results). The HBsAg test, which has a detection limit below 1 ng/mL, is the more sensitive of the two tests to be performed. The highest concentration of HBsAg found in donor samples is approximately 1 mg/mL.[8] Therefore, the washing station must be able to decrease contamination by a factor of at least one million.

Concern for ease of human interaction with the system was a major design factor. The system, once installed in a laboratory, must be operated by medical technologists with limited computer experience. For this reason, the software needed to be designed with simple operator interfaces and limited access to low-level sampler functions. Provisions were required for easy addition of laboratory specific requirements such as blanks and controls. In addition, routines for maintenance fuctions such as flushing, priming and calibration must be easy to perform. All racks and fixtures had to be accessible and removable for cleaning and maintenance and include features that would facilitate ease of use (e.g. position labeling, rack and fixture locators).

Data management capabiltites also needed to be incoroprated into the sampler. The system should read the bar coded labels on each test plate and sample tube and track the location of each sample in the test plates. From this, a load list can be printed or electronically transferred to another computer for integration with test results.

SYSTEM DESCRIPTION

The system chosen for this application was the Tecan Sampler 505. The sampler (Figure 1) is a dedicated liquid handling system comprised of a stainless steel probe which is manipulated by a cartesian type (x-y-z) robot arm over a 12 x 24 inch work surface. The probe incorporates a sensor which allows the sampler to detect the presence of liquid. This feature allows the sampler to deal with the variable liquid levels which are found in sample tubes. Pipetting and dispensing are done by two independently programmable syringes which aspirate or dispense liquid through the probe or draw liquid from an external reservoir. Each syringe is connected to the probe and reservoir through a three way valve by flexible tubing. The probe tip is coated externally with Teflon to minimize adherence of contaminants. The sampler is controlled by an IBM PC which communicates with it through a standard RS-232 interface.

The probe washing station is located on the sampler work surface. It consists of a vertical column with a blind cavity in the middle. Washing is achieved by positioning the probe tip in the chamber and dispensing liquid through it. This flushes the inside of the probe and the liquid flowing up the cavity and around the probe cleans the outside. The wash liquid then flows out the top of the wash station and

FIGURE 1. The Tecan Sampler 505 dispensing plasma samples from donor tubes into Abbott test plates.

exits through a drain surrounding the column. A plexiglass cover was built around the wash station to contain splashing and minimize the diffusion pathway for aerosols.

Fixtures were fabricated to position objects on the sampler work surface. A flat rack was constructed to hold the two test plates and one intermediate dilution plate, as the 441:1 dilution was accomplished in two steps. A standard 96 well microtiter plate was used to hold the intermediate dilution. Sample tubes were held in a circular carousel that accomodates a maximum of 60 tubes. The carousel was mounted on an indexing rotor which sequentially presents tubes to be sampled. A fixture was also constructed to hold the sample diluent container. A separate container, located to one side of the sampler, held the deionized water which was used as the wash solution. Waste was collected in a bottle which was located on the floor containing sodium hypochlorite solution for disinfection.

A laser bar code scanner (Laserscan 6000, Symbols Technologies, Bohemia, NY) was mounted so that it could scan labels on donor tubes as the carousel indexed. One position on the carousel was modified to allow the test plates to be positioned for scanning before each run. The scanner was triggered through a programmable switch closure which is built into the sampler.

The software to operate the sampler was written in RatBas[9], a language preprocessor which allows structured program development. The preprocessor translates the structured code into standard BASIC. Menu driven routines are used to set up the sampler, process samples, prime the liquid handling system and shut down the sampler at the completion of processing. Once a routine is started, the only controls that the operater can operate are a pause key which stops the routine and allows it to be restarted at the point where it was stopped and an abort key which ends the routine, parks the probe in a safe position and reinitializes the sampler. The only other operator intervention necessary is to manually input a bar code if it could not be read by the scanner. This structure provides a user-friendly interface while, at the same time, it reduces the risk of user induced errors.

A theoretical model of the washing process was developed to predict the factors which govern the wash process. The most significant factors were found to be the amount of sample aspirated into the probe and the volume of wash solution used to flush the probe. Since the amount of sample to be aspirated was determined by the test procedure, studies were performed to identify the quantitative relationship between wash volume and contaminant washout. Based on these studies, a wash volume of at least 6 mL was indicated to maintain 1 part/million or less of carry-over. To minimize contamination on the outside of the probe while sampling, the liquid level sensing capability was exploited to allow the probe to find the liquid. The rate at which the liquid level fell was calculated from the rate of aspiration and the diameter of the tube; the probe was moved downward at this rate as fluid was aspirated. Air gaps of 7 uL were taken before and after each sampling or dispensing step which helped to eliminate drops hanging on the probe tip and maintained separation of liquids to avoid mixing with wash solution and diluent which will introduce inaccuracies.

The sampling procedure involves, first, picking 200 uL of diluent from the diluent reservoir and the 210 uL of sample from the donor tube. Two hundred uL of sample are dispensed into the HBsAg plate and the remaining 10 uL are dispensed with the diluent into the intermediate dilution plate. The probe is then washed. Next, the sampler picks up

200 uL of diluent and then 10 uL of diluted sample from the intermediate dilution plate. Diluent and sample are then dispensed into the HTLV-III plate and the probe is washed again before proceeding to the next sample.

The probe is washed twice during each sampling cycle. A first wash of 2.5 mL is done after HBsAg sample is pipetted and the first 20:1 dilution is made for the HTLV-III assay. This wash removes undiluted sample from the probe tip which was found to cause substantial error in the final dilution. A second wash of 4.5 mL is done after the final HTLV-III dilution is made. Thus, a total 7 mL was sufficient to prevent any detectable sample carry-over. Dividing the wash also allowed a 5 mL syringe to be used for washing rather than a 10 mL syringe. The smaller syringe could be run at a considerably faster speed thus enhancing throughput.

TESTING AND VERIFICATION

Dilution accuracy was measured suing a fluorescent tracer, 4-methylumbelliferone (Sigma Chemical Co., St. Louis, MO) diluted in plasma. The tracer was diluted 441:1 by the sampler. As a standard, a series of 2:1 serial dilutions was performed manually using the same starting material. The samples' fluorescence was read on a Dynatech MicroFluor Autoreader (Dynatech Laboratories, Inc., Alexandria, VA). Because there was a nonlinear relationship between tracer concentration and fluorescence (due to changes in protein concentration and pH), a second order polynomial was fit to the standard curve. The coefficients of the polynomial were used to compute the dilutions from the experimental measurements.

In addition, volumetric accuracy was measured using colored standards from a pipette volume calibration kit (Medical Laboratory Automation Inc., Mt. Vernon, NY). The standards were diluted in diluent included in the calibration kit and the results were read on a colorimeter (Quantumatic, Abbott Laboratories, Chicago, IL). Dilutions were also done using plasma as a diluent to better simulate the properties of actual donor samples. Results were compared to a calibration curve derived from reagent standards included in the kit.

Mean dilution values and coefficients of variation (CV) were determined for both the sampler and a skilled technologist performing the 441:1 dilution manually. Dilutions performed manually had a mean value of 384:1 with a CV of 33% using plasma as a diluent. Dilutions performed by the sampler using plasma as diluent had a mean value of 400:1 with a CV of 13%. These results indicated that the sampler was performing the dilutions adequately.

Verification that the wash station sufficiently eliminated sample carry-over was accomplished by testing 30 known HBsAg positive samples, all with concentrations of approximately 0.1 mg/mL, which were randomly dispersed among 30 known negative samples. The Abbott HBsAg test was then conducted on all 60 samples. The test correctly classified all 60 reactions (no false positive reactions due to carryover). A test was also run to detect carry-over in subsequent sample tubes. The sampler was programmed to take plasma from a known positive sample and then from a known negative sample with a wash in between. Samples from the known negative tube were tested; no detectable contamination was found.

IMPLEMENTATION

The sampler was installed at the American Red Cross Penn-Jersey Region in Philadelphia, PA. This facility processes about 1200 samples/day. The system was installed without the data management capability since the software to interface the sampler to other test instruments and computers was not available. The sampler processed samples in parallel with the routine manual method to verify that the automated method performed comparably. Once 800 samples were processed without discrepancy, the sampler became the routine method for preparing samples for the Center's HBsAg and HTLV-III testing.

The sampler processes donor samples at a rate of between 130 and 140 samples/hour. To date, over 3000 samples have been run. No adverse effect on the quality of the test results for either HBsAg or HTLV-III, as indicated by the percent of false positive results, has been noted.

DISCUSSION

Several advantages were achieved by automating the sample preparation procedure for these tests. The system consolidated two manual tasks into a single automated one that could run unattended for 20 minutes at a time, enabling a technologist to perform other work. Cost savings are realized by washing the probe, thus eliminating the need for disposable pipette tips. Increased accuracy and uniformity in sample preparation should improve the quality of the test results and produce additional cost savings since fewer retests may be required and less reagents wasted. Another advantage to such an automated system is its flexibility. The sampler can be reconfigured with relatively minor hardware and software modifications to accomodate changes in test requirements. This is an option not often found in more expensive dedicated pieces of laboratory equipment.

Table 1 shows parameters used to calculate return on investment for the Tecan Sampler. These figures are based on current estimated operating expenses for a typical Red Cross laboratory performing these tests. These figures do not reflect labor saved on data management tasks, potential cost savings associated with increased accuracy (reduced

TABLE 1. Estimated Return On Investment.

Disposables Savings (2 pipette tips, 1 dilution tube eliminated; 1 microplate added)	$ 0.052 / sample
Labor Savings (Dispensing samples @ $12.50 / man-hour including fringe benefits)	$ 0.041 / sample
Total Savings	$ 0.093 / sample
Tecan Sampler Cost (all options)	$ 30,000.00 / unit
Service Costs (10% sampler cost / year)	$ 3,000.00 / year
Processing volume to achieve Return on Investment of 2.5 years	165,000 samples / year

retesting rate) and less tangible advantages associated with positive sample identification (reduced chance of substandard product being released or good quality product being withheld). Although these savings are substantial, no data exists on which to base our calculations.

By using the sampler in conjunction with devices which exist for doing other phases of the testing, the laboratory can adopt a modular approach to automation. A larger laboratory requiring more throughput than a single sampler produces can buy additional units. Additional uints also provide a back-up capability should a unit experience mechanical failure. This approach also makes automation an affordable alternative for a broader range of testing facilities whose processing needs cannot justify the purchase of large, more expensive, dedicated instruments.

ACKNOWLEDGEMENTS

We wish to thank the laboratory staff at the American Red Cross Penn-Jersey Region for their help and cooperation and Ms. Rebecca Akins for performing the HBsAg assays.

REFERENCES

1. Wasmuth, E. H. et al., in Advances in Laboratory Automation - Robotics, Hawk, G.L. and J.R. Strimaitis, eds., Zymark Corporation, Hopkinton, MA, 1984, p. 209.

2. Brodheim, E., W. Ying and R.L. Hirsch., Vox Sang., 40, 175 (1981).

3. Severns, M. L. and G. L. Hawk., in Robotics and Artificial Intelligence, Brady M. et al., eds., Springer-Verlag, Berlin, 1984, p. 633.

4. Severns, M. L. and J. E. Brennan., in Advances in Laboratory Automation - Robotics, Hawk, G.L. and J.R. Strimaitis, eds., Zymark Corporation, Hopkinton, MA, 1984, p. 323.

5. Zenie, F. H., in Advances in Laboratory Automation - Robotics, Hawk, G.L. and J.R. Strimaitis, eds., Zymark Corporation, Hopkinton, MA, 1984, p. 1.

6. Brennan, J. E. and M. L. Severns., Proc. 37th ACEMB., 26, 253 (1984).

7. Dodd, R. Y., Personal Communication.

8. Nath, N. and Dodd, R. Y., in CRC Handbook Series in Clinical Laboratory Science, Section D: Blood Banking, Volume 2, T.J. Greenwalt and E. A. Steane, eds., Chemical Rubber Company., Boca Raton, FL, 1981, p. 301.

9. Sharpe, W. F. and B. Weaver., RatBas - A Software Tool for Users of BASIC, Wells Fargo Investment Advisors, San Francisco, CA, 1982.

A ROBOT FOR PERFORMING RADIOIODINATIONS

William M. Hurni, Willian J. Miller,
Edward H. Wasmuth, William J. McAleer
Virus and Cell Biology Research
Merck Sharp & Dohme Research Laboratories
West Point, Pennsylvania 19486

ABSTRACT

A robotic system for radioiodination of proteins has been developed by Merck, Sharp & Dohme Research Laboratories and Zymark Corporation. The robot radioiodinates the protein of interest, separates the free from the bound ^{125}I, makes appropriate dilutions so that the free and bound peaks can be located using a gamma counter and performs a complete cleanup after the procedure. The robot runs unattended during the iodination procedure so that operator exposure is minimized. Operator time and attention is also reduced.

INTRODUCTION

Radioiodination of proteins is a common technique used to provide tracers for radioimmunoassays and to follow biological events. Presently, radioiodination of proteins such as antibodies is most frequently performed using the chloramine T method.[1] This procedure requires precise additions of small amounts of reagents and accurate timing of the reaction to insure uniform results. The operator is exposed not only to low-level gamma radiation but also potential inhalation of the volatilized ^{125}I during the course of the procedure. The cleanup after labeling again risks low level exposure to gamma radiation as well as the potential for accidental spills of radioactive liquids. The iodination procedure, which includes labeling of the protein of interest, separation of free from bound ^{125}I, cleanup and various safety tests, requires a significant amount of operator time and attention such that only one labeling can be performed in a day. Not only has exposure to gamma radiation and inhalation of volatilized ^{125}I become a consideration but also throughput limits were becoming apparent. With the evolution of electronic technology over the past 10 years and the drastic reduction in the price for this technology, a robot for performing iodinations became a reasonable solution to the problem of reducing exposure and expanding throughput with the additional benefit of providing greater uniformity.

METHODS

The manual method of chloramine T radioiodination of proteins as originally described by Hunter and Greenwood[1] has many variations. The method used in our laboratories is described as follows: Ten uL of 0.1N NaOH containing 1 mci of ^{125}I are added to a reaction vial. Ten uL of a freshly made solution containing 2.5 mg/mL of chloramine T in 0.5M PO_4 buffer pH 7.5 are added. Ten uL of sample containing 5 mg/mL protein are added to the reaction vial and allowed to react for 15 seconds. Ten uL of a 2.8 mg/mL solution sodium bisulfite are added followed by 200 uL of a solution containing 2 mg of Bovine Serum Albumin (BSA) and 0.2 mg

KI. The contents of the reaction vial are then charged to a 14 mL Sepharose (Pharmacia, Inc., Piscataway, N.J.) column to separate the free from the bound ^{125}I. An appropriate dilution of the column fractions is made and counted on a gamma counter to locate the free and bound peaks. This is followed by a cleanup procedure in which radioactive liquids and contaminated solid waste are segregated and placed in appropriate containers for disposal. Samples of urine both before and after the iodination procedure, as well as samples from the air monitoring devices for the iodination hood and the operator breathing zone are collected and tested to be certain they are within acceptable limits. To protect the operator and to minimize the release of ^{125}I into the atmosphere, this procedure is carried out in a dual hood system. The larger hood is a 30" X 60" fume hood. Located inside the fume hood is a second smaller plexiglass box with an integral charcoal filter before an exhaust fan. The actual iodination reaction is carried out in this plexiglass box. It was our objective to develop a robot that would perform the above procedures and that would be flexible so that the procedure could be modified for different requirements for labeling.

The iodination robot built by Zymark Corporation to our specifications is shown in Figure 1.

ROBOT OPERATION

The initial setup for operation includes packing the column which separates the free from the bound ^{125}I with appropriate bed material for the protein being labeled and making sure that all the lines are filled

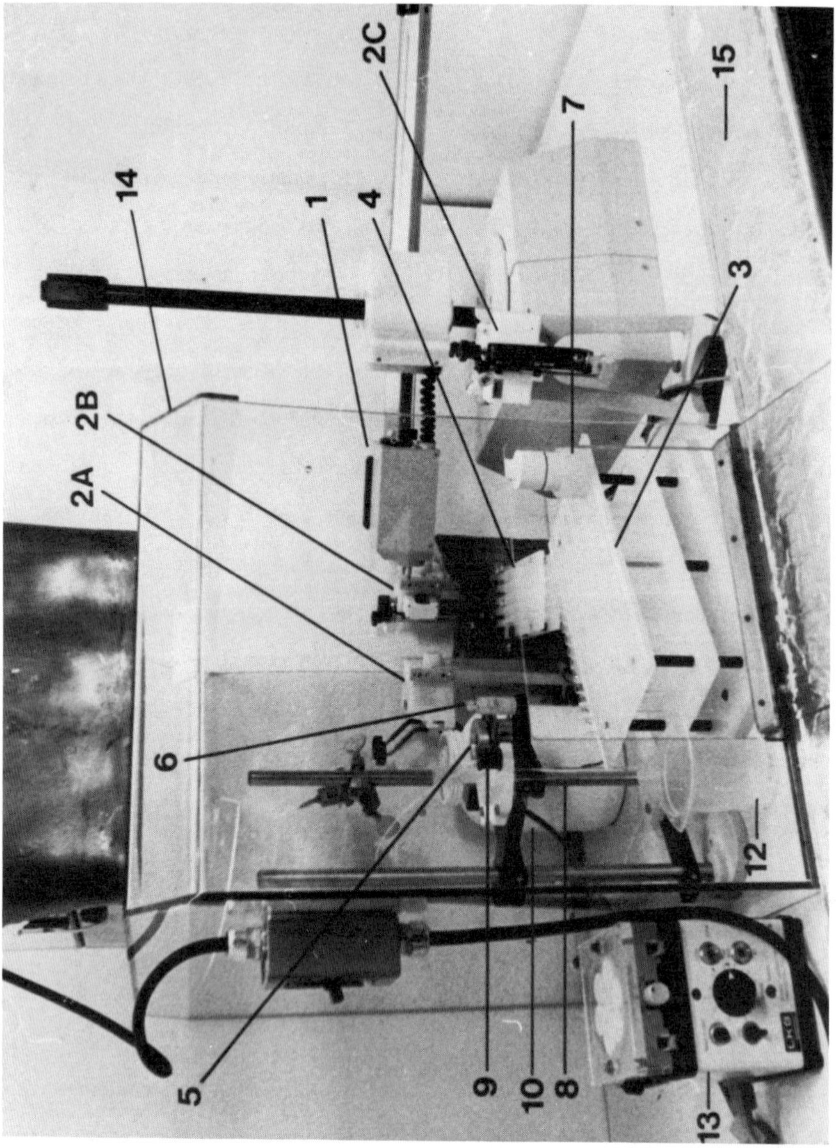
1
2A
2B
2C
3
4
5
6
7
8
9
10
12
13
14
15

FIGURE 1. Robotic Layout for Antigen and Antibody Radioiodinations.

1) The robot arm is controlled by a computer (not shown).
2a) "Gripper hand" to manipulate a lead pig and the iodine vial, to collect the column fractions, and perform all of the cleanup operations.
2b) Dedicated ^{125}I syringe hand to withdraw ^{125}I from its storage container and place it in the reaction vial (Note the lead shield around the lower portion of the "hand").
2c) General purpose syringe hand to withdraw and dispense liquids using disposable plastic tips.
3) Tube rack which precisely locates the test tubes for the collection of fractions, performing dilutions and holding reagents used during the iodination. At the far end of this rack is a holder used for locating the lead pig (Note that the lead pig is in place).
4) Pipette tip rack used to hold disposable pipette tips used.
5) Iodine vial holder to locate and hold the storage vial of ^{125}I.
6) Reaction vial holder.
7) Lead pig holder.
8) Column station for separation column used to separate free ^{125}I from bound ^{125}I (Note that the column is in place).
9) Station to hold and locate the tube coming from the bottom of the column. This remote dispenser carrying the column effluent is moved from tube to tube in the tube rack during the course of collecting fractions by the robot arm with the gripper hand attached.
10) Liquid radioactive waste container into which all radioactive liquid waste is placed during the cleanup operation.
11) 1000 mL plastic disposable beaker into which all contaminated solids such as pipette tips and empty tubes will be placed during the cleanup operation (Note that this beaker is not in place in the photograph. It is located in the space beneath the reaction vial holder, 6).
12) 250 mL plastic disposable beaker to catch any liquid waste coming from the column, used primarily during setup of the column for an iodination.
13) Dual channel peristaltic pump used to control the flow rate of the column and to add eluant to the top of the column.
14) Small plexiglass hood. The charcoal filter is located behind the lead shield on the top of the hood. Above that is the blower which draws air through the charcoal filter.
15) Primary fume hood.

with liquid. The tip rack is filled with disposable pipette tips. The tube rack is filled with the required tubes. Three sets of 18 tubes are used. Into each of the second set of 18 tubes are placed 500 uL of water. Three other tubes are present which contain the sodium bisulfite, BSA, KI and the chloramine T used in the iodination reaction. The chloramine T solution is prepared just prior to use. Ten uL of the material to be labeled are placaed in the reaction vial and the reaction vial placed in its holder. Both hoods are turned on and the sliding plexiglass door on the front of the large fume hood is pulled down.

The program is started. The robot attaches the gripper hand, removes the top from the lead pig and places it on a clear space on the tube rack. It then transfers the vial containing the ^{125}I to the vial holder and returns the gripper hand. It attaches the ^{125}I syringe hand and transfers 10 uL of ^{125}I into the reaction vial. The robot then returns the ^{125}I syringe hand and attaches the general purpose syringe hand. It picks up a disposable pipette tip and subsequently transfers 10 uL of chloramine T solution, 10 uL of sodium bisulfite solution and with a fresh pipette tip 200 uL of BSA KI solution to the reaction vial. Finally, the contents of the reaction vial are loaded to the top of the column. Subsequently, the robot returns the general purpose syringe hand to its holder and attaches the gripper hand. Using the gripper hand, the robot picks up the remote dispenser attached to the output of the column and moves it over the top of the first tube in the first group of 18 tubes. The robot acts as a fraction collector for eighteen 1 mL fractions at a rate of one fraction every 10 minutes. After the 18 fractions have been collected, the robot attaches the general purpose syringe hand and, using separate disposable tips, removes 10 uL from each fraction and places it in one of the second group of 18 tubes containing 500 uL of water, thus effecting a 1:50 dilution of each column fraction. Using the same general purpose syringe hand, the robot removes 10 uL from each of the 1:50 dilution of the fractions and places them in the third set of 18 tubes. These are the tubes used to locate the free and bound ^{125}I peaks. At this point the robot stops and the tubes are manually placed into a gamma counter. The desired fractions are combined and diluted for use. The robot is instructed to proceed with the cleanup. This consists of emptying the liquid from all remaining tubes into the liquid radioactive waste container and then placing the tube in a 1000 mL disposable beaker. The ^{125}I is returned to the lead pig and the top replaced. This is all accomplished using the gripper hand attached to the robot arm. The operator then removes the liquid and solid waste container and the column to a suitable bulk disposal container.

RESULTS

Since the installation of the robot in our laboratory, 50 iodinations have been performed. The substrates for iodination have been both antigens and antibodies. The robot has performed each of these iodinations without mishap, and the resultant products have paralleled the products that have been made by the manual procedure prior to the introduction of the robot. Table 1 shows the specific activity of a single biological entity that is routinely labeled in our laboratory.

Table 1. Iodination Results.

Date Labeled	Specific Activity, CPM/ug
3/16/84	768,800
4/24/84	1,269,050
6/14/84	661,980
7/21/84	1,193,159
8/21/84	954,725
9/24/84	984,200
robot introduced	
11/7/84	367,364
12/28/84	881,500
1/31/85	992,300
3/8/85	1,122,536
4/9/85	1,716,083
5/17/85	1,368,456

While the number of iodinations of this particular entity using the robot is small, the indication is that the product is substantially the same whether done manually or by the robot.

DISCUSSION

The main objective for the design and implementation of a radioiodination robot was operator protection. A secondary objective was greater throughput by virtue of the robot's ability to run unattended. In order to meet thses objectives, a system that is well contained and reliable is required. In order to meet this criterian, a simple, compact robot system is needed.

Containment is of the utmost importance when performing chloramine T radioiodination. The conventional solution to this problem has been to utilize two hoods, one within the other. The inner hood is equipped with a blower and a charcoal filter which is located over the area where the iodination reaction is to occur. The charcoal filter traps most of the volatilized ^{125}I. The remaining volatile ^{125}I will be withdrawn by the second hood and vented to the atmosphere. This vented air stream is monitored to be certain that it does not contain amounts of ^{125}I greater than that allowed by the Nuclear Regulatory Commission. In order to meet these specifications, we limited the size of our robot such that it would fit in our 30" X 60" iodination hood. A smaller second hood with a blower and charcoal filter was constructed to fit over that portion of the robot where the iodination reaction is performed.

Reliability is insured by keeping the robot operation as simple as possible. The operating format was structured so as to keep the number of stations and operations to a minimum, even at the expense of greater operator input in preparing the robot for an iodination.

It was found that the Zymate Laboratory Automation System (Zymark Corporation) would fit within our iodination hood and was capable of performing most of the required steps in the iodination format. This system consists of a computer which controls a robot arm. The arm can move on the X, Y and Z axis as specified by a computer program. A variety of "hands" is available which can be attached to the robot arm to perform required tasks such as gripping and transferring liquids. This modular approach enabled us to assemble a robotic system by selecting various standard laboratory stations and arranging them in a configuration which would fit within our standard fume hood. By using this approach it was possible to keep the need for custom designed subsystems to a minimum.

The Zymate robot performs its tasks by knowing the precise location of all the stations it must attend during the course of the iodination procedure. A precise location for all but one of the stations was attained by bolting the stations down to a baseboard on which all the robot components, except for the controller, are mounted. The radioactive iodine station required a custom design. Since the primary objective was to have minimal operator exposure to gamma radiation, it was necessary to have the robot begin with the iodine vial within a lead pig. In order for the robot to know the precise location of the bottom of the vial so that it could withdraw 10 uL from a total volume of 30 uL, a special holder was required which would compensate for the substantial variation in the manufacture of the pig and the vial. To overcome this limitiation, it was necessary to remove the vial from the pig, thus eliminating the pig as a source of variability and place it in a special springloaded holder which would allow the vial to move up and down over a range of movement sufficient to cover the manufacturing tolerance of the vial. The robot then can find the bottom of the vial over the range of movement provided by the cushion using the dedicated ^{125}I syringe hand. A solenoid operated lock holds the vial in place so that the needle used to pick up the ^{125}I can be withdrawn through the rubber septum on the top of the vial, thereby leaving the vial behind.

It is important that the robot perform those operations which place a person at risk during a conventional radioiodination procedure. In addition, the robot should carry out those routine tasks required which consume an operator's time and attention such as running the column to separate the free from the bound ^{125}I, performing appropriate dilutions for counting and, finally, doing a thorough cleanup which includes the segregation of radioactive liquids and contaminated solids into separate containers for disposal. Consequently, operator time required for an iodination is reduced, thus freeing up time for other work. Also, iodinations can be performed at night unattended.

In summary, the employment of a robot for performing radioiodination has been demonstrated to be a reasonable and effective solution to operator exposure to radiation and nuclide inhalation. The robot has also demonstrated its ability to reduce manpower requirements for the radioiodination of biological substrates.

REFERENCES

1. Hunter, W.M. and Greenwood, F.C., Nature, 194, 495 (1962).

Zymark and Zymate are registered trademarks of Zymark Corporation.

DEVELOPMENT OF AN AUTOMATED URINE ANALYSIS SCHEME FOR DETERMINATION OF ppb LEVELS OF As and Se VIA HYDRIDE/ATOMIC ABSORPTION

Linda Lester, Tim Lincoln and Haig Donoian
Xerox Corporation
800 Phillips Road
Webster, NY 14580

ABSTRACT

The routine analysis of urine for ppb levels of arsenic and selenium at Xerox Corporation has historically proven to be an extremely labor intensive activity. For this reason a fully automated system utilizing a Zymate System interfaced to an Atomic Absorption (Perkin-Elmer Model 5000 with automated MHS-20 hydride system) was developed. This paper is intended to describe robot operation, bench & module setup, experimental design and gives a detailed description of statistical data used for method verification along with labor efficiencies realized after system start up.

INTRODUCTION

One aspect of the Xerox Corporation Employee Health and Safety program is a routine monitoring of urine for arsenic and selenium. Urinalysis for arsenic and selenium in the parts-per-billion level is a recognized screening for possible work place exposure to these elements. This paper deals with an internal response, made by the corporate analytical lab, to the challenge of providing a benchmarked service with respect to turnaround, cost and quality.

Specifically, this project dealt with the development of a sophisticated robotic scheme requiring the interface of an automated Perkin-Elmer Model 5000 Atomic Absorpiton/MHS-20 hydride generator with a Zymate System (Figure 1).

With the complete robotic protocol, sixteen urine samples can be analyzed within a sixty hour cycle time. Forty-two hours of the cycle represent elapsed time required for controlled furnace drying and ashing of the samples.

The total capital cost required for automation has been approximately $40,000. The atomic absorption instrumentation, not included in this estimate, was previously used for other manual lab applications prior to the automation project.

The approach taken to staff this project centered primarily on a team of contributors representing both Zymark and Xerox. From Zymark, Dan Reasoner (Sales Representative) and Brian Lightbody (Applications) participated, and representing Xerox lab personnel were Linda Lester, Haig Donoian, Ken Waehner and Tim Lincoln. The project took approximately six months to complete, with an extimated twelve man-months of activity.

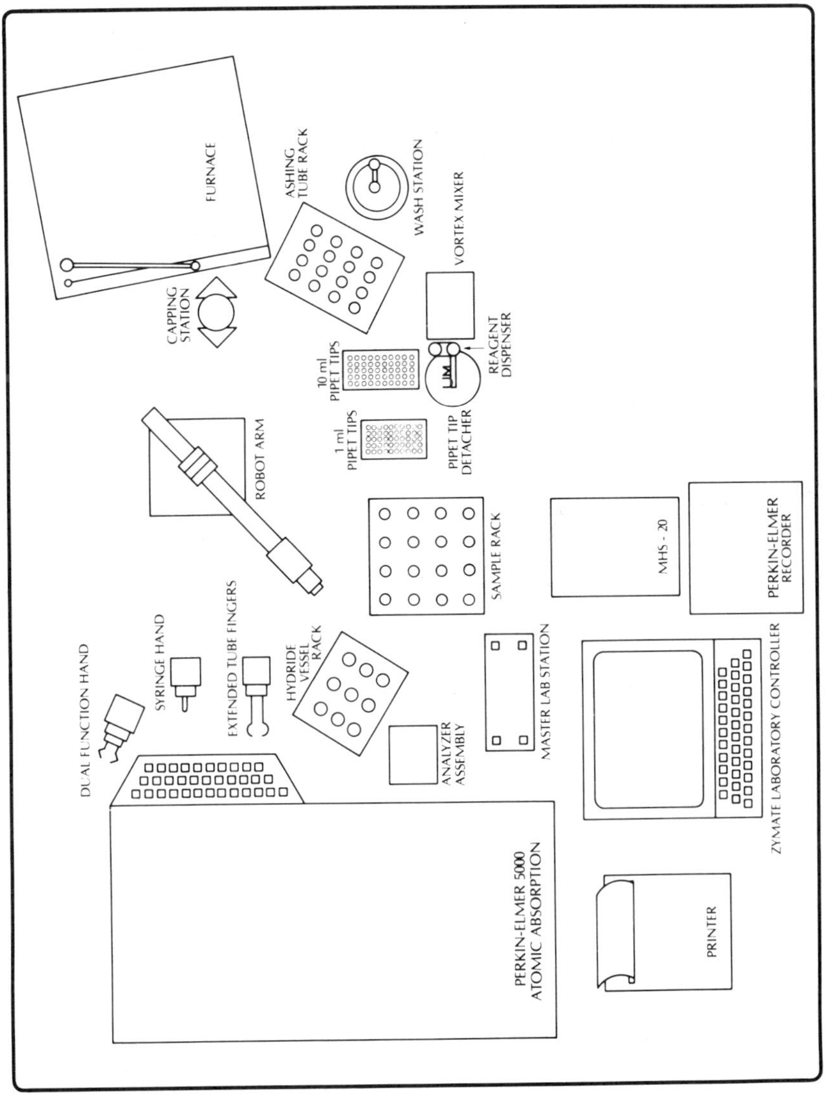

FIGURE 1. Zymate System layout for monitoring arsenic and selenium in urine.

The efficiency realized from this automation project is best illustrated in Figure 2. Opportunities for potential time savings is strictly dependent on sample volume per unit submission. Given the nature of the monitor program at Xerox, the mean number of samples per submission is approximately fifteen. This has resulted in a corresponding time savings of approximately 80% compared to an entirely manual protocol. Additionally, the development of this method provided a number of safety and logistical benefits involving operator handling of the urine.

The automation of urinalysis for arsenic and selenium stands out as a success at Xerox for two primary reasons:

1) Good business: achieving cost savings while increasing quality and quantity of service.

2) Team approach: development of the robotic protocol demonstrated some very obvious cost and timetable related benefits.

OVERVIEW

The method incorporates a slow evaporation of an aliquot of the urine sample, in the presence of magnesium nitrate[1] in a programmable furnace. This is followed by a slow increase of the furnace temperature to above the melting/decomposition point of the magnesium nitrate, which acts as a fusion flux. In this dry ashing step the organo arsenic (As) and selenium (Se) species present in the urine are converted to acid-soluble inorganic salts.

The As and Se salts are next dissolved in a measured quantity of acid, and the resulting solution is measured by an automated hydride

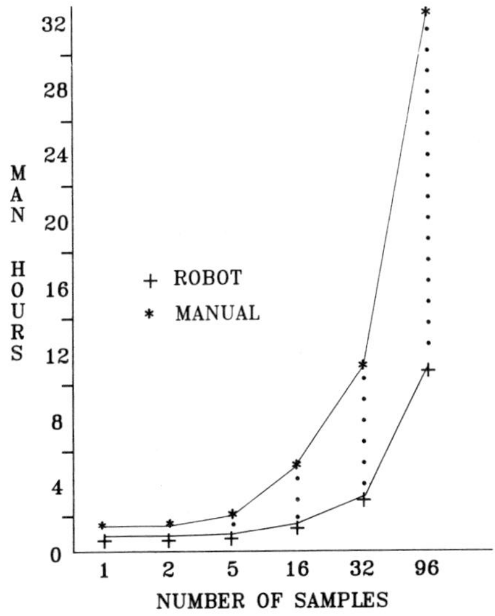

FIGURE 2. Urinalysis As/Se efficiency graph.

generation apparatus (a Perkin-Elmer Model MHS-20) fitted to an atomic absorption spectrophotometer (Perkin-Elmer Model 5000). By means of the hydride generation system a small aliquot of the digest solution is place into an enclosed vessel containing HCl. This charge is automatically purged of dissolved oxygen. An addition of sodium borohydride solution reduces arsenic and selenium ions to their respective gaseous hydrides, AsH_3 and H_2Se. These volatile hydrides are collected in the generating vessel headspace for a prescribed number of seconds and then swept with argon gas into a heated (900°C) quartz tube placed in the optical path of the spectrophotometer. Depending on whether an As or Se EDL lamp is in operation, the quantity of As or Se is measured, respectively. Because the process involves a preconcentration of the analytes, the detection limit of As and Se are in the order of one nanogram absolute. Control samples are carried along with each sample cluster in order to monitor the recoveries of As and Se.

The automation of the above sample preparation, along with the instrument calibration and analyses is accomplished by employing a Zymate System together with its several ancilliary accessories.

METHOD SUMMARY

Urine samples are received from the customer weekly in 125 mL Nalgene bottles and stored in a refrigerator until they are ready to be analyzed. The analyst removes the samples from the refrigerator and places them into a rack that has a capacity for sixteen bottles. The program is initialized to commence the urine preparation. A method summary is listed below to outline the sequence of events.

Urine Sample Preparation

1) Load urine bottles into rack.
2) Uncap bottle.
3) Pipet urine into ashing tube.
4) Add $Mg(NO_3)_2$.
5) Cap urine bottle and return to rack.
6) Repeat for each sample bottle in the rack.
7) Load ashing tubes into furnace.
8) Dry and ash urine for total of 42 hours.
9) Remove tubes from furnace and return to rack.
10) Digest ashed sample with HCl.
11) Vortex to complete digestion.

Presentation of Sample to AA for As/Se Analyses

1) Transfer aliquot (of samples) to hydride vessel.
2) Add HCl (to react with $NaBH_4$).

3) Position vessel in hydride generator.
4) Start MHS-20.
5) Print data.
6) Remove generator vessel.
7) Wash generator vessel.

METHOD DETAIL

Sample Preparation

The robot attaches a dual function hand to obtain a bottle of urine from the indexed sample rack. The bottle is placed in the capping station to be uncapped (Figure 3). It verfies that the cap has been removed by inserting it in an infrared light sensor. In addition, an electrical switch confirms the presence or absence of a sample bottle.

The dual function hand turns 180° to receive a 10 mL pipet tip that is used to transfer 16.0 mL of urine to a rack of indexed ashing tubes

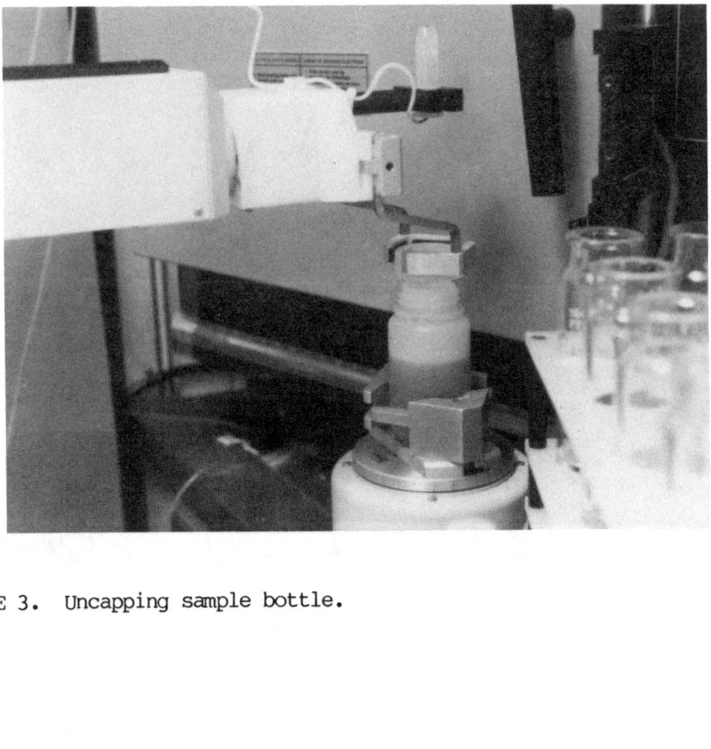

FIGURE 3. Uncapping sample bottle.

(Figure 4). Using the 10 mL syringe from the master laboratory station (MLS), 8.0 mL of urine are transferred twice. To insure that urine does not leak from the pipet tip the 10 mL syringe from the MLS is constantly drawing air during the transfer. Due to the foaming of urine when it is dispensed, another 10 mL of air is used to purge the pipet tip. The pipet tip is then removed by a 10 mL detacher and the hand rotates to its original position, where the cap is still in place. The sample bottle is then capped and returned to its rack. The ashing tube with urine is moved to the reagent dispensing station where 5 mL of 60% aqueous solution of $Mg(NO_3)_2$ is added using a 10 mL syringe from the MLS. This cycle is repeated until the last sample is completed. The dual function hand is parked and an extended tube fingers hand, designed

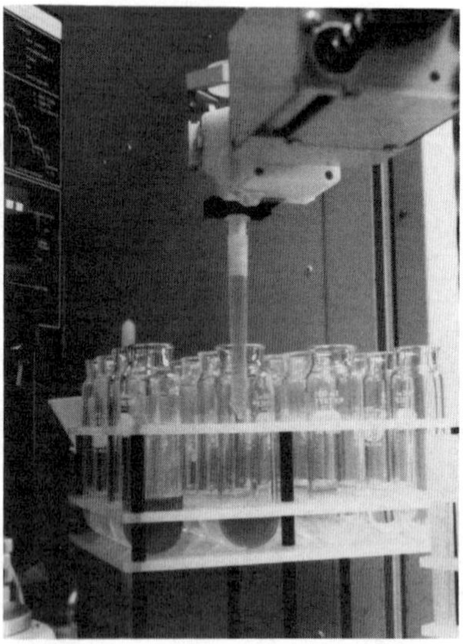

FIGURE 4. Transferring sample aliquot.

and manufactured by Zymark, is attached. This hand is used to load and remove ashing tubes to and from the furnace (Figure 5). The furnace door is automatically opened and closed using a pneumatic piston operated by the power and event controller. The robot hand turns the programmed furnace on, and the urine is dried and ashed for a total of forty-two hours. This preparation enables a single sample to be prepared for both arsenic and selenium.

Calibration

After the samples are removed from the furnace, the atomic absorption unit is calibrated for the first element (selenium). A 100 ug/L arsenic and selenium standard is uncapped and the cap is stored by the parking station of the dual function hand until the calibration of instument is

FIGURE 5. Loading furnace for ashing.

completed. The 1 mL syringe hand and the 1 mL pipet tips are used for the standard aliquots of 50 uL, 100 uL, and 200 uL, respectively. Before the aliquot is taken, 10 uL of air is drawn. This is done to prevent leaking of the aliquot and to ensure complete delivery. After the syringe hand dispenses the aliquot in the hydride generating vessel, (Figure 6) the extended tube fingers hand is attached and the aliquot is diluted with 0.9 mL of 6N HCl and 9 mL of H_2O at the reagent dispensing station. The vessel is then taken to the analyzer assembly (Figure 7), where a lift operated by the power and event controller secures the vessel in place. A proximity sensor was installed to verify the presence of a generator or vessel. The MHS-20 controller hard wired to power and event controller is activated. Sodium borohydride is dispensed in the generating vessel, and the reaction takes place. The

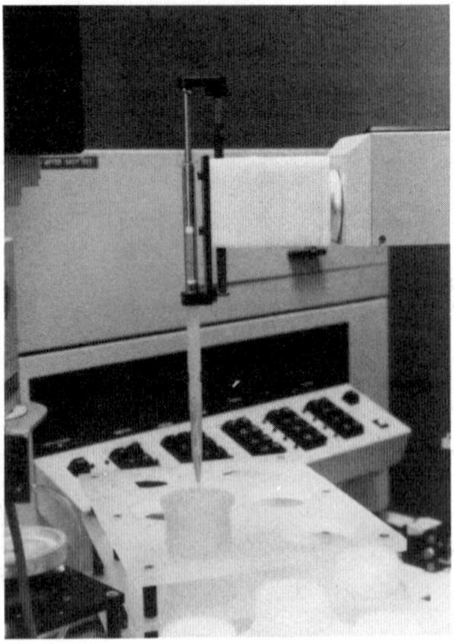

FIGURE 6. Dispensing aliquot to hydride generator vessel.

absorbance is printed out on the Perkin-Elmer PRS-10 printer and plotted on the Perkin-Elmer 056 recorder. A timer is used in the program so that the vessel is removed from the analyzer assembly upon completion of the reaction. The vessel is washed at the washing station (Figure 8) and returned to its original rack. A blank sample of acid and H_2O is analyzed after each aliquot has been analyzed in duplicate. The standard is then capped and returned to its rack. It should be noted that the hydride generator analyzer assembly and vessels were modified by Zymark. The tabs on the vessels were removed and a pneumatic lift was added to the analyzer assembly to ensure a tight seal. The vessel is no longer secured in place by hand. The robot places the vessel at the gasket and the lift is raised to ensure a tight fit.

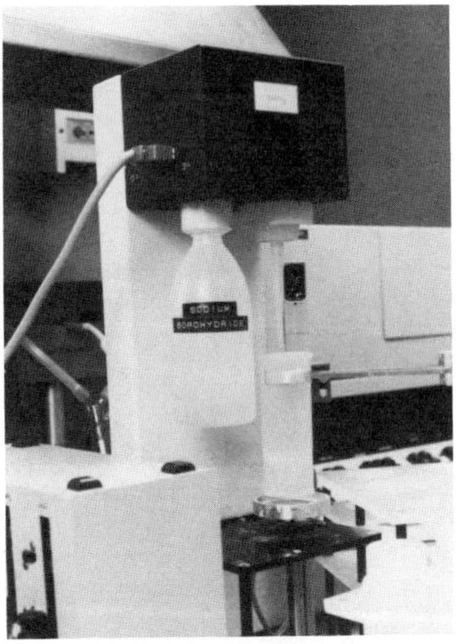

FIGURE 7. Hydride generating vessel at analyzer assembly.

FIGURE 8. Washing station.

Analysis

The dual function hand takes the ashed urine to the reagent dispensing station and adds 25 mL of 6N HCl to digest the sample. The tube is then placed in a vortexer for 45 seconds to complete the digestion. The ashing tube is then returned to its rack and the cycle is repeated until all ashed samples are digested.

The 1 mL syringe hand and 1 mL pipet tips are used to deliver an aliquot of 0.05 to 0.7 mL to the generating vessel. Once the aliquot is taken, the hand is programmed to hit the sides of the ashing tubes to remove liquid on the outside of the pipet tip. The aliquot is then dispensed into the generating vessel and the same procedure is followed as for the calibration of the instrument. Two blanks are analyzed after each sample has been tested in duplicate to verify that there is a steady base line. The cycle is repeated until all samples are completed. Upon completion of samples, the calibration of the instrument is repeated.

The analyst sets up the instrument for the next element, arsenic, and initiates the program to begin. The arsenic program is the same as selenium except that the 25 mL addition of 6N HCl is omitted. The samples are vortexed before they are analyzed to ensure homogeneity after long standing. The data generated is gathered by the analyst, reviewed, and calculated. Usually one element is analyzed during the day and the other at night. To prevent the EDL lamps and the furnace on the AA from burning out, timers are set before leaving to turn off the power supply to these devices.

Procedural Review

The robotic configuration and process of the complex system is reviewed in the flow chart (Figure 9). The flow chart gives a step-by-step detailed description of the complete robotic protocol including sample preparation, analysis and necessary human intervention. Also shown are several fail-safe verification devices which eliminated some problems that were encountered during the initial system debugging trials.

VERIFICATIONS

A rigorous statistical study has not yet been attempted; however, recoveries from spiked aqueous solution were made. Precision and recovery studies on urine samples compared to manual procedures were

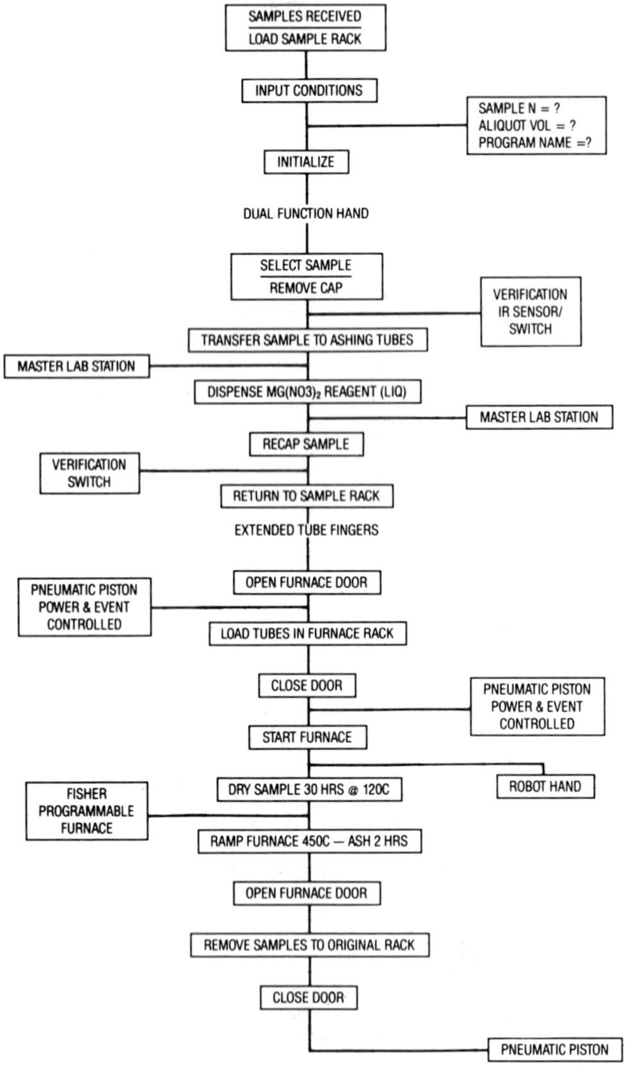
SAMPLES RECEIVED
LOAD SAMPLE RACK
INPUT CONDITIONS
SAMPLE N = ?
ALIQUOT VOL = ?
PROGRAM NAME =?
INITIALIZE
DUAL FUNCTION HAND
SELECT SAMPLE
REMOVE CAP
VERIFICATION
IR SENSOR/
SWITCH
TRANSFER SAMPLE TO ASHING TUBES
MASTER LAB STATION
DISPENSE MG(NO3)2 REAGENT (LIQ)
MASTER LAB STATION
RECAP SAMPLE
VERIFICATION
SWITCH
RETURN TO SAMPLE RACK
EXTENDED TUBE FINGERS
OPEN FURNACE DOOR
PNEUMATIC PISTON
POWER & EVENT
CONTROLLED
LOAD TUBES IN FURNACE RACK
CLOSE DOOR
PNEUMATIC PISTON
POWER & EVENT
CONTROLLED
START FURNACE
ROBOT HAND
DRY SAMPLE 30 HRS @ 120C
FISHER
PROGRAMMABLE
FURNACE
RAMP FURNACE 450C — ASH 2 HRS
OPEN FURNACE DOOR
REMOVE SAMPLES TO ORIGINAL RACK
CLOSE DOOR
PNEUMATIC PISTON

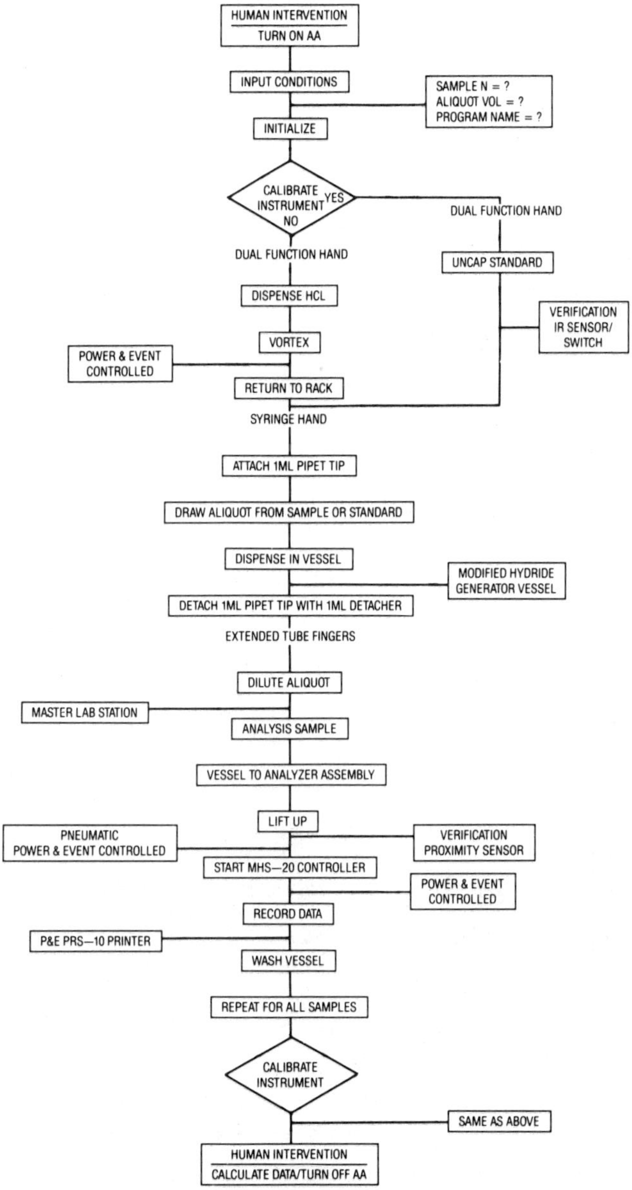

FIGURE 9. Procedural review flow chart for determining arsenic and selenium in urine.

performed as well.

Aqueous Spiked Samples

Commercial standard concentrates (Spex Industries, Inc.) of Se^{+4} and As^{+3} were used to prepare spiked samples at three levels of concentration. Analyses for Se and As were performed both robotically and manually. The results are presented in Tables 1 and 2 for As and Se, respectively. The data represents single sample preparations and the mean of at least two hydride generation determinaitons.

TABLE 1. Inorganic Arsenic in Water - Robot Versus Manual

Spike	Manual		Robot	
(ug/L)	Found (ug/L)	Recovery (%)	Found (ug/L)	Recovery (%)
30.0	29.7	99	26.4	88
50.0	40.7	81	44.9	90
70.0	66.3	95	66.6	95

Robot vs Manual: Corr. Coef. = 0.97

Slope = 1.05

Very good recoveries of As were obtained both robotically and manually, on the average of about 91%. Linear regression of robotic versus manual results gave a slope of 1.05 vs. unity with a correlation coefficient of 0.97. (Table 1)

In the case of Se (Table 2) the recoveries were again very good except that the manual determinations showed somewhat lower, but acceptable, recoveries. Linear regression of robotic versus manual results indicated a slope of 1.03 with a correlation coefficient of 0.99.

TABLE 2. Inorganic Selenium in Water - Robot Versus Manual.

Spike (ug/L)	MANUAL Found (ug/L)	MANUAL Recovery (%)	ROBOT Found (ug/L)	ROBOT Recovery (%)
24.6	16.2	66	20.7	84
42.5	37.2	88	45.6	107
58.0	49.9	86	54.7	94

Robot vs Manual: Corr. Coef. = 0.99

Slope = 1.03

Spiked Urine Samples

A large volume of a stock of blended urine was used to prepare three concentration levels of As and Se. The arsenic spikings were 50% inorganic As^{+3} (Spex) and 50% organo arsenic as cacodylic acid (dimethyl arsenic acid, soduim salt, trihydrate, J.T. Baker Chemical Co.). Selenium spikes were totally inorganic (Spex Industries). A total of three 4-liter samples were prepared. Table 3 describes the constituents of each "control". From each of these urine controls, one liter was

TABLE 3. Spiked Urine Controls.

Control	As[a] Added (ug/L)	Se Added (ug/L)	Total As (ug/L)	Total Se (ug/L)
Stock	0	0	38.1[b]	19.8[c]
1	0	8.3	38.1	28.1
2	20.0	25.0	58.1	44.8
3	40.0	41.7	78.1	61.5

a) 50% as inorganic and 50% organic As.
b) Average of 7 determinations; 4 manual, 3 robotic.
c) Average of 8 determinations; 4 manual, 4 robotic.

retained for these studies, and the other three liters were shared with the Xerox Medical Department. They are submitted as blind, coded controls accompanying every cluster of urine samples delivered for urine analysis. The fractional amounts of Se added were a result of an independent determination of a "10 ppm" solution used to spike the urine which analyzed to be 8.3 ppm.

The arsenic results (Table 4) showed excellent precision and accuracy. The relative standard deviations (RSD) were low, in the order of 3-6%, and the recoveries essentially 100% for both manual and the robotic determinations. Linear regression of robotic versus manual results gave a slope of 0.96 and a correlation coefficient of 0.9995.

The selenium results (Table 5) reflect the fugacious nature of this element when measured in the parts-per-billion level. The RSD for the manually obtained data ranged from 15 to 40%, and for the robotic data, from 5 to 20%. The rcoveries were about 50% and 78% for the manual and robotic determinations, respectiviely. For selenium these would be fair to good recoveries in trace analysis. Interestingly, the robot again out-performed the human element in this regard (Table 2 and Table 5). As expected, the linear regression of robotic versus manual data yielded a slope of 2.0 versus unity with a correlation coefficient of 0.90.

CONCLUSIONS AND FUTURE PLANS

Much "hands-on" work up-front has culminated in a new "hands-off" finished procedure for analyzing reliably trace (ppb) levels of arsenic and selenium in urine. The present robotic protocol requires about 10%

TABLE 4. Determination of Arsenic in Spiked Urine.

		MANUAL				ROBOT			
Sample	Nom. As Conc. (ug/L)	No. of Repl.	Mean Conc. (ug/L)	RSD (%)	Recovery (%)	No. of Repl.	Mean Conc. (ug/L)	RSD (%)	Recovery (%)
Stock	38.1	4	37.8	3.8	--	2	38.4	3.7	--
Control 1	38.1	12	38.1	5.4	100	7	38.9	5.3	102
Control 2	58.1	16	57.4	5.9	98	8	58.1	4.5	100
Control 3	78.1	14	78.3	3.1	100	8	77.3	2.7	99

Robot Versus Manual: Corr. Coef. = 0.999
Slope = 0.96

TABLE 5. Determination of Selenium in Spiked Urine.

Sample	Nom. Se Conc. (ug/L)	MANUAL				ROBOT			
		No. of Repl.	Mean Conc. (ug/L)	RSD (%)	Recovery (%)	No. of Repl.	Mean Conc. (ug/L)	RSD (%)	Recovery (%)
Stock	19.8	4	18.5	40.3	--	4	21.0	20.5	--
Control 1	28.1	14	15.7	18.2	59	7	23.1	15.4	87
Control 2	44.8	14	21.6	27.2	50	7	31.6	4.8	73
Control 3	61.5	14	27.2	17.0	45	8	43.6	12.1	73

Robot Versus Manual: Corr. Coef. = 0.90
Slope = 2.00

human intervention over a four-day span to complete the sample preparation and analytic measurements of sixteen urine samples for the two analytes.

Future activities planned:

- The urine analysis program will be further serialized to increase sample output rates. For example, significant amount of time would be gained by having a second aliquot prepared for hydride generation while one is being analyzed.
- An HP150 PC will be interfaced with the spectrophotometer and the Zymate System. This will enable automated computations, decision making, data storage and retrieval. One can envision this system to select proper EDL lamp and other spectrophotometer parameters, establish the calibration curve, compute sample data, make decisions such as required for lower or higher aliquot repeats, store the data and output the results in a finished report format.

The benefits realized with the present configuration and further with the planned improvements are evident. Reduction and/or elimination of human error have been realized. The shake-down and debugging of the robotic manipulation trouble spots have brought the reliability of the system to a point where the analyst now has more time to tackle other technical problems. In addition, the hardware configuration and software program were designed such that they are easily amenable in whole or in part to analyses of other matrices for arsenic and selenium. Indeed any element measurable by hydride generation atomic absorption can be monitored by this robotic protocol.

REFERENCES

1. J. P. Buchet, R. Lauwerys, and H. Roels, Int. Arch. Occup. Environ. Health, 46, 11-29 (1980).

Zymark and Zymate are registered trademarks of Zymark Corporation.
Nalgene is a registered trademark of Nalge Co., Inc.

ROBOTIC SAMPLE RETRIEVAL FROM PHARMACEUTICAL DISSOLUTION TESTERS

B. J. Compton, W. Zazulak and O. Hinsvark
Pennwalt Pharmaceutical Division
755 Jefferson Road
Rochester, NY 14623

ABSTRACT

Four six-spindle (24 vessel) dissolution apparatus and a robotic arm are arranged in a star configuration to allow automated sampling from the dissolution apparatus. Actual sampling utilizes a custom robotic hand with a replaceable low volume in-line filter. Since the filter is used for all sample withdrawals during a study and only replaced between studies, sample cross-over is avoided by using a precise sampling sequence of filter back-flushing and cleaning. The sampling duty cycle is approximately 40 seconds allowing sampling of 24 vessels at half hour intervals. Samples are presently archived in sealed autosampler vials for off-line analysis. The system has been extensively studied for formulations of capsules, tablets, and suspensions and is used routinely in their testing.

INTRODUCTION

Dissolution testing, in the context of the pharmaceutical industry, is a means of quantitating "in vitro" release rates of drugs from formulations into stimulated biological fluids. This type of testing is used in many phases of drug development and manufacturing. For instance, the test is used in the development phase when a new formulation of a drug is being characterized. On the other hand, it is used as a final product control to determine if product specifications are met. The test is known to be difficult to implement because of the requirement to control a multitude of system variables.[1,2,3]

While it is not the expressed objective of the test, in some instances "in vitro-in vivo" correlation of release rates to bioavailability data has been demonstrated.[4] In general, the test is utilized with the acceptance that a multitude of test and sample variables are being studied. One of the objectives of conducting automated dissolution testing is to control test variables.

A variety of commercial automated dissolution testers utilize multi-channel continuous pumping systems with in-line filters. Quantitation is usually accomplished with the aid of UV-VIS absorption measurements, the spectrometer being multiplexed to the system. Alternatively, dissolution samplers use continuous or discrete sampling[5,6,7] for on- or off-line presentation to HPLC systems.

The system described here uses discrete robotic sampling for on- and off-line analysis. It is unique because the probe design can accommodate a variety of sampling methods, such as the use of individual tip filters[8] or an integrated in-line filter, and uses a sharp terminal probe for introducing samples into septa-sealed vials.

We use the system in the off-line analysis mode with samples being stored in a number of different vials or test tubes. We favor off-line analysis because our daily operation uses HPLC, GC, UV-VIS spectroscopy and continuous flow post-dissolution chromophore generating systems with the analysis being conducted in three different laboratories. The system is also used with a multitude of formulations in their development and post-development stage.

We further use the system in a mode which incorporates a single in-line filter for all sampling. This can be accomplished because of the use of an unique sampling and filter back-flushing sequence.[9]

Since many of our development products are controlled-release suspensions (ie. based on the Pennkinetic system), we tend to run multi-point dissolution tests from three to twenty-four hours. Thus, the system has been automated only to the extent of sampling.

EXPERIMENTAL

Equipment and dissolution parameters are listed in Tables 1 and 2, respectively. The robotic system from Zymark Corporation (Hopkinton, MA) was purchased specifically for the application described. The Van-Kel Industries, Inc. dissolution systems were also specifically obtained for this application, because they contain some design features, such as accessibility to each vessel, which makes them desirable for robotic sampling.

The robotic arm and controller comprise an integral unit which mimics a

TABLE 1. Robotic Assisted Dissolution Testing Equipment

4, 6-spindle dissolution systems (Vander Kamp 600) with water baths and controllers
1 Zymark robotic arm and controller (Z110 and Z120)
1 Zymark master lab station (Z510)
1 Zymark power and event controller (Z830)
1 Zymark printer (Z820)
4 Zymark racks to hold 50 Vials (large) with needle guides (No. 38320)
4 Zymark racks to hold 50 Vials (small) with needle guides (custom racks)
2 Zymark custom racks to hold test tubes
1 In-house designed custom robotic sampling probe
In house designed and manufactured 45-um Luer lock type filter
17-guage 3.5 inch long stainless steel needle

TABLE 2. Summary of Dissolution Parameters

Parameters generally follow USP XX recommendations.[10]
Bath Temperatures 37°C $\pm$0.5°C.
Type I and II configurations
Paddle stir rate 50 or 100 RPM.
Dissolution media 500, 800, 900 or 1000 mL.
Samples were drawn 2 cm above the top of paddles.
Suspensions were added by disposable syringe, weighed by difference.
Capsules were added intact, manually in glass coils.
Sampling intervals were 0.5, 1.0, 3.0, 6.0, 12.0 and 14.0 hrs to a resolution of 3 seconds. An option for 45 min single point is also available.
Dissolution media varies with the application.
Manual sampling involved pulling samples through cotton syringe tip filters.
Robotic sampling involved between 6.5 and 10 mL pull up, 2 mL filter clearance, between 0.5 - 5.5 mL sample placement and wash-out with 20 mL, 70% aq. MeOH.
Sample placement and wash-out with 20 mL, 70% aq. MeOH.
Sampling filters were 45 u blanks assembled in-house from blanks by Micron Separations, Inc. (Honeoye Falls, NY).

human arm in movement. The arm is placed centrally to all four dissolution baths, each system being at a tangent to a 30 in. diameter circle. The robotic sampling probe (Figure 1) placed at the end of the arm was designed with sampling flexibility and the dissolution baths in mind. Of the innumberable sampling modes available, the simplest was chosen, an in-line Luer lock filter terminated by one 3.5 inch 17-guage stainless steel needle. In this mode the probe is connected with teflon tubing to a master laboratory station, which contains programmable syringes used in sample aspiration and dispensing.

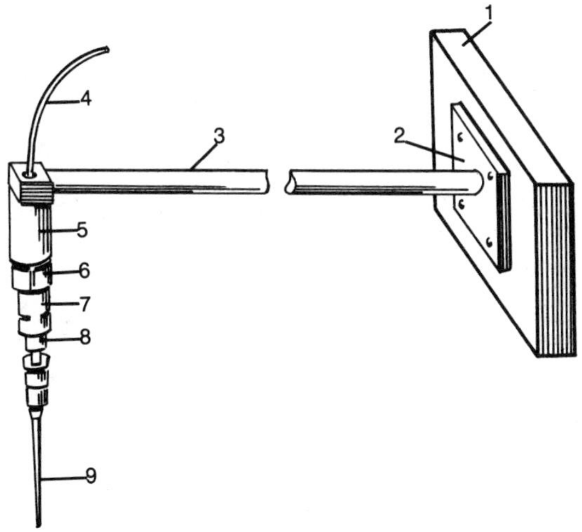

FIGURE 1. Schematic representation of the custom sampling probe. 1) blank hand, 2) base plate, 3) extension rod, 4) Teflon tube terminating at the master lab station, 5) extension tube, 6) and 7) threaded male luer lock connectors, 8) replaceable in-line filter, 9) hypodermic needle. Parts 3, 5, 8 and 9 are easily modified to suit different sampling requirements.

Accessories to the system are the power and event controller which presently is used to signal, via a buzzer, appropriate sample introduction times and a printer which documents exact sampling intervals and program execution. Sample racks, an integral part of the system, are used to hold samples dispensed from the sample probe. The racks hold either small or large autosampler vials and, with self-sealing septa, limit sample evaporation during extended dissolution runs and aid in cleaning the exterior of the probe needle. Additional racks to hold test tubes are also available.

Sample sizes vary from the placement of 0.5 mL into small vials, 2.0 mL into large vials and 5.5 mL into test tubes. At an appropriate pre-programmed time, the sample arm moves into position and places the sample probe needle through a hole in the vessel cover to a prescribed depth. An aliquot is pulled through the probe needle and in-line filter into the tubing connecting the probe to the master lab station. The connecting tubing holds a maximum of 10.0 mL. With the probe still in place in the vessel, 2.0 mLs of sample are flushed back through the filter into the vessel to clear the filter element anad ensure no sample particulates are transferred in the subsequent sample delivery step. The probe is then removed from the vessel and the sample is transferred by piercing the vials' septa with the probe needle. The remaining sample, which is used in the sample uptake to clear the sampling line, is dispensed to a waste container and the sampling system is washed with 20 mLs of 70% aq. MeOH.

The above sampling sequence requires less than 40 seconds per sampling cycle for completion and thus can be utilized to service four six-vessel baths with 17 minute sampling cycles. Our sampling intervals are generally 30, 60, 180, 360, 720 and 1440 min. with most tests terminating at 180 or 360 min. The system also will conduct a 45 minute single point run.

VALIDATION STUDIES

The validation studies used to authenticate the reliability of the robotic method used Pennkinetic controlled release suspensions and capsules, and were designed to show:

1. The robotic system produces samples indistinguishable from samples collected manually.
2. The probe system contains adequate sample wash-out to prevent vessel-to-vessel sample carry-over.
3. Overall vessel-to-vessel and bath-to-bath reproducibility is acceptable for routine usage.

Specific experimental designs are described below (actual robotic start-up procedures and options are presented in Appendix A).

Study (A) Sample carry-over (vessel-to-vessel) and Manual vs. Robotic Crossover: Samples of a suspension containing drugs A and B were dissoluted in alternating dissolution vessels (3 vessels for each formulation) in Tank 1. All analyses were conducted using HPLC.

Study (B) Vessel-to-vessel and bath-to-bath reproducibility and Manual vs. Robotic Crossover: Samples of one lot of drug B capsules were dissoluted using Baths 1-4, representing 24 vessels. Analysis of the samples was conducted as in Study (A).

All data was manually collated and summarized as detailed in the discussion.

DISCUSSION

Results from Study (A), the manual-to-robot crossover and vessel-to-vessel carry-over experiment, are summarized in Table 3. A paired T-test statistical analysis of the manual vs. robot data indicated no difference between the samples generated by either method. Also, the HPLC system used to assay the samples quantitated drugs A and B simultaneously. No A in B nor "vice versa" was found, indicating no vessel-to-vessel carry-over.

The four bath, 24 vessel capsule study, Study (B), produced results summarized in Table 4. Application of a paired T-test statistical analysis for the manual versus robot data indicated no difference between the samples generated in either method. Also, vessel-to-vessel and bath-to-bath reproducibility was found to be low with a relative standard deviation of less than 4%, with few exceptions. Tank 4 in this study showed a systematic error tending towards low dissolution rates and was found to be due to low bath temperature (34°C rather than 37°C).

TABLE 3. Study (A): Suspensions of Drug A and B.

Drug A.

	% LABEL RELEASED (RSD)		
TIME (Hrs)	Robotic Sampling	Manual Sampling	RME%[2]
0.5	41.0 (3.6)	41.5 (2.1)	1.2
1.0	45.4 (1.3)	46.4 (1.6)	2.1
3.0	56.1 (0.98)	58.5 (2.9)	4.1
		Absolute mean	2.5

Drug B.

0.5	48.5 (5.7)	48.3 (1.3)	-0.4
1.0	56.0 (1.1)	57.7 (0.2)	2.3
3.0	69.7 (1.6)	73.2 (0.2)	4.8
		Absolute mean	2.5

1. Means and relative standard deviation of (RSD) three vessels (n=3).
2. Relative mean error % = $\text{mean}_{\text{manual}} - \text{mean}_{\text{robot}}$ X 100 $\text{mean}_{\text{manual}}$

A detailed study of the variability of the system was not required because the variances found in the system are acceptable, generally less than 4% relative standard deviation (RSD). A simple "analysis of variance" not directly related to method validation is presented below to show where variability in the system is thought to originate.

TABLE 4. STUDY (B): Drug B capsules 3 hour Dissolution, 4 bath (24 vessel).

% LABEL RELEASED (RSD)

	TANK NUMBER	MANUAL SAMPLING	ROBOT SAMPLING	RME%
TIME 0.5 HRS				
	1	50.8 (3.9)	51.7 (5.3)	-2.0
	2	51.8 (4.4)	52.6 (3.3)	-2.0
	3	52.0 (0.8)	51.1 (1.7)	-2.0
	4	48.1 (2.7)	49.0 (1.9)	-2.0
	means of means	50.7 (3.5)	51.1 (2.9)	absolute mean 2.0
TIME 1.0 HRS				
	1	62.4 (2.7)	63.4 (3.9)	-2.0
	2	61.5 (3.8)	62.8 (3.5)	-2.0
	3	61.5 (2.7)	62.2 (1.8)	1.0
	4	58.9 (1.7)	59.7 (1.6)	1.0
	means of means	61.1 (2.5)	62.0 (2.6)	absolute mean 1.5
TIME 3.0 HRS				
	1	76.5 (4.7)	76.0 (1.6)	1.0
	2	75.2 (1.1)	75.0 (0.8)	1.0
	3	74.7 (0.7)	75.3 (0.9)	-1.0
	4	73.0 (1.6)	73.5 (1.1)	-1.0
	means of means	74.9 (1.9)	74.9 (1.4)	absolute mean 1.0

1. Means of six vessels per tank (n=6) and relative standard deviation (RSD).

2. Relative Standard Deviation of means.

ANALYSIS OF VARIANCE

Variance for the dissolution method (σ^2_m) is calculated from

$$\sigma^2_m = \sigma^2_{HPLC} + \sigma^2_{Diss} + \sigma^2_{Sampling} + \sigma^2_{Sample}$$

σ^2_{HPLC} is variance associated with the HPLC method, σ^2_{Diss} is variance associated with the actual release of drug of interest in the dissolution apparatus, $\sigma^2_{Sampling}$ is variance originating from the sampling method, and σ^2_{Sample} is uncertainty about the sample introduced into the apparatus. Variance contributions are as follows:

σ^2HPLC - Estimates of this variance are typically in the range of RSD of 1-2% unless care is taken in the analysis, in which case values as low as 0.3% are attainable.

σ^2Sample - A RSD of less than 3% has been found during sampling of Pennkinetic suspensions in potency and content uniformity studies. Values of approximately 3% can be associated with capsule following USPXX guidelines.

σ^2Diss - The variance associated with dissolution apparatus is not known and is determined here.

σ^2Sampling - Likewise the variance associated with dissolution sampling is not known.

σ^2m- During the course of this work standard deviations of less than 5% within tanks (six vessels in one tank) and between tanks were found as shown in Tables 3 and 4.

Substituting variance values from above, it is found to a first approximation that:

$$(\sigma_{Diss} + \sigma_{Sampling})^{1/2} = 3\%$$

In other words, the variance associated with well maintained and calibrated dissolution apparatus is low, as is the variance associated with the sampling methods (manual and robotic) used here. The study of the capsules does indicate that the system is not immune to systematic errors. As mentioned previously, Tank 4 showed a systematically low dissolution rate attributed to a low bath temperature. Steps are being taken to minimize the occurrence of such errors, while overall system variance will be minimized using periodic system calibrations.

SYSTEM PERFORMANCE, LONG TERM

The system has been utilized on a daily basis, with two operators, for one year. During this time certain system limitation were recorded. The sampling method has difficulty with formulations containing excipients that gel or otherwise slow the filtration sequence such that the filter back-flush step is not properly executed. This problem is solved by slowing the syringe fill rate of the system. An additional problem has been associated with formulations which apparently are not fully wetted by the dissolution media and consequently float. These tend to coat the outside of the probe needle and can potentially cause vessel-to-vessel carry-over. The actual amount of material transferred has been too low to detect. This potential problem has been addressed by introducing an external needle wash line to clean the needle exterior simultaneously with the system wash-out.

SYSTEM COST-RECOVERY STUDY

The introduction of robotic sampling of dissolution tests which sometimes require samples drawn at 12 and 24 hrs. has revolutionized our test facilities. We estimate a general 400% increase in utilization of personnel and 200% in equipment. Since dissolution testing, in our facility, is distinct from the actual analysis function by GC, HPLC or

UV-Vis and has been upgraded independent of the analysis laboratories, we are now capable of generating considerable more samples than can be analyzed.

From Appendix B, we calculated that the system has a cost-recovery of 126 days. This is conservative since the calculations do not consider the increase in the reliability of the testing steps and less need for sample retesting.

ACKNOWLEDGEMENT

We acknowledge the continued support of the Department of Pharmaceutical Development, Pennwalt Pharmaceutical Division, and Stefanie Viggiani for preparation of this manuscript.

REFERENCES

1. D.C. Cox and W. B. Furman. J. Pharm. Sci., 73, 670-675, (1984).

2. D. C. Cox, C. E. Wells, W. B. Furman, T. S. Savage and A. C. King, J. Pharm. Sci., 71, 395-398, (1982).

3. D. C. Cox and W. B. Furman, J. Pharm. Sci., 71, 451 (1982).

4. L. P. Amsel, O. N. Hinsvark, K. Rotenberg and J. L. Sheumaker, Pharm. Tech., April 1984.

5. Vander Kamp 10475, VAN-KEL Industries, Inc., Edison, NJ.

6. Model 27 Dissoette Hanson Research Corp., Northridge, CA.

7. Application Study #3/84, Perkin-Elmer corp., Norwalk, CT.

8. Application Note "Automated Dissolution Testing", Zymark Corp., Hopkinton, MA.

9. U.S. Patent Pending.

10. USPXX, United States Pharmacopoeial Convention, Inc., Rockville, MD, 1980, p. 959.

APPENDIX A. Robotic Start-up Procedure and Options:
Program Dissolution

Vessels are initially prepared for dissolution testing as per SOP or USPXX recommendations. The program DISSOLUTION is edited to include:

1. Tank Number (1 - 4).

2. Run Duration (3, 6, 12, or 24 hrs. Note: 0.5 and 1 hr samples are taken).

3. Dissolution Media Volume (500, 800, 900, or 1,000 mL).

4. Vial size (large or small) or test tube.

The program is then executed. Samples to be tested are manually fed into each vessel; the time intervals are indicated by the sounding of a buzzer from the power and event controller.

APPENDIX B. Cost-Recovery Evaluation of the Dissolution System

		FORMULA	PRESENT METHOD	ZYMATE SYSTEM
A.	Number of Samples per Sample Group	Input	24	24
B.	Total Time per Sample Group (Hrs)	Input	6	6
C.	Number of Sample Groups per day	Input	1	2
D.	Operating Hrs per Day	B*C	6	12
E.	Technician Cost per Hr inc. Fringe	Input	15	15
F.	Technician Hrs per Operating Day	Input	20	4
G.	Technician Hrs per Sample Group	F/C	20	2
H.	Technician Cost per Sample Group	E*G	300	30
I.	Instrumentation Cost	Input	32,000	60,000
J.	Est. User Set-Up & Programming Cost	Input	1,000	2,000
K.	Total Investment	I+J	33,000	62,000
L.	Annualized Investment	K/5	6,600	12,400
M.	Investment Cost per Day	L/250	26.40	49.60
N.	Investment Cost per Sample Group	M/C	26.40	24.80
O.	Total Cost per Sample Group	H+N	326.40	79.60
P.	Zymate System Saving per Sample Group: 0(Present Method) - 0(Zymate System Method)			246.80
Q.	Number of Sample Groups to recover Investment	K/P		251.00
R.	Days to Recover Investment	Q/C		126.00

Zymark and Zymate are registered trademarks of Zymark Corporation. Vander Kamp is a registered trademark of Van Kel Industries, Ltd. Pennkinetic is a registered trademark of Pennwalt Corporation.

LABORATORY ROBOTICS FOR TABLET CONTENT UNIFORMITY TESTING

P. Walsh, H. Abdou, R. Barnes, B. Cohen
Quality Control Division
E.R. Squibb & Sons, Inc.
New Brunswick, NJ 08903

ABSTRACT

Content uniformity is an important aspect of pharmaceutical products. The repetitive and intensive labor involved, as well as the requirement for identical treatment of all samples in testing for uniformity, makes this an ideal application for a robotic system. A Zymate Laboratory Automation System was installed in the Quality Control laboratories to perform sample preparation on some of Squibb's solid dosage forms. The primary components of the system are a controller, a robotic arm, and several laboratory stations that include a balance, capper, vortex mixing unit, HPLC autosampler-vial carousel turntable, ultrasonic bath, and a solvent dispensing master lab station. The analytical determination for the active ingredient is subsequently conducted by HPLC.

The precision of the robotic method was affirmed by determining content uniformity (10 tablets) on seven lots of the same product and comparing the results with those obtained using the approved manual preparative technique. The relative standard deviation was comparable for the two techniques, ranging from 1.4 to 3.9% for the robotic system and 2.0 to 3.2% for the manual procedure. The averages of the seven lots analyzed were 99.6 and 99.2% of theory for the respective robotic and manual preparations showing that the accuracy of the two methods is similar.

INTRODUCTION

Procedures for instrumental methods of analysis consist of four major steps: sample preparation, instrumental measurement, data handling, and report generation. Advancements in the area of microprocessor technology have resulted in a new generation of analytical instruments that are designed for unattended operation. Laboratory computers have revolutionized the areas of data acquisition, data processing, and report generation. And now sample preparation, which involves physical manipulation, is also benefiting from advances in microelectronics. The development of microprocessors and sophisticated mechanical arms has bridged the gap between the microcomputer and the physical world. Laboratory robots can perform a number of routine laboratory operations, thereby bringing the analytical laboratory one step closer to being totally automated.

Automated analysis came into existence in clinical laboratories during the 1950's with Skeggs' adaption of classical methodology into automated procedures through his development of segmented flow systems.[1] These autoanalyzers became increasingly sophisticated over the years, evolving into systems which can process untreated samples, conduct the analysis, and generate an analytical report.[2] This type of instrumentation is commonly referred to as hard or dedicated automation, because the system is optimized for a single procedure and would require major modifications for alternative use. Hard automation is well-suited for operations such as a clinical laboratory, which processes large numbers of samples requiring identical test conditions.

The pharmaceutical quality control laboratory differs from the clinical laboratory due to the variety of samples submitted for testing which may have specific requirements for sample preparation or analytical measurement.[3] In contrast to the hard automation of autoanalyzers, laboratory robots are flexible automation systems capable of being programmed to perform specific steps appropriate for different samples. Variables common to robotic procedures include: the choice of solvents, volume of solvent delivered, and secondary dilution ratios.

A procedure typically performed in pharmaceutical quality control laboratories is the determination of the uniformity of dosage units. Compendial requirements for content uniformity of tablets are met if the amount of active ingredient in 10 individual tablets lies within the range of 85.0 to 115.0 percent of the label claim and the relative standard deviation is less than or equal to 6.0 percent.[4] In this work we describe the validation of a commercially available laboratory robot for the tablet content uniformity sample preparation of two potencies of anti-hypertensive tablets.

EXPERIMENTAL

Robotic Equipment

The primary components of the system are a controller, a robotic arm and several laboratory stations. The spacial relationship between the components is shown in Figure 1. All components were obtained from Zymark Corporation, Hopkinton, MA except where indicated.

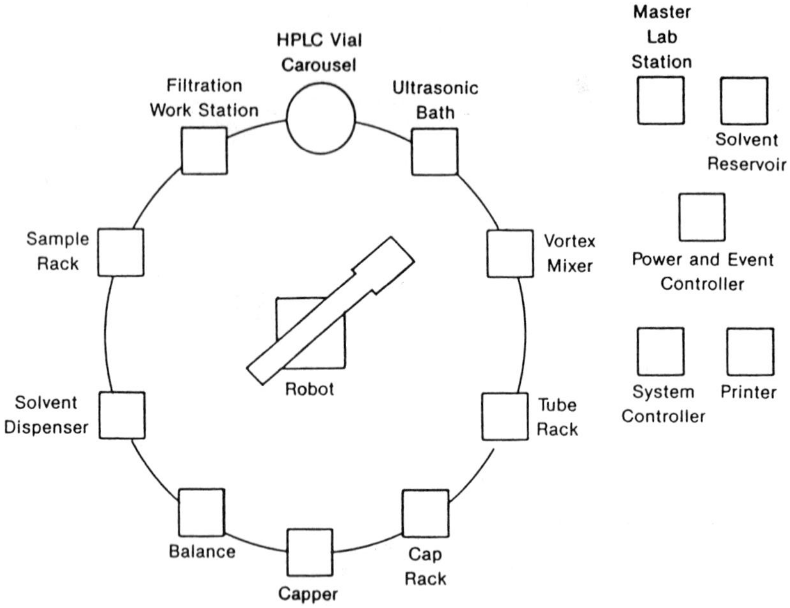

FIGURE 1: A schematic representation of the robotics system for the preparation of tablets for content uniformity testing.

1. Controller - The controller (Model Z100) uses microprocessor technology and EasyLab software to control the robot and laboratory stations as an integrated system. Programming is dictionary based, where the user names the locations and operating commands and stores these names in the computer's dictionary. These names are then recalled and built into sample preparation procedures.

2. Robotic Arm - The robotic arm transfers samples from station-to-station according to user programmed procedures. A dual purpose hand allowing the use of gripper fingers or a remote dispenser nozzle was used in this procedure. Rigid gripper fingers were selected to better grasp the caps and glassware.

3. Laboratory Stations - The laboratory stations are placed within reach of the robotic arm. The active stations are electronically connected to the controller. These stations include a Mettler AE-160 balance with RS-232 interface, (Mettler Instrument Corporation, Highstown, NJ), capper (Model Z410), vortex unit (Model Z620), autosampler vial carousel turntable, an ultrasonic bath (Model T-28, L&R Manufacturing Company, Kearny, NJ) and solvent dispensing master lab station (Model Z510). Other lab stations include a centrifuge tube rack, test tube rack, syringe barrel rack, and syringe barrel holding and shucking station.

High Performance Liquid Chromatography System

The HPLC system consisted of a Model 590 pump, a WISP Model 710B injector and a Model 481 variable wavelength detector (Waters Associates, Milford, MA). The column was 30 cm x 4.6 mm internal diameter, stainless steel, packed with octadecylsilane chemically bonded to 10 um diameter porous silica (ES Industries, Marlton, NJ). The mobile phase consisted of 55% methanol, 45% water and 0.05% o-phosphoric acid. HPLC grade methanol was obtained from J.T. Baker Company, Phillipsburg, NJ. HPLC grade o-phosphoric acid, 85%, was obtained from Fisher Scientific Company, Springfield, NJ. Water was purified using a Milli-RO reverse-osmosis system in combination with a Milli-Q deionization system (Millipore, Bedford, MA).

The Robotic Procedure

The procedure was designed to follow the equivalent steps of the existing manual preparation procedure. The primary steps in the robotic procedure are:

1. Transfer the tablet to a previously weighed tube.
2. Weigh the tablet (for information purposes).
3. Add a nominal 25.0 mL of methanol. Weigh and convert to volume using density.
4. Cap the tube.
5. Disintegrate the tablet and dissolve the active ingredient using an ultrasonic bath and a vortex mixer.
6. If required by the potency of the tablet being tested, dilute an aliquot of the suspension with methanol in the following fashion. Transfer a nominal 5.0 mL aliquot to a tared tube. Weigh and add the equivalent weight of methanol from the dispenser. Weigh again and determine the weight of methanol added (for information, used to verify the 1:1 dilution).
7. Filter into an HPLC autosampler vial.

Validation of the Balance

A tube is tared on the balance by the robot. Then, a class S calibration weight is manually placed in the tube. The combined weight of tube plus calibration weight is then determined and transmitted to the controller. The observed weight obtained by difference is compared to the class S calibration weight. Table 1 gives typical results.

TABLE 1. Weighing Validation.

Observed Weight (g)	Class S Calibration Weight (g)	Difference From Calibration Weight (g)	Acceptance Criteria (g)
0.1004	0.1002	+ 0.0002	± 0.001
1.0004	1.0000	+ 0.0004	± 0.001
9.9991	9.9996	- 0.0005	± 0.001

Validation of Delivery of Methanol

Delivery of methanol was validated in the following manner. The master lab station was programmed to deliver 25.0 mL of methanol to ten preweighed tubes. The weight of methanol was determined by difference and converted to volume using density. Typical results are given in Table 2.

TABLE 2. Validation of Methanol Delivery.

24.95 mL	25.00 mL
25.03	25.02
25.02	25.01
25.00	25.02
24.99	25.02

Avg. = 25.01 mL
RSD = 0.09%

Validation of Secondary Dilution

Typical results of a ten tablet run requiring a 1:1 secondary dilution are shown in Table 3 where weight of aliquot, weight of methanol added and the difference are listed. Slightly less methanol than required for a 1:1 dilution was found to be added. The average % error was -0.32% with a range of -0.15 to -0.47% and considered acceptable.

TABLE 3. Validation of Secondary Dilution Step.

Tablet Number	Weight of Sample Aliquot (g)	Weight of Methanol Diluent (g)	Δ*	% Error**
1	3.8349	3.8274	- 0.0075	- 0.20
2	3.8751	3.8619	- 0.0132	- 0.34
3	3.7408	3.7231	- 0.0177	- 0.47
4	3.8860	3.8802	- 0.0058	- 0.15
5	3.8778	3.8633	- 0.0145	- 0.37
6	3.8499	3.8372	- 0.0127	- 0.33
7	3.8546	3.8468	- 0.0078	- 0.20
8	3.8551	3.8444	- 0.0107	- 0.28
9	3.8447	3.8283	- 0.0164	- 0.43
10	3.8419	3.8265	- 0.0154	- 0.40
Average	3.8461	3.8339	- 0.0122	- 0.32

* Δ = Weight of methanol diluent - weight of sample aliquot.

** $\% \text{ Error} = \frac{\Delta}{\text{Weight of aliquot}} \times 100$

Robotic System Operation

The system is prepared by the analyst in the following fashion. The solvent reservoir is filled with degassed methanol. Centrifuge tubes and caps, test tubes, pipette tips, syringe-filter-needle assemblies and autosampler vials are loaded into their respective racks. Single tablets are placed into test tubes. Ten tablets are processed serially to the point of being placed in the ultrasonic bath. All ten samples are simultaneously ultrasonicated for 15 minutes. Each sample is then filtered into an HPLC autosampler vial, the septum of which has been prepunctured to prevent build up of back pressure. Another set of ten tablets can then be processed.

Analysis by High Performance Liquid Chromatography

At the completion of the preparative stage, the analyst places the carousel of vials into a Waters WISP autosampler. Typical chromatograms for standard and sample are presented in Figure 2. The amount of anti-hypertensive per tablet is calculated as follows:

$$\frac{A \times C/D \times P/100}{B \times 1/F \times G} = \text{mg anti-hypertensive/tablet}$$

Where:

A = peak response of sample.

B = peak response of standard.

C = weight of standard (mg).

D = dilution of standard (mL).

F = actual volume of methanol used to dissolve individual tablet (mL).

P = purity of standard (%).

G = secondary dilution factor (where applicable).

RESULTS

Anti-hypertensive Tablets (25 mg/Tablet)

Because of tablet variability, it was desirable to first compare the robotic preparation procedure against the manual procedure using a common grind of sample. A comparison of the two preparative procedures was then made using the content uniformity test.

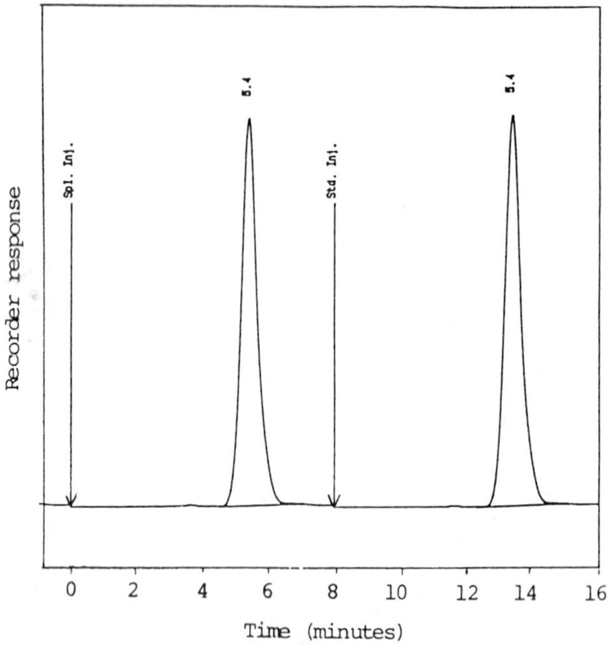

FIGURE 2: HPLC chromatogram of sample and standard preparations of anti-hypertensive. Retention time equals 5.4 minutes.

Common Grind: Twenty tablets were ground. Portions (0.100 g), accurately weighed, were placed in ten centrifuge tubes. Five of the tubes were processed through the manual procedure. The other five tubes were processed by the robot with a suitable modification in the program to start with the solution step. The ten samples were then analyzed by HPLC.

Results are shown in Table 4. The averages obtained by the two methods are similar, 25.1 mg/t for manual and 25.9 mg/t for robotic. The relative standard deviation for the five assays for the robotic preparation agrees closely with that for the manual preparation, 1.0% versus 0.7%, respectively.

TABLE 4. Comparison of Robotic and Manual Preparations Using a Common Grind, Anti-hypertensive Tablets (25 mg/Tablet).

Manual Preparation	Robotic Preparation
25.3	25.8
24.9	26.0
25.1	25.6
25.3	26.3
25.0	25.8
Avg. = 25.1 mg/t	Avg. = 25.9 mg/t
RSD = 0.7%	RSD = 1.0%

Content Uniformity: Seven lots were analyzed for content uniformity using both robotic preparation and manual preparation. Results are shown in Table 5.

One manual determination and two robotic determinations were performed on each lot. For the manual preparation, the relative standard deviation on ten tablet assays ranged from 2.0 to 3.2%. The RSD on ten tablet assays using the robotic preparation ranged from 1.4 to 3.9%, indicating good precision.

TABLE 5. Comparison of Robotic and Manual Preparation Content Uniformity Testing, Anti-hypertensive Tablets (25 mg/Tablet)

	Manual			Robotic		
Lot	Range (% of Theory)	Average (Mg/T)	% $RSD_{(10)}$	Range (% of Theory)	Average (Mg/T)	% $RSD_{(10)}$
1	96.1-106.0	25.2	3.2	98.0-107.3	25.4	2.5
				97.0-101.6	24.8	1.4
2	94.2-100.2	24.3	2.0	93.4-103.2	24.9	2.9
				94.1-102.4	24.4	2.4
3	94.1-102.4	24.8	2.5	98.2-104.9	25.1	2.4
				97.6-102.8	25.0	1.6
4	94.1-103.2	24.5	2.5	91.5-103.8	24.5	3.9
				94.6-100.9	24.7	2.0
5	96.9-102.5	24.8	2.0	95.9-105.4	24.7	3.1
				95.8-102.7	24.9	1.9
6	96.0-101.8	24.7	2.0	96.1-106.9	25.0	3.2
				95.2-105.4	24.9	3.2
7	96.4-104.4	25.2	3.0	93.3-104.5	25.1	3.1
				99.6-109.2	26.0	2.7
	Grand Average =	24.8		Grand Average =	24.9	

Compendial Requirements
(Uniformity of 10 tablets): 1) No value outside the range of 85.0-115.0% of label claim.

2) $RSD_{(10)} \leq 6.0\%$

The averages of the seven lots analyzed were 99.6 and 99.2% of theory for the respective robotic and manual preparations showing that the accuracy of the two methods is similar.

Anti-hypertensive Tablets (50 mg/Tablet)

Five lots were analyzed using the robotic procedure incorporating the secondary dilution step. One manual and two robotic determinations were performed on each lot. Results are shown in Table 6. For the manual

TABLE 6. Comparison of Robotic and Manual Preparation Content Uniformity Testing, Anti-hypertensive Tablets (50 mg/Tablet)

	Manual			Robotic		
Lot	Range (% of Theory)	Average (Mg/T)	% $RSD_{(10)}$	Range (% of Theory)	Average (Mg/T)	% $RSD_{(10)}$
1	96.1-105.6	50.4	2.8	98.2-102.4	49.9	1.4
				98.7-106.6	51.2	2.4
2	95.1-106.0	49.4	3.1	95.8-104.3	49.5	2.9
				97.1-106.6	50.3	3.4
3	101.2-102.6	50.1	2.1	97.2-101.1	49.6	1.1
				97.3-104.4	50.6	2.3
4	98.6-104.8	51.1	1.6	98.5-105.3	50.9	2.1
				100.0-107.7	51.6	2.4
5	98.2-105.8	51.0	2.1	94.7-105.0	49.9	3.3
				95.7-105.8	51.4	2.9
	Grand Average =	50.4		Grand Average =	50.5	

Compendial Requirements (Uniformity of 10 Tablets):

1) No value outside the range of 85.0-115.0% of label claim.

2) $RSD_{(10)} \leq 6.0\%$

preparation, the relative standard deviation ranged from 1.6 to 3.1%. The RSD on ten tablet assays for the robotic preparation ranged from 1.1 to 3.4%, indicating good precision. The averages of the five lots were both 101% of theory showing that the accuracy of the two methods is equivalent.

SUMMARY

Robotic sample preparation using the Zymate System has been successfully applied to the content uniformity test for two potencies of a tablet formulation. Validation of the system included calibration of the balance and solvent delivery system, comparison of results obtained from analysis of a common grind prepared for assay using both the robotic and manual methods, and finally the replicate analyses of several lots for content uniformity using both robotic and manual methods. The results of this study show that the robotic method of sample preparation can replace the current manual method as the primary preparation procedure. The robotic arm is slower in its physical movement than the human hand. However, it is designed for unattended operation without interruption and is utilized during the off-hours. We anticipate that a significant increase in laboratory efficiency will be realized for content uniformity testing of tablet products.

Other efforts in the robotic area will involve the use of the HPLC interface and the expansion of the robotic methodology to the sample preparation for the content uniformity of other products.

ACKNOWLEDGEMENTS

The authors are grateful to Dr. E. Gusmano and Dr. C Papastephanou for their continued support throughout this study, to Dr. G. Hassert for his helpful discussions and to Mr. A. Barcia for his technical assistance.

REFERENCES

1. Skeggs, L.J., Amer. J. Clin. Path., 28, 311 (1957).

2. Burns, D.A., Pharm. Tech., 5, 52 (1981).

3. Leiper, K.J., Trends in Anal. Chem., 40, 4 (1985).

4. The United States Pharmacopeia, 21st rev., The National Formulary, 16th ed., Mack Publishing Co., Easton, 1984, p.1277.

ZYMATE LABORATORY AUTOMATION SYSTEM IN A CONTACT LENS PRODUCT RESEARCH AND DEVELOPMENT LABORATORY

Marlene A. Hall, Anthony J. Dziabo,
and Richard M. Kiral,
Allergan Pharmaceuticals, Inc.
2525 Dupont Drive
Irvine, CA 92715

ABSTRACT

Contact Lens Products Research and Development laboratory at Allergan Pharmaceuticals has a new helper for assisting in repetitive laboratory tasks, a Zymark robot. Laboratory testing is required by the FDA for certifying the use of each new contact lens care solution with a particular class of contact lens. Testing consists of modeling the repeated use of a contact lens care solution by adding an aliquot of a solution to each contact lens in a 10 mL vial, allowing the lens to soak for a defined length of time, removing the used solution by aspiration, and adding a fresh aliquot of solution for the next testing cycle. Each study involves an average of 30 contact lenses and, in many cases, two or three different solutions may be used in one study during a 24-hour period. Solutions requiring heating can be tested using a capping station to bottle each lens and solution. The bottle can then be transferred to a hot water bath. When tablets are part of the test, they are delivered to certain vials by a specially designed funnel. Coordination of simultaneous studies can be carried out by four clocks, seven timers, and an infinite number of counters that are selected by the programmer. A printer can be programmed to automatically record cycle name and number, counter numbers, time of day, and any other pertinent information.

INTRODUCTION

Contact lens care products are essential for the safe and successful wear of contact lenses. These care solutions include lens cleaners, lens disinfectants (both chemical and heat), deposit removers, lens conditioning agents, soaking solutions, inserting solutions, and rinsing solutions. All of these solutions can be used in a contact lens care regimen in a myriad of combinations. A typical soft lens care regimen for one day's wear might include use of a disinfectant, a protein deposit remover and a saline rinsing solution.

Contact lenses and contact lens care solutions must be approved by the FDA before they can be marketed. This premarket approval process includes verification that the care product and intended contact lens are compatible. To substantiate compatibility, various tests are performed according to established FDA guidelines (Figure 1). The solution/lens combination must be proven to be compatible with the physiology of the eye and provide stable optical correction. In addition, it must also be proven that none of the care solution components (disinfectants, preservatives, surfactants, etc.) adsorb onto the intended contact lens or concentrate in the lens, thereby causing a delayed toxicological effect.

For approval of a disinfection product, a typical contact lens compatibility testing program might require 20 lenses of the same lens type treated for 30 consecutive treatment cycles in a chemical disinfection system, or 20 lenses for 90 consecutive treatment cycles in a heat disinfection system. In this example, one treatment cycle would

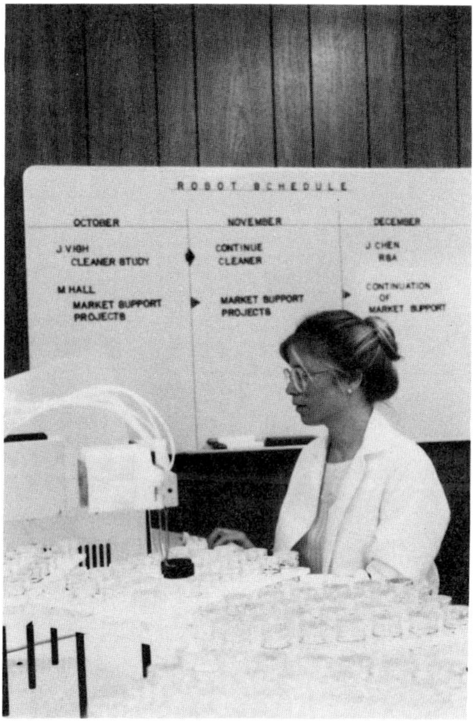

FIGURE 1. Various studies must be completed to meet FDA standards for contact lens products.

consist of the soaking time as labeled on the package insert (or planned package insert) for the solution or testing regimen. Compatibility of the lenses with a given care product would be proven during the testing period by measuring optical and physical parameters of the lens and uptake of preservatives or other chemicals into the lenses.

OBJECTIVES

The compatibility studies are tedious, repetitive and time-consuming. Before Allergan purchased the Zymate Laboratory Automation System, lens

TABLE 1. Human Worker versus Laboratory Automation Completing Comparison Study.

COMPARISON STUDY	
COMPLETED BY HUMAN	COMPLETED BY ROBOT
30 WORKING DAYS	15 DAYS TOTAL
ONE CYCLE PER DAY	2 CYCLES PER DAY
OR	INCLUDING WEEKENDS
ABOUT 1 1/2 MONTHS	AND HOLIDAYS

treatments could only be performed when a technician was on duty to change solutions, generally within the 40 hours available per work week. For example, a study consisting of lenses soaking in a chemical disinfection solution for ten hours each treatment cycle would take 30 working days, at one cycle per day for a laboratory technician (Table 1). With a laboratory robot, the study would take approximately 15 days at two cycles every day, including weekends and holidays. The objective of the Zymate System at Allergan Pharmaceuticals Contact Lens Research and Development laboratory is to automate the lens cycling, freeing the laboratory technician from this time-consuming task, thereby increasing laboratory productivity[1] while increasing cost savings.[2] Additionally, the robot ensures each sample receives identical processing, decreasing errors caused by human variability.[3]

The robot is not able to take the optical and physical measurements upon which the studies are based. These must be completed by the technician. The optical parameters include lens diameter, wet weight, and base curve. Physical parameters include lens discoloration, clarity,

TABLE 2. Documentation Provided By Zymate Printer.

DOCUMENTATION PROVIDED BY ZYMATE PRINTER
* NAME OF STUDY
* CYCLE NUMBER
* TIME CYCLE PERFORMED
* REMINDERS TO MEASURE LENS OPTICAL/PHYSICAL PARAMETERS

deposits and tearing rates. Parameters are examined at approximately four checkpoints during the compatibility testing (Table 2). Lens preservative uptake analysis is also required for a study report. We are presently working on methods for analyzing preservative uptake of lenses using the robot interfaced with an HPLC, but at present this analysis is being performed in another laboratory.

The most repetitive portion of the compatibility studies, the changing of solutions, can be easily performed by the Zymate System (Figure 2). A pair of robot gripper fingers (Figure 3) removes the cover from a vial containing a contact lens held in a custom designed rack. The hand then rotates by 90 degrees to allow access to the vial by the aspiration finger, a Teflon tube. The aspiration finger moves from side to side in the upper and middle portions of solution in the vial to move the contact lens out of the way, allowing the finger to then move safely down into the vial without pinning the lens to the bottom. The used solution is thereby aspirated without harming the lens. The aspiration of used solution is accomplished by a pump connected to the robot through the power and event controller. Fresh solution, pumped through a Teflon tube leading from the solution's holding tank to a master

FIGURE 2. Zymate Laboratory Automation System at Allergan Pharmaceuticals.

FIGURE 3. The Aspirator hand for contact lens product research.

laboratory station, is then pumped through the Teflon tubing from the master laboratory station to the dispensing robot finger. Finally, the robot rotates its wrist to the initial hand position, where the lid to the vial has been held in its cover gripper during the aspiration and filling operations, and replaces the lid on the vial.

Another way to configure this type of study is to set up a filling station. In this case the vial would be picked up in the cover gripper and delivered to the filling station for cover removal, aspiration, filling, replacing the cover and delivery back to its rack. This is a more time-consuming method but does not require the Teflon tubing to be suspended from the robot arm, risking entanglement.

Each master lab station has 3 syringes to deliver solutions to vials. Additional master lab stations can be connected to the system if more syringes are required. Our robot station at Allergan Pharmaceuticals is set up to deliver nine solutions (Figure 4). These stations may contain such solutions as a cleaner, a disinfectant, a soaking solution, a saline solution, and a diluent for the protein removal tablet. A typical compatibility study involving one lens care regimen might use three solutions during a 24 hour period. Between treatment cycles, the lens is usually soaked for 30 minutes in an artificial tear solution to simulate wearing time. Some solutions require multiple rinsing of the lens in distilled water before the next solution can be added. Furthermore, several compatibility studies using different regimens and solutions may be served by the robot during a 24 hour period. This requires a large number of many solutions to be available. The Zymate

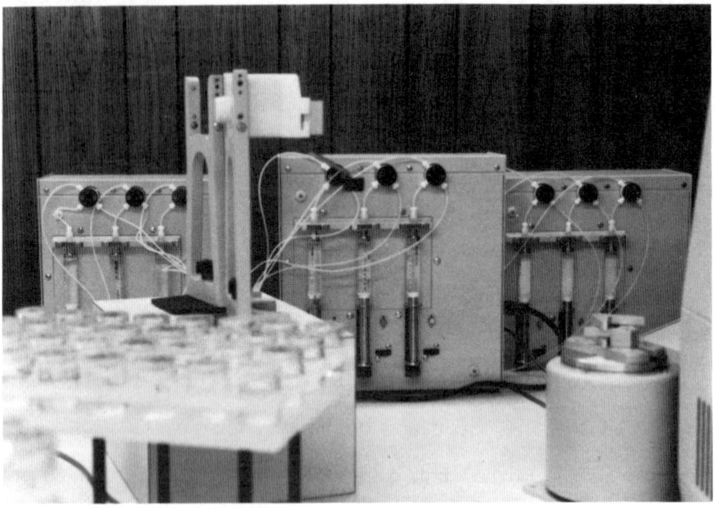

FIGURE 4. Multiple master laboratory stations for delivering nine solutions.

System has the equipment available to distribute these large amounts. After a study, the Teflon tubing that contains a particular solution can be flushed out and replaced by another solution if necessary.

An arrangement was devised to deliver tablets 1/4" in diameter to the vials. Tablets were put into sample test tubes held in a custom rack for robot delivery to the vials. However, the tablets sometimes missed the vial if they happened to be sitting tilted at an angle inside the test tube. At these times the tablet tended to land outside the vial. A funnel was designed to sit on top of the vial after the vial's lid was removed (Figure 5). The funnel insured that the tablet went into the vial every time. A custom set of gripper fingers were designed and attached to a second hand for picking up the test tubes and funnel.

FIGURE 5. Hand 2, the funnel and tube hand for pouring tablets.

Hand 1 and hand 2 had to be exchanged twice during the service of each vial. The robotic arm automatically changes the hands as programmed.

A much more efficient arrangement for tablet delivery was implemented when we began testing tablets that require a completely dry atmosphere. Zymark engineers, in keeping with Zymark's philosophy of support of the system through design and manufacture of custom accessories and supplies,[1] designed and built a chamber to contain the tablets in an environment kept dry by periodically flushing the chamber with dry nitrogen. The dry environment could also be maintained by adding hydroscopic pebbles to channels around the outside of the plate containing the tablets. To deliver a tablet from the chamber to a vial, the lid to the chamber is opened by the robot. A nozzle attached to the robot hand vacuums up an indexed tablet from the plate. The nozzle suction is provided by a vacuum pump turned on by the power and event controller. The tablet is then taken by the robot to the appropriate vial. When the nozzle is above the opened vial, the power and event controller signals the vacuum pump to turn off and the tablet drops from the nozzle into the vial. The nozzle is fitted onto the first hand, which has previously removed the lid from the vial, aspirated used solution, filled the vial with fresh diluent, and now delivers the tablet and replaces the vial lid. In this new system no time is lost in the exchange of hands as in the first tablet delivery system which involved picking up and placing the small test tubes and funnel by means of multiple hand exchanges.

Occasionally, we must test a solution that requires heating the lenses. In this type of study the robot carries a screw-cap vial to the capping station. The vial is uncapped, solution aspirated, fresh solution added, cap replaced, and the vial carried to a hot water bath for the required length of time.

FUTURE STUDIES

The future of the Zymate System at Allergan will see a variety of new tasks in addition to the compatibility study cycling. This is made possible by the flexibility of the physical environment and also programmability of the robotic system, in which the instrument is not permanently fixed to any one task but can be changed as needed to perform a new procedure.[2] With LC equipment, methods and robot interface, the robot will be able to carry out chemical uptake studies.

In addition, the robot will interface with a Mettler titrator for automatic sample delivery.

RESULTS AND CONCLUSIONS

The Zymate System has proven to be time and cost effective in its use at Allergan Pharmaceuticals Contact Lens Research and Development laboratories. The physical environment of the robot is versatile, enabling the robot to change functions in the laboratory as needed. The programming necessary to set up the robot for studies is simple to learn and use. The automation of studies provides for an increase in the precision of tasks, removes the laboratory technician from repetitive tasks, and increases productivity.

REFERENCES

1. Zenie, F. H., Amer Lab, 51, Feb (1985).

2. Papas, A. N., Alpert, M. Y., Marchese, S. M., Fitzgerald, J. W., and Delanye, M. F., Anal Chem, 57, 1408 (1985).

3. Hawk, G. L., Little, J. N., Zenie, F. H., Amer Lab, 96, June (1982).

INTERACTION BETWEEN A ROBOTIC SYSTEM AND LIQUID CHROMATOGRAPH - HPLC CONTROL, COMMUNICATION AND RESPONSE

Kevin J. Halloran and Helena Franze
Boehringer Ingelheim Pharmaceutical Corporation
Danbury, CT 06810

ABSTRACT

The system discussed is an automated content uniformity analysis of multiple dosages of a pharmaceutical tablet. A Zymate System prepares a sample solution and directly injects it into a liquid chromatograph. A series of interactions have been programmed into the system that allow the System to determine the status of the analysis. The System tracts baseline response, identifies peak retention times, estimates peak response and determines the reproducibility of multiple injections. Also, the System evaluates sample response to determine if the sample meets specifications.

INTRODUCTION

The benefits of laboratory automation have been discussed time and again. The more efficient the robotic system is, the more effective will be its role in the laboratory. The robotic application discussed herein is a Zymate System that is used to conduct content uniformity determinations for multiple potencies of a pharmaceutical tablet.

Samples are prepared by the robot for HPLC analysis. Some form of communication between the robot and the HPLC is, of course, desirable. Shutting down the HPLC after the analysis is complete would conserve solvents, recorder paper, UV lamp life, pump seals, etc. A means of transmitting data signals from the HPLC to the robot would be invaluable. By examining the output from the HPLC detector, the robot can evaluate the status of the anaylsis and track the chromatography to insure that a stable baseline is maintained. The robot could detect system malfunctions such as a pump shutdown or detector lamp instability, and subsequently cease to inject samples. Ideally, the robot could shut down the system, thereby conserving solvents, samples and consumable items (i.e. pipette tips). To be complete, an error message could automatically be printed, flagging the malfunction.

A more sophisticated means of communication could give the robot access to chromatographic peak responses. Based on these peak responses, the robot could determine the reproducibility of multiple injections. Ultimately, the robot could compare sample and standard peak responses to determine if the smaple meets specifications. If a sample is judged to be out of specification, the robot could respond accordingly:

A) The sample could be reinjected.

B) The standard could be reinjected.

C) Another sample could be prepared.

D) The sample could be flagged with a message sent to the printer.

Criteria for USP content uniformity testing of tablets calls for the testing of ten tablets. All ten tablets must fall within 85% to 115% of label with relative standard deviation not greater than 6%. If one tablet is out of specification or if the RSD is >6% then twenty more tablets must be tested. Conceivably a robotic system could detect outliers and automatically test an additional twenty tablets.

In an effort to set up a communication link between the Zymate System and the HPLC, we explored several possibilities. One option was to use the Zymate computer interface linked to a microcomputer accessing a chromatographic integrator. As with many large laboratories, our chromatographs are linked to a main laboratory automation system. In our opinion, it would be counter productive to implement a separate data handling system for the robot. In light of the current FDA interest in computer validation, complete validation of a separate system would be necessary. The man-hours spent on validation, along with the actual cost of hardware (computer interface, microcomputer, integrator, etc.) and software (external or internal) makes this option impractical in our situation. The second option explored was to provide a link between the robot and the laboratory computer (Hewlett Packard 3357). Our final conclusion was that development of any elaborate communication link would require a substantial commitment of time and money. Such an endeavor might best be left to Zymark and Hewlett Packard. The third option was to link the robot to the HPLC via the analog input on the Zymate injector station and "simulate" the actions of an integrator. The actual data collection would still be performed by the Hewlett Packard 3357. The robot would estimate peak responses and use these approximations as a basis for the decisions described above. The advantages of this option are:

1) no additional hardware is needed

2) the programming is simple relative to programming a micro or mini computer

3) since the main laboratory computer is still collecting the data upon which the final results are generated, no additional computer validation is necessary

The obvious drawback to this approach is that the Zymate HPLC interface is not an integrator and therefore is not as accurate. Also, the HPLC interface does not have peak identification capabilities. The data generated by the HPLC interface is used only as a means of evaluating the analysis. The data generated by the HPLC interface need only approximate the real data sufficiently enough to be a reliable base for this evaluation.

While planning the communication programming, the following objectives were set as goals:

1. The robot could have the ability to turn any component of the HPLC on and off.

2. The robot would have the ability to switch mobile phase when needed.

3. In the event of an HPLC malfunction, the robot would not continue to prepare and inject samples.

4. The robot would monitor the chromatographic baseline and respond to an unstable baseline by either rezeroing the detector, pausing the injections until a stable baseline is reached or, if a stable baseline cannot be reached, shutting down the system.

5. The communication programs must be able to handle a multiple levels of active concentration. This system will prep samples of three potencies (0.1, 0.2 and 0.3 mg of active/tablet).

6. The robot would determine if the system meets suitability criteria.

THEORY

Data acquisition by the Z310 is performed by a series of eight registers. Each register collects output from the detector over a specific time interval. The signal collection is not continuous over the entire time interval. Instead, each register accumulates output from the detector at discreet instances during the interval. The size of the time interval, or window, is the product of the number of these instances and the time elapse, or time base, between each reading. There are limitations to the use of the Z310 as a data collector. The Z310 is not an integrator; therefore, it does not have inherent peak detection capabilities. Second, there are limitations to the quantity of signal that can be accumulated during each window (no more than 32,999 response counts), the size of the time base (not less than 0.1 seconds) and the quantity of readings per window (not more than 100 readings). We found that a signal response greater than 32,999 is transformed into a negative number. Conditions must be maintained such that the size of the response in any one window does not approach this limit.

Since the Z310 is not capable of identifying a peak automatically, its usefulness would be severely limited unless a method of peak identification could be developed. Chromatographic parameters, such as column condition and mobile phase variability, may alter the chromatography from run to run. In fact, the chromatography may change over the course of one run. For this reason the data acquistion parameters must be flexible and change in response to the changing chromatography. Also, a means of calculating a total peak response that is flexible to changes in chromatography must be included. The data acquisition programs must therefore include a means of identifying a peak, tracking a peak from injection to injection, and deriving a total peak response that simulates the actual peak area determined by the laboratory computer.

Data Acquisition Window Setup

Setup of the data acquisition windows are automaticlaly done by injecting two standards prior to the actual analysis. These injections (called the "scope" injections) approximate the peak retention time and zero the windows in on the peak.

In this particular anlaysis, the active peak elutes at anywhere from 1.75 to 3 minutes. Prior to one minute there is a very large solvent front. The first window starts one minute after the injection and each window is 30 seconds wide (number of readings = 100, time base = 0.3 sec). The windows occur in succession; the second window begins where the first window ends (1.5 min.), the third window begins at 2 minutes, etc. The eight windows span the range of from one minute to five minutes after the injection (Figure 1).

After the first chromatogram is complete, the window with the largest response is identified. The robot then starts a second scope injection with the windows set as follows (Figure 2):

1. The windows are ten seconds wide and occur in succession.
2. The start of window number four is set equal to the start time of the window with the largest response in the first scope injection.

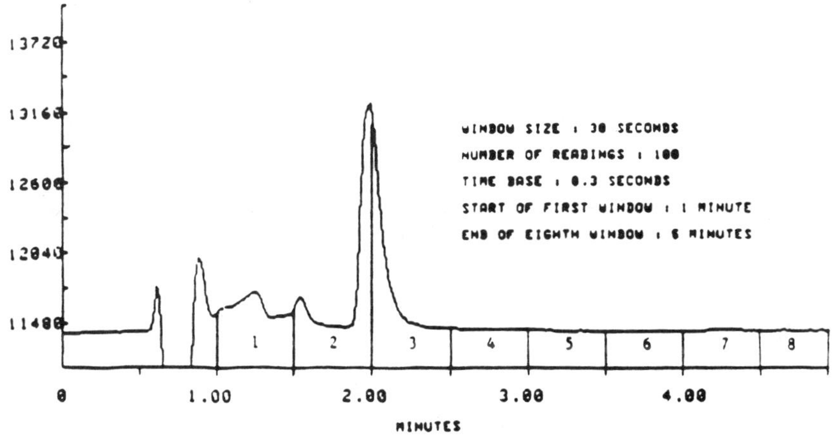

FIGURE 1. First "scope" injection, data acquisition windows.

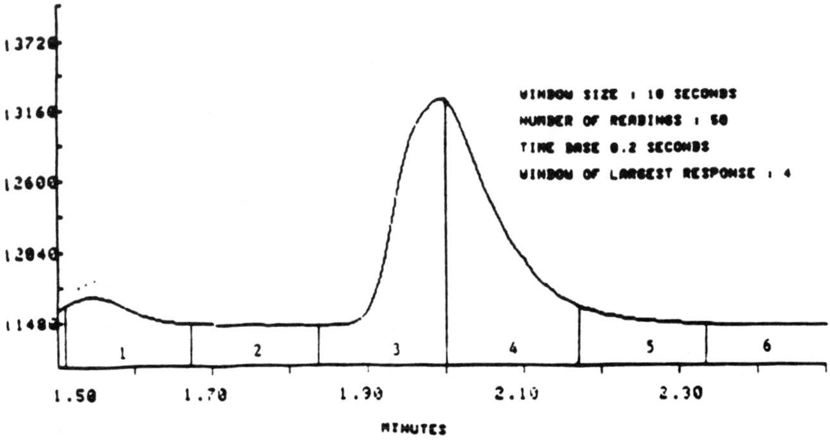

FIGURE 2. Second "scope" injection, data acquisition windows.

After this chromatogram is complete, the window with the largest response is again identified. The final windows are set up as follows (Figure 3):

1. The start of window number four is set equal to the start time of the window with the largest response in the second scope injection.
2. If the first samples to be analysed are 0.1 mg active/tablet or 0.2 mg active/tablet then each window is set five seconds wide (number of readings = 50, time base = 0.1 sec).
3. If the first samples to be analysed are 0.3 mg active/tablet then the windows are 7.5 seconds wide (number of readings = 75, time base = 0.1 sec).
4. The windows occur in succession.

There are different size windows for the higher potency samples to insure that no individual window response approaches the 32999 response limit.

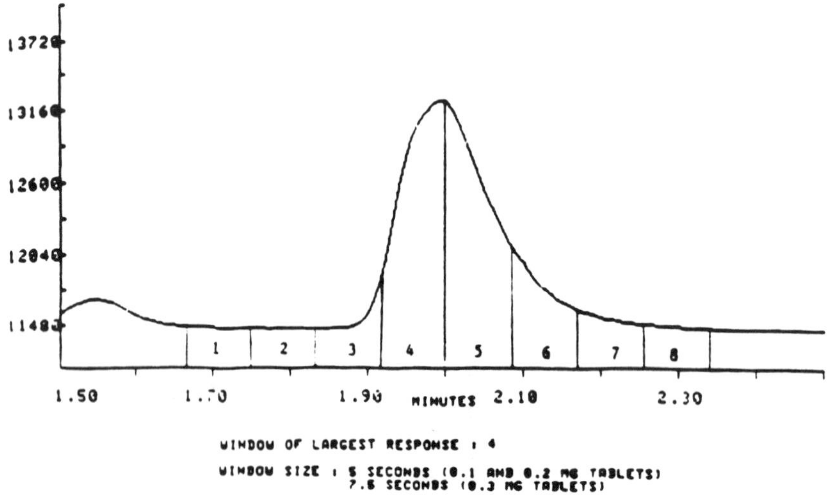

FIGURE 3. Standard injection, data acquisition windows.

Data Acquisition

There are three functions included in the final data collection program:

1. The program identifies baseline drift.

2. The program detects shifts in peak retention time and shifts the windows accordingly.

3. The program calculats a total peak response that simulates the actual peak area derived by the lab computer.

This simulated response will be the basis that the robot will use in its evaluation of the analysis.

Response to retention time drift is accomplished by determining the window with the largest response and, if that window is not window number four, shifting all windows such that window number four for the next injection begins at the time at which that window began. The program, therefore, responds to changes in retention time in the injection subsequent to the injection where the change was detected. In a case where the peak retention time is near the end of one window and the start of the next, the windows may shift back and forth from one injection to the next. If no window response is greater than 10% of the lowest response, the chromatography is not representative and the windows do not shift. If a radical shift occurs, a message is sent to the printer flagging that injection.

Several methods of approximating peak response were examined. A simple, yet reasonably effective, calculation was developed. The program has already identified the window of largest response. Each window prior to the window of largest response is then examined and the window with the lowest response prior to the peak is found. This process is repeated with the windows appearing after the window of largest response. Any window that appears before the first low window or after the second low window is disregarded (Figure 4). The two low windows are then compared to determine the background response. Data stored in all windows between and including the two low windows are summed. The background response is then multiplied by the number of windows included in that summation. The difference between these two values is the calculated peak response.

Baseline Monitor

Baseline monitoring is a vital function of the system. By monitoring the baseline, we are insured that in the event where conditions change

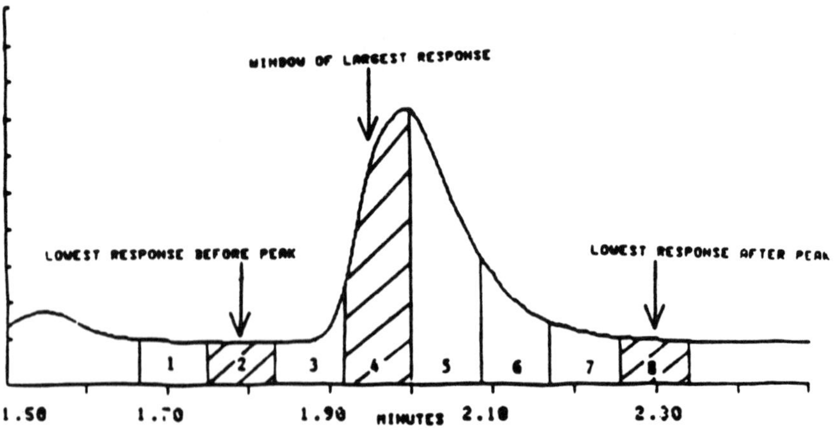

FIGURE 4. Data acquisition window response breakdown.

drastically, such as a pump shutdown, the robot will cease to inject and, if needed, shut down the entire system.

After the liquid chromatograph has been turned on and stabilized for a period of time, the detector is zeroed by the robot. The baseline is then monitored by one five second window. This value is stored in the variable LAST.BASELINE. After ten minutes another baseline reading is taken and stored in the variable BASELINE. If the BASELINE value differs from the LAST.BASELINE by less than ten percent, then the system is ready for injection. If the difference is greater than ten percent, the sytem is stabilized for an additional fifteen minutes and the process is repeated. After each baseline evaluation, the value stored in BASELINE is transferred to LAST.BASELINE.

When the system has met baseline suitability requirements, the system proceeds with the scope injections. Another baseline evaluaiton is performed prior to the start of the actual analysis. After each injection, the low window response is stored in BASELINE, and a baseline evaluation is performed. In effect, the baseline is checked after each injection by comparison to the baseline value from the injection directly prior to it.

The following conditions were set as guidelines for baseline acceptability:

1. If the difference between LAST.BASELINE and BASELINE is less than ten percent, the baseline is stable.

2. If the difference in baseline is greater than twenty percent, there is a serious drift and the system shuts down. In this case, an error message is sent to the printer.

3. If the change in baseline is greater than ten percent but less than twenty percent, a warning message is printed. In this case, LAST.BASELINE is not replaced by BASELINE and another sample is injected. If this new BASELINE differs from LAST.BASELINE by less than 15% and the peak response is consistent, the analysis continues. If the difference is greater than 15% and the peak response is not consistent, the system shuts down and an error message is printed.

EXPERIMENTAL

The Zymate System includes the following stations:

1. Robot arm and controller
2. Master lab station
3. Power and event controller
4. Vortex station
5. Analytical balance, Mettler AE-160

The liquid chromatograph includes the following:

1. Detector, Kratos Spectroflow 773
2. Pump, Perkin Elmer Series 10
3. Recorder, Perkin Elmer R100
4. Injector, Zymate Z310
5. Solvent selector, Autochrom 101

The data is collected and processed by a Hewlett Packard 3357 laboratory information system. A layout of the system can be seen in Figure 5.

The liquid chromatograph is integrated with the Zymate robot as seen in Figure 6. The detector power is controlled by the power and event controller via the A.C. outlet. The detector can be zeroed by a switch closure on the power and event controller. The pump and recorder chart drive are started and stopped by power and event controller switch closures. HPLC interface switch closures control the solvent selection valve and create a chart mark on the chart paper at each injection. A switch closure also initializes data collection by the laboratory computer.

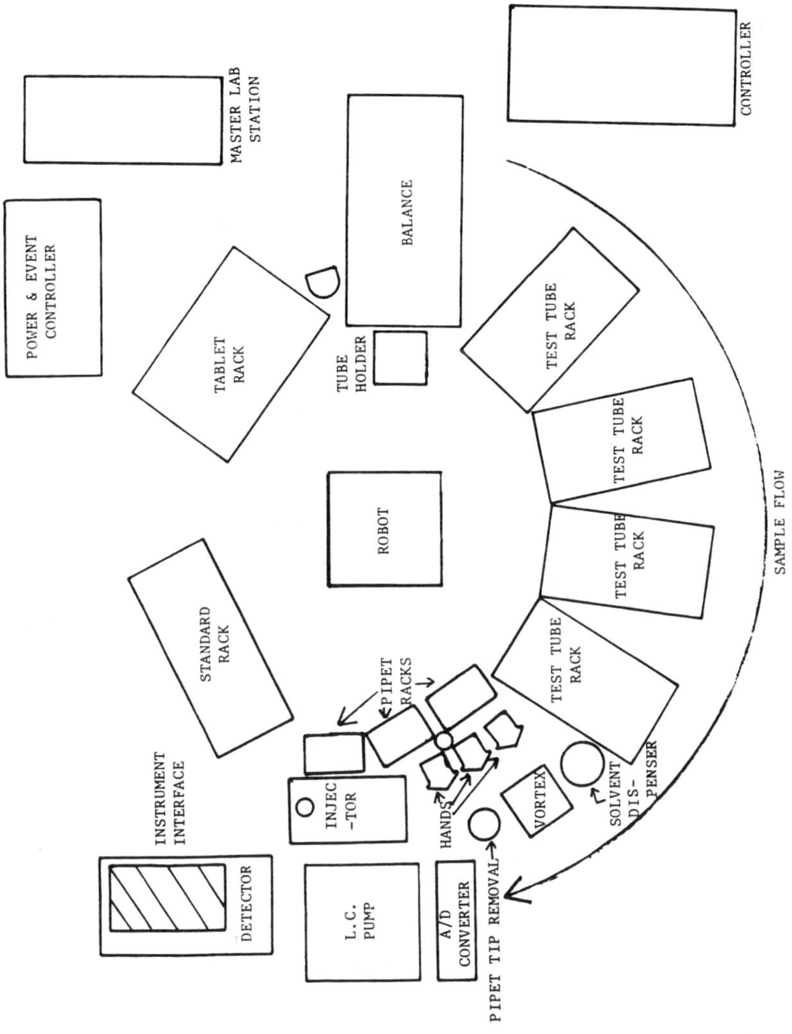

FIGURE 5. Content uniformity system configuration.

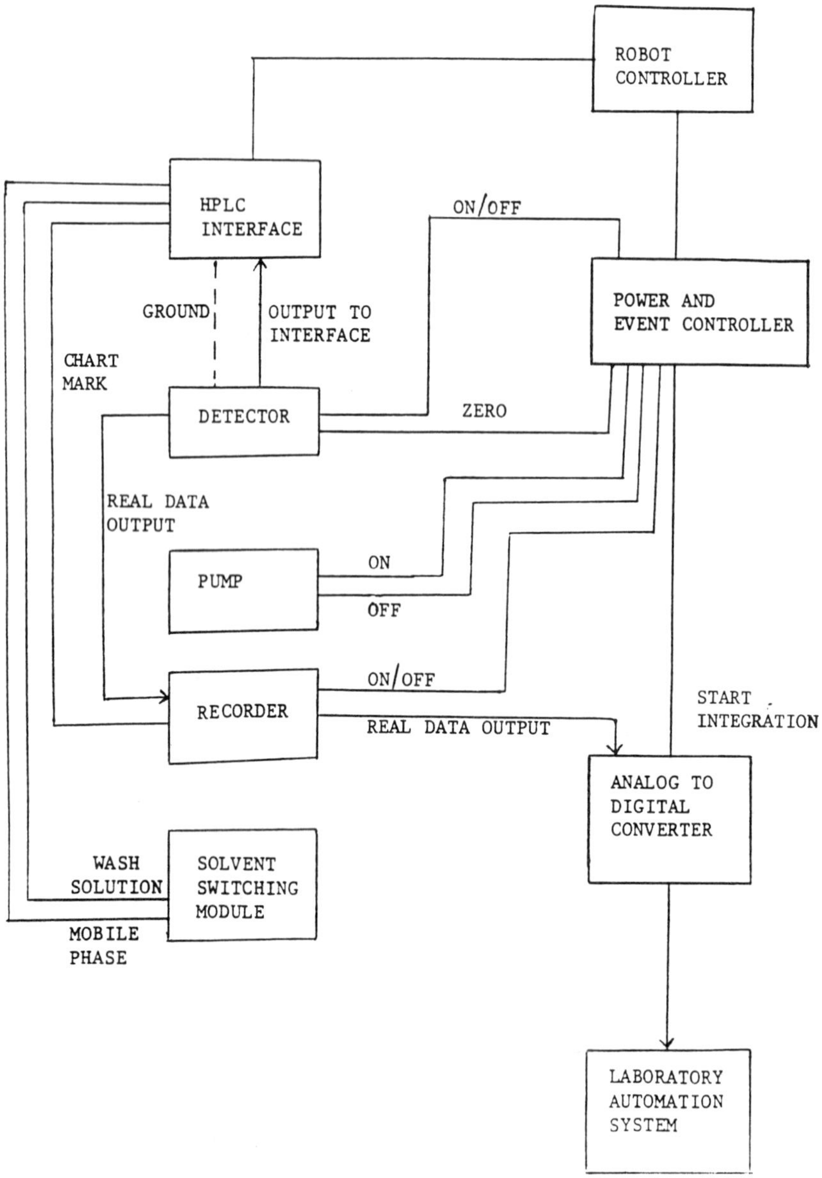

FIGURE 6. Equipment interface for HPLC and a Zymate System.

Output from the detector is channeled in two directions. The detector is attached to the recorder, which in turn, is connected to an analog to digital converter. Real data collection and processing is performed by the laboartory computer. The detector output is also channeled to the analog input on the Z310 HPLC interface. Data collected by the interface is then sent to the controller for use in internal decisions. To eliminate external noise, the cable from the detector to the interface is as short as possible. The cable is also shielded and a diode is used to drain off any charge accumulating at the analog input.

The chromatographic conditions used in this analysis (Table 1) are less than ideal. The flow rate of 3.0 mL/min results in a backpressure of over 3000 psi. An attenuation of 0.030 AUF at a wavelength of 220 nm challenges the sensitivity of the detector. Even at this low attenuation the peak size of the 0.1 mg/tablet potency is small (about 7000 area counts). Finally, the peak elutes on the tail of the solvent front. These conditions make data collection by the robot more difficult than would be the case in the majority of routine analyses.

The robot is programmed to analyse all three potencies of tablets during the same run. Before initializing the analyses the analyst is stepped through a series of instructions by the robot. The majority of the instructions concern the proper setup of the system. The analyst also enters sample information such as lot number, potency and number of tablets per lot. The robot is programmed to bracket each sample lot with two sets of standard injections. Since the robot has associated each lot of tablets with the proper potency, the robot will inject the

TABLE 1. Chromatographic Conditions.

column:	C8 (15 cm x 4.6 mm x 5 micron)
mobile phase:	aqueous/acetonitrile/methanol (16:3:1 v/v/v)
aqueous phase:	water/triethylamine (200:1 v/v) adjusted to pH of 5.8 with phosphoric acid
pump flow rate:	3.0 mL/minute
detector wavelength:	220 nm
detector attenuation:	0.030 AUF
injection size:	50 microliters
analysis time:	5 minutes

proper concentration standard. For example, if four lots of samples are to be run (two of potency one, one of potency two and one of potency three) the analysis would be performed as seen in Figure 7.

DATA WINDOW SETUP
↓
STANDARD SUITABILITY
↓
LOT #1 POTENCY #1
↓
TWO POTENCY #1 STANDARDS
↓
LOT #2 POTENCY #1
↓
TWO POTENCY #1 STANDARDS
↓
TWO POTENCY #2 STANDARDS
↓
LOT #1 POTENCY #2
↓
TWO POTENCY #2 STANDARDS
↓
TWO POTENCY #3 STANDARDS
↓
LOT #1 POTENCY #3
↓
TWO POTENCY #3 STANDARDS
↓
SYSTEM SHUTDOWN

FIGURE 7. Flow diagram for multipotency tablet analysis.

Our laboratory has adopted a policy for evaluating system suitablity. At the beginning of each HPLC run two standard preparations are injected repeatedly. The first standard, termed primary standard, is injected twice. The second standard, termed secondary standard, is then injected twice. These injections are then repeated for a total of eight injections (4 primary and 4 secondary). Each peak response is then normalized for standard weight. The system meets suitability requirements if the relative standard deviation between all normalized peak responses is not greater than two percent. The robot has been programmed to determine an approximate response for each peak, normalize each response for weight and calculate the relative standard deviation between the normalized responses.

As a side note, error messages have been carefully added to the sytem in an attempt to make the system more user friendly. For instance, if a tablet is dropped during the tablet pour, the robot will immediately abort that sample preparation; send a message to the printer identifying the malfunction, lot number and tablet number; and immediately go on to the next sample. This insures that now erroneous sample preparation is injected into the liquid chromatograph. Also, any sample that the robot determines to be out of specification by comparison to the standard peak responses is flagged by a warning message sent to the printer identifying the lot and tablet number.

At the end of the analysis, or in the event of a serious malfunction, the robot executes a system shutdown routine. The robot turns off the detector and recorder and switches the solvent selection valve to pump a system wash solution. A debubbling device traps any air that may enter the system during the solvent switching. After the system is flushed for thirty minutes, the pump is shut down. The mobile phase reservoirs are sealed to insure that airbound particulates cannot enter the system and that volatile components of the mobile phase do not evaporate.

After each lot is analysed, a summary report is dumped to the printer. This report lists the tablet weights, mean tablet weight, standard deviation, high and low values. In the event of a system malfunction, a message will flag the problem. If the system can continue, as in the case of a dropped tablet, it will. If the problem is in a vital step, such as the malfunction of the balance door opener or shutdown of the HPLC pump, the run will be aborted and the sytem will shut down.

RESULTS

Ideally, a plot of the peak response as calculated by the robot verses the computer generated peak area should produce a linear relationship. In Figures 8 and 9 this data is plotted for the 0.2 mg/tablet and 0.3 mg/tablet potencies. Although the relationships are not exactly linear, a significant linearity is demonstrated (correlation coefficients of 0.97 and 0.95 respectively). Once again, it should be stressed that the robot peak response calculation is a model and only approximates the actual peak areas. There does appear to be a bias between the two responses; the computer generated peak areas are consistently larger in value. This is misleading since the two responses are not of the same units. The magnitude of the robotic responses are not important; the change in response as a reflection of the change in actual peak areas is

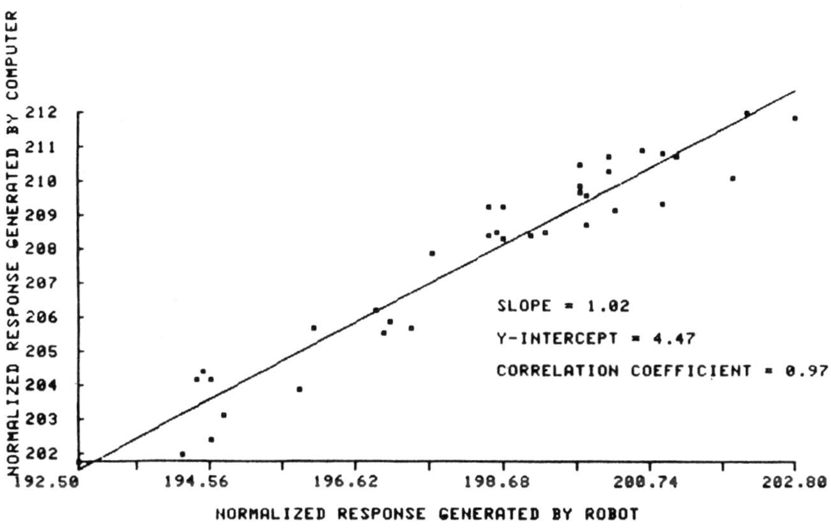

FIGURE 8. Comparison of normalized peak response as generated by the computer and by the robot for 0.2 mg active/tablet potency.

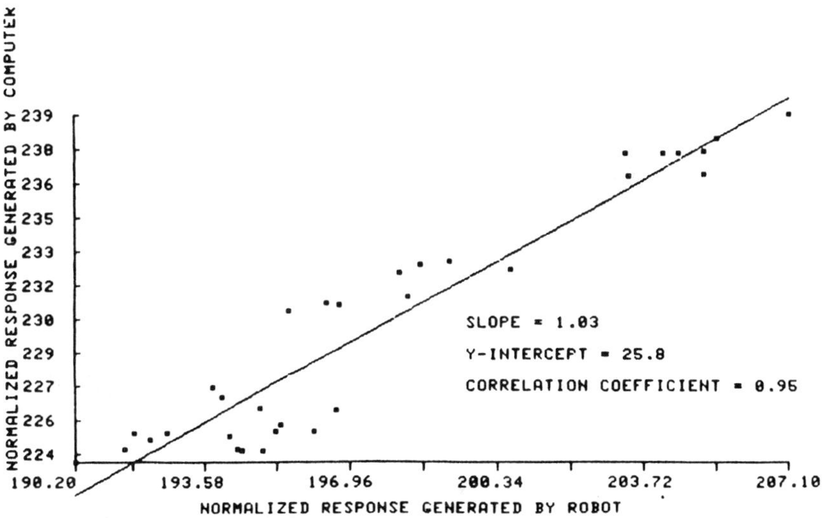

FIGURE 9. Comparison of normalized peak response as generated by the computer and by the robot for 0.3 mg active/tablet potency.

the important parameter. This bias is constant for each potency and can be eliminated by a simple calculation.

To determine if the robot peak responses are representative, standard suitability results based on these responses were generated. Figure 10 illustrates typical standard suitability patterns as calculated by the computer and by the robot. Clearly, the peak areas generated by the computer are mimicked by the robot. In Figure 11 the robot response pattern has been superimposed over the computer response pattern. Although the response patterns are not exactly identical, the robot response pattern does closely simulate the changes in peak area from injection to injection.

In Tables 2, 3 and 4 actual standard suitability results as generated by the robot and by the computer are compared. To illustrate the responsiveness of the system over a wide variability in reproducibility, standards were purposely prepared that would produce a wide range of relative standard deviations. Typical standard suitability RSD's are normally less than 1.25%. In the case of the 0.2 and 0.3 mg/tablet potencies, the robot normally reproduced the actual standard suitability to within 0.1% RSD.

The last data pair in Table 2 should be highlighted as an extreme test of the system. In this case an injection problem causing partial injection volumes resulted in standard RSD of 19.3%. The robot calculated an RSD of 18.9%, demonstrating the system's response to extremely irreproducible injections. Data pair #8 in Table 3 is an

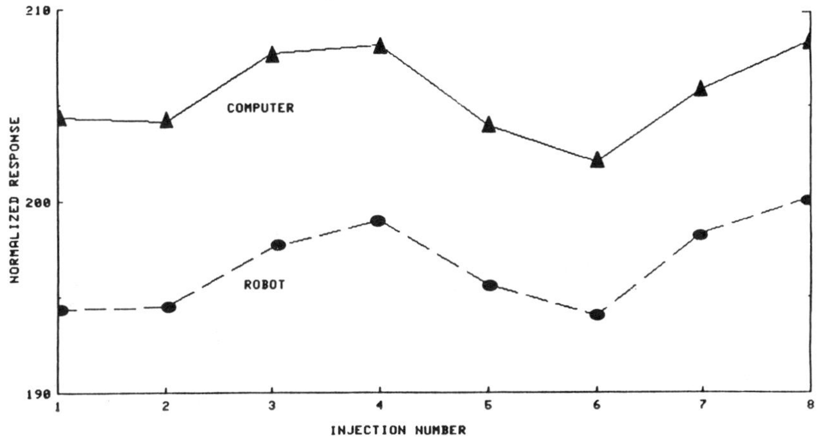

FIGURE 10. Normalized standard response patterns as generated by the computer and by the robot.

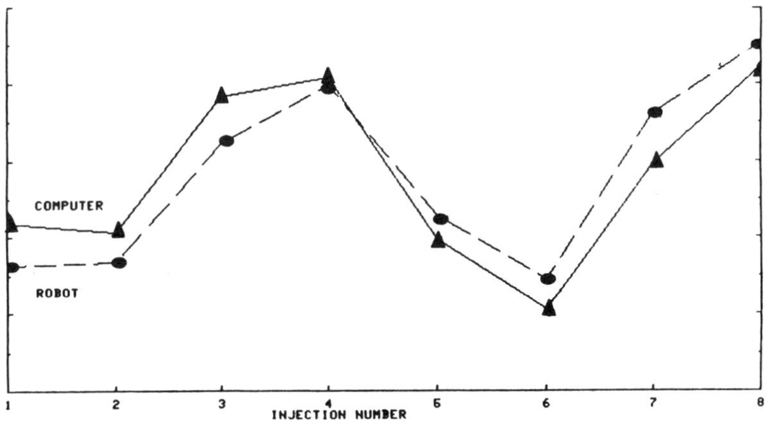

FIGURE 11. Robot response pattern superimposed over computer response pattern.

TABLE 2. Comparison of Relative Standard Deviation Between Response of Standards as Generated by the Robot and the Laboratory Computer System. Tablet Potency = 0.3 mg Active per Tablet.

Robot	Computer	Difference	% Difference
0.71	0.63	0.08	12.7
0.76	0.77	0.01	1.3
0.84	0.82	0.02	2.4
1.14	1.22	0.08	6.6
1.34	1.46	0.12	8.2
1.52	1.50	0.02	1.3
1.96	1.85	0.11	5.9
2.78	2.86	0.08	2.8
2.84	2.95	0.11	3.7
18.94	19.31	0.37	1.9

TABLE 3. Comparison of Relative Standard Deviation Between Response of Standards as Generated by the Robot and the Laboratory Computer System: Tablet Potency = 0.2 mg Active per Tablet.

Robot	Computer	Difference	% Difference
0.33	0.25	0.08	32.0
0.37	0.45	0.08	17.7
0.55	0.50	0.05	10.0
0.71	0.71	0.00	0.0
0.98	0.91	0.07	7.7
1.16	1.13	0.03	2.6
1.38	1.31	0.07	5.3
2.07	1.87	0.20	10.7
3.02	2.79	0.23	8.2
3.09	2.82	0.27	9.6

TABLE 4. Comparison of Relative Standard Deviation Between Response of Standards as Generated by the Rboot and the Laboratory Computer System: Tablet Potency = 0.1 mg Active per Tablet.

Robot	Computer	Differences	% Differences
0.36	0.33	0.03	9.1
0.45	0.38	0.07	18.4
0.68	0.51	0.17	33.3
1.25	1.01	0.24	23.8
1.92	1.73	0.19	11.0
2.16	2.02	0.14	6.9
3.22	2.85	0.37	13.0

example of an instance where the robot would reject a good standard suitability as being greater than 2%.

It should be pointed out that the results for the 0.1 mg/tablet potency, as seen in Table 4, were more varied than for the other potencies. We believe that the limit of sensitivity of the HPLC interface is being challenged by this low a signal. We are currently investigating ways of increasing the signal to noise ratio for this potency.

DISCUSSION

The HPLC interface has proven to be a suitable means of communication between the robot and the liquid chromatograph. Although it is not accurate enough for generation of actual test data, it will generate approximate data for use in evaluating the status of the analysis. One advantage of this method of feedback over other methods is that no additional equipment is necessary. Also, since the data manipulation is done by the robot controller software, the programming is much less complex to develop. Finally, since the actual data is collected by the laboratory computer and the robot generated data is only used for internal decisions, formal computer validation is not necessary.

We are currently investigating new additions to our robotic content uniformity system. One future plan is to fully automate the data collection programs so that the robot will set the most suitable window parameters based on the scope injection data. We also plan to expand the robot control of the liquid chromatograph to include flow rates, attenuation, wavelength and column selection. Our ultimate plans are for the sytem to have the ability to run completely different analyses in series.

ACKNOWLEDGEMENTS

The help and encouragement of Zymark Corporation, and especially Mr. Art Martin, are greatly appreciated. The help of Ms. Nirmal Aggarwal, Mr. James Hannigan and Mr. Gregory Wilmes of Boehringer Ingelheim is also recognized.

MULTI-PRODUCT ROBOTIC SAMPLE PREPARATION
IN THE PHARMACEUTICAL QUALITY ASSURANCE LABORATORY

C. Hatfield, E. Halloran, J. Habarta,
S. Romano, W. Mason
Ortho Pharmaceutical Corporation
Raritan, NJ 08869

ABSTRACT

A Zymark robotic arm and its modules were designed and purchased with multi-product sample preparation in mind. This paper disucsses the equipment, unit operations, validation and programming techniuqes utilized for a suppository sample preparation application. Future applications for a variety of product types is also briefly discussed.

INTRODUCTION

As a Quality Assurance Laboratory in the pharmaceutical industry, we are charged with performing assays on products in an accurate and timely manner. To help us perform our function, we rely on automation. The first steps toward automation were sample introduction into the chromatographs (autoinjectors) and data reduction and reporting (laboratory data system). The most labor intensive portion of an assay is sample preparation, and this is where the Zymate System arm has contributed to our laboratory.

The system described in this paper is our second Zymate System. The first Zymate System is used to prepare content uniformity assays for oral contraceptive tablets. This entails two preparations (one for HPLC samples and one for GC samples) which are similar in unit operations.[1] For our second system, we decided to set up the arm and the components so we could perform many of our solid dosage form preparations on the same table. These dosage forms include suppositories, large tablets and granulations. Keeping in mind all potential applications, we programmed the system to perform a suppository assay first since this was the most intricate procedure and incorporated elements of all future assays we will be performing. This paper describes the equipment, unit operations and the validation of the suppository assay and will include some discussion of the planned future applications.

EQUIPMENT

All the modules purchased or designed for this table were done so with many applications in mind. The modules include: (Figure 1)

Controller (not shown)
Power and Event Controller (A)
Mettler AE 160 Balance with Interface (B)
2 Master Lab Stations (C)
2 Vortex Mixers (D)
Linear Shaker (E)
Sonicator (F)
Centrifuge (G)
Capping Station (H)
Evaporation Station (I)
2 Custom Cap Dispensers (J)
2 Custom Cap Seats (K)
Custom Dispensing and Aspirating Station (L)
2 Custom Transfer Racks (M)

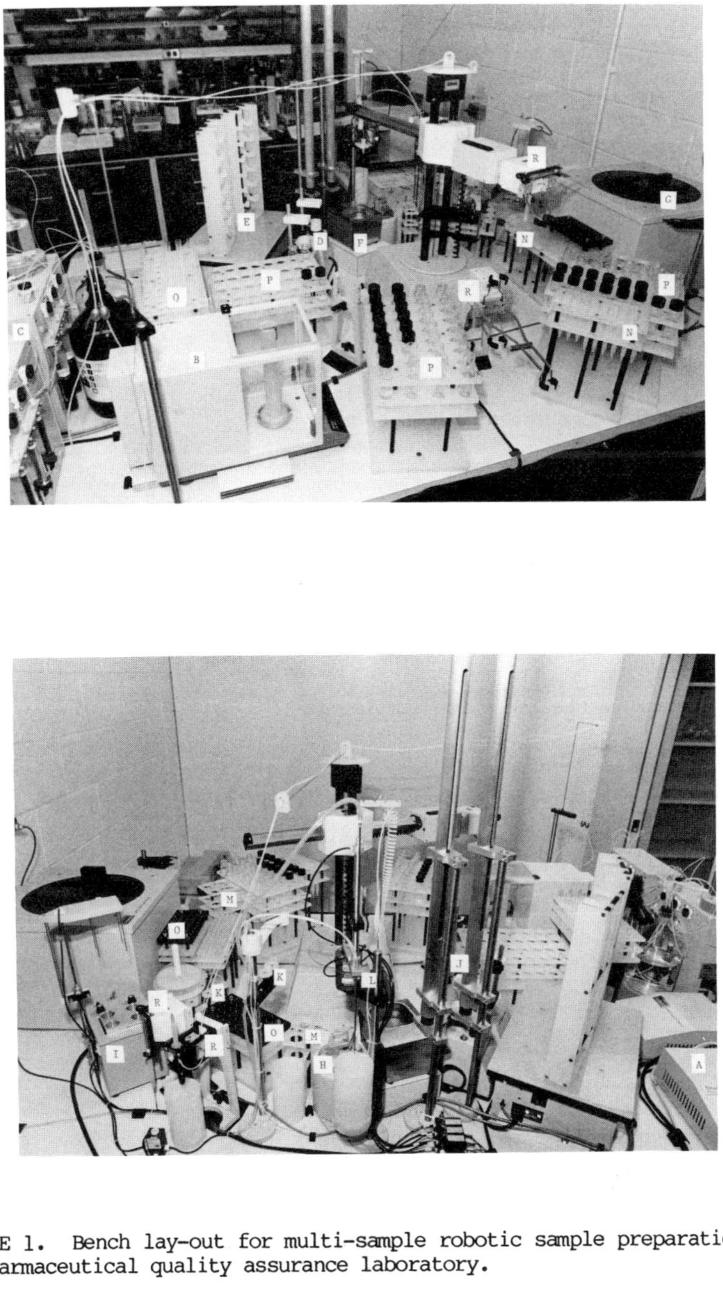

FIGURE 1. Bench lay-out for multi-sample robotic sample preparation in the parmaceutical quality assurance laboratory.

2 Pipet Tip Racks (N)
2 Dense Packed Vial Racks with Templates (O)
3 Dense Packed 50 mL Centrifuge Tube Racks (P)
Sample Test Tube Rack (Q)
4 Hands (R)
LC Injector Station (not visible)

This section describes the custom cap dispensers, the evaporation and extraction stations and the custom hands that are being used. Further comments about custom pieces will be made in the section describing the unit operations.

All sample preparations are performed in 50 mL centrifuge tubes. The samples are weighed, vortexed, shaken, etc. in these tubes. Normally, the tubes would be placed in their racks with the caps on. The arm would then take a tube to the capping station, remove the cap, park the cap and proceed with sample preparation. To save time, cap dispensers were designed. Using a cap dispenser, the tubes are placed in a rack uncapped, the sample is weighed into the tube and solvent is dispensed before the cap is placed on the tube.

The cap dispensers consist of a long metal tube with a platform at the end. Caps are placed in the top of the tube and are gravity fed to the bottom. The capping fingers remove a cap and the space is filled with a new cap. We employ two dispensers which hold approximately forty caps each. If an assay requires a subdilution into a second centrifuge tube, there must be approximately double the number of tubes and caps as samples. The cap for the tube used for the subdilution is taken from the second dispenser.

We employ a unique system of pneumatics and a peristaltic pump to perform our evaporation and extraction functions. Compressed air is routed through a series of two multiport valves whose operation is controlled through the power and event controller. For the evaporation station, not only is the air used to evaporate the sample controlled through these valves, but also the placement of the manifold over the samples is controlled pneumatically. The manifold is attached to a linear actuator which moves the manifold down over the samples as air is applied.

The dispensing and aspiration station utilized two pneumatic actuators and a peristaltic pump. The dispensing and aspirating station is essentially the capping station where the tube sits while liquids are dispensed into or aspirated out of the sample. Three cannulas are attached to a rotary actuator which moves the cannulas over or clear from the tube. Two of these cannulas are for dispensing and are connected by tubing to the master lab station (see Figure 2). The third cannula is used for aspiration of solvent out of the sample and this is attached to a linear actuator and a peristaltic pump. After the rotary actuator brings the cannulae around over the sample tube, the linear actuator is activated which brings the aspiration cannula into the sample (see Figure 3). The other end of this cannula is connected to tubing which goes to a peristaltic pump. The pump is then activated and the solvent is aspirated out of the sample, the tube can then be capped and the sample continues in the preparation scheme.

The system employs four hands. The first is a vibration hand with three

FIGURE 2. Rotary actuator for dispensing solvent into sample tube.

FIGURE 3. Linear actuator for aspirating solvent from the sample tube.

sets of fingers (see Figure 4). In the center of the flange block is a set of pouring fingers set ten inches away from the block. The distance between the fingers and the block provides two advantages. First, we can access the furthest position in the sample tube rack from any arc, while still employing two other standard sets of fingers on the same hand (the other fingers do not hit the rack preventing access). Secondly, the length of the fingers allows us to employ a high pulse rate in the vibration hand which translates to a smooth, fine pouring action at the end of the fingers.

The second hand employed is a multifunctional hand consisting of a 1 mL syringe with a 3 inch long 1 mL pipet tip adapter and a set of heavy duty capping fingers. Next is a modified liquid distribution hand (see

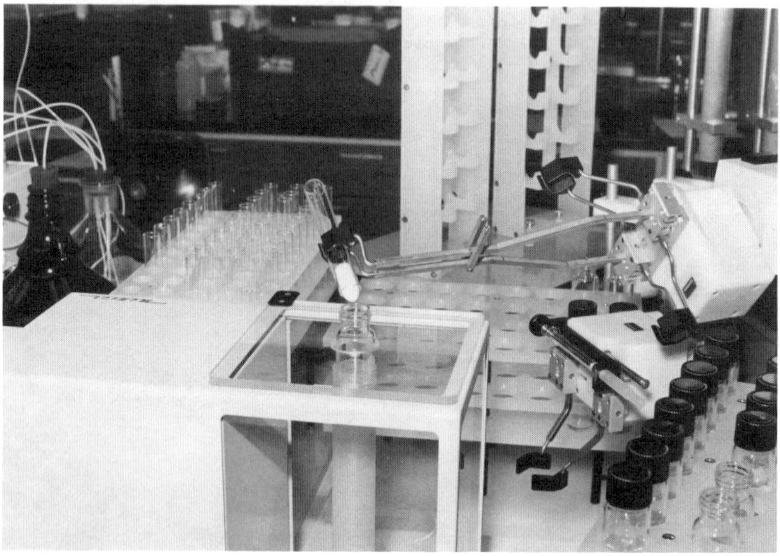

FIGURE 4. Vibration hand for pouring suppository sample with 3 sets of fingers.

Figure 5). Here the mounting plate has been extended to match the reach of the other hands.

In this fashion, the racks can be placed so that the innermost and the outermost positions of each rack can be accessed by every hand. Attached to this mounting plate are two pipet tip adapters. One is a rubber 5 mL pipet tip adapter for pipetting. A sample is drawn into the pipet tip and an accurate amount is delivered to a second tube. To accurately deliver a specified amount to this second tube, a tight seal must be maintained between the pipet tip and the adapter. Rubber was used instead of the usual hard plastic to compensate for minor differences in pipet tips which can cause a loose seal. One hundred eighty degrees to this adapter is a 1 mL pipet tip adapter which has been lengthened so that with a pipet tip attached, it can reach to the bottom of 50 mL centrifuge tube if necessary.

The fourth hand employed is a high torque syringe hand (see Figure 6). The syringe is equipped with a cannula whose total length is 122 cm so it can reach the bottom of a centrifuge tube. Since this syringe is used to fill previously capped LC or GC vials, the cannula is heavily reinforced to prevent flexing. Two-thirds of its length is plastic and the upper portion of the metal has a reinforced shank. The end of the cannula is pointed and a small vent tube has been added to relieve pressure while filling capped vials.

UNIT OPERATIONS

As previously mentioned, the first application transferred to the

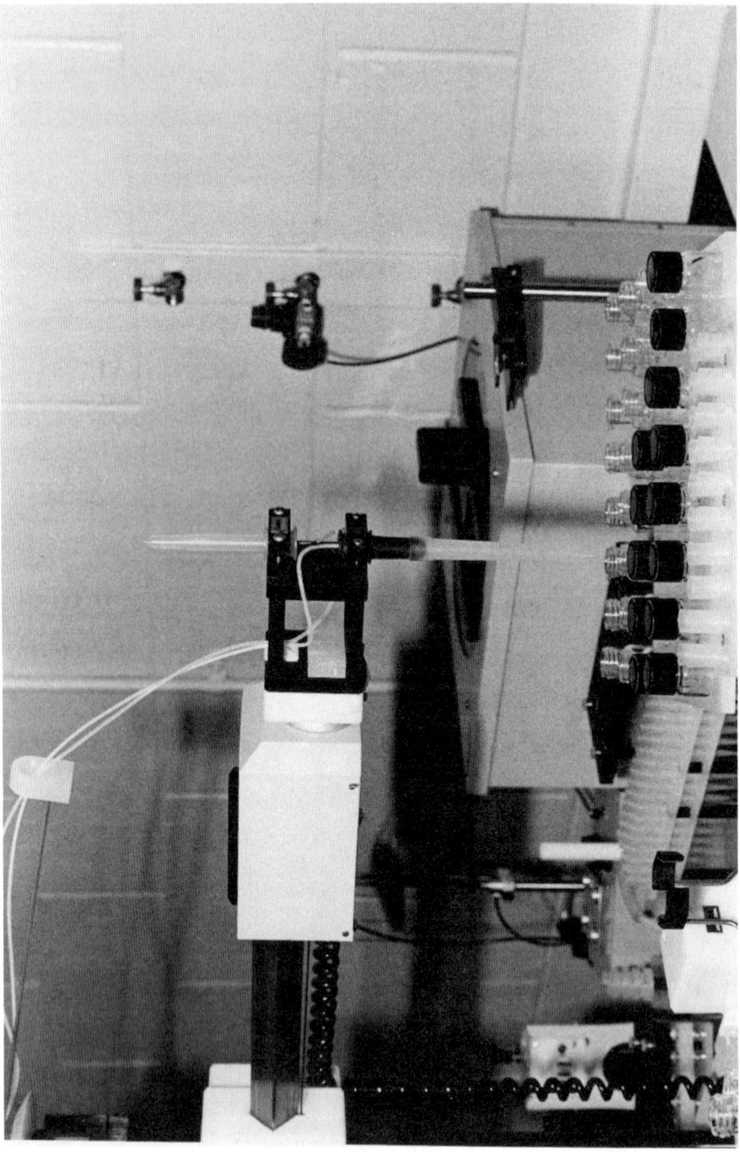

FIGURE 5. Modified liquid distribution hand.

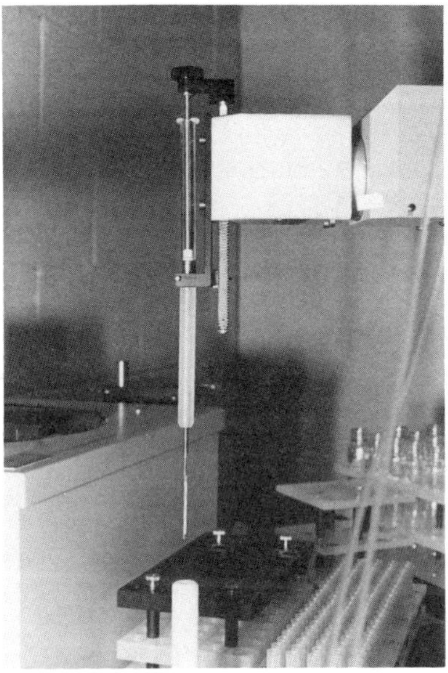

FIGURE 6. High torque syringe hand.

robotic arm was our MONISTAT Suppository assay. Table 1 lists the unit operations performed for a single suppository content uniformity assay.

The first operation performed is weighing the suppository. The arm takes a 50 mL centrifuge tube, places it in the balance and waits for the balance to tare. A sample tube (16 x 100 mm test tube containing a suppository) is then picked up with the pouring fingers and the sample is poured with vibration into the tube in the balance. The sample tube is replaced into its original rack. After the weight of the sample is recorded, the centrifuge tube is moved to the capping/extractiion station. Here the dispensing cannulae move over the tube and pentane is added. The arm retrieves a cap from the cap dispenser, caps the tube,

TABLE 1. Unit Operations for MONISTAT Suppository Assay

- WEIGH SUPPOSITORY
- DISPERSE SUPPOSITORY
  - Add 30 mL Pentane
  - Vortex and Centrifuge
  - Withdraw Pentane
- EXTRACT EXCIPIENT MATRIX (repeat 4 times)
  - Add 20 mL Pentane
  - Vortex and Shake
  - Centrifuge and Withdraw Pentane
- EVAPORATE RESIDUAL PENTANE
- RECONSTITUTE SAMPLE
  - Add 40 mL Methanol
  - Sonicate and Shake
  - Sonicate and Vortex
- PIPET SAMPLE ALIQUOT TO SECOND TUBE
- EVAPORATE TO DRYNESS
- ADD INTERNAL STANDARD
  - Vortex and Sonicate
- FILL CAPPED VIALS USING SYRINGE HAND

and places it into the vortex to disperse the sample. After dispersion, the tube is moved to the centrifuge and spun down (a balancing tube has previously been placed in position 3 of the centrifuge). The sample is returned to the capping/extraction station where the cap is removed and held while the pentane is aspirated out (using the peristaltic pump) and fresh pentane is added. This begins the Extract Excipient Matrix operation.

In this particular assay, the active ingredient is insoluble in pentane while the excipient matrix is soluble. Repeated washings with pentane are performed to isolate the active ingredient. Since the active and residual excpients are packed in the bottom of the tube when centrifuged, the sample is not only vortexed, but also placed on the linear shaker to insure dispersion after a fresh addition of pentane. The linear shaker has been fitted with towers to accommodate 50 mL centrifuge tubes. The packing effect caused by the centrifugation also allows us to withdraw "used" pentane without fear of removing some of the active. The aspiration cannula has been set so its depth in the sample tube remains above the solid material and always leaves some pentane behind.

Once the excipient matrix has been removed (four extractions after initial dispersion), the residual pentane must also be removed. The hand parks the cap in a cap seat which has been designed to hold a cap in either a normal or inverted position. Since the cap will be reused on the same sample, the cap is parked in an inverted position while the tube is in the evaporation station. The hand retrieves the tube from the capping station and places it into the evaporation station. The pneumatic valve is actuated and the manifold comes down into the sample. The valve is switched and air starts to flow over the sample.

When the residual pentane is evaporatead, the tube is retrieved from the evaporation station and placed again in the capping station. Methanol is added from one of the cannulae on the rotary actuator to dissolve the active ingredient. The cap is retrieved from its parked position and replaced on the tube. The sample then goes through a series of manipulations to insure dissolution of the active. The tube is held in the sonicator, shaken on the linear shaker, sonicated again and finally placed in the vortex mixer.

At this point a quantitative aliquot must be taken, dried and reconstituted to prepare for the chromatography. The sample tube has been taken from the vortex mixer, placed in the capping station and had its cap removed and parked. The arm then retrieves a second 50 mL centrifuge tube from another rack and places it in a small (2 position) transfer rack next to the capping station. The arm parks the multifunctional hand and picks up the liquid distruibution hand. In this case the volulme to be transferred is 2 mL, so the pipet tip adapter picks up a 5 mL pipet tip from its rack. It should be mentioned that the line from the pipet tip adapter back to the master lab station is filled with liquid to allow more reproducible delivery by reducing the total volume of air in the system. The pipet tip is inserted into the sample solution and a portion of the sample is drawn up into the tip. This is expelled and another portion is drawn up. We withdraw 0.5 mL more than is needed into the pipet tip. The tip then moves over the second tube and the accurate amount is delivered into the tube. The hand moves over a waste container, expels the rest of the liquid in the tip, shucks the tip at a shucker and parks the hand where the end of the adapter is flushed with solvent. This flushing is performed not only to prevent any cross-contamination that may occur, but also to fill the line totally with solvent again so each delivery begins with the same amount of air in the system.

The second tube containing the sample aliquot is placed into the evaporation station. While the sample is drying, the arm recaps the original sample tube and places it in its original rack in its original position. After the aliquot is dried, the tube is moved from the evaporation station into the capping station where an aliquot of internal solution is added via a movable remote dispenser. The tube is capped with a new cap from the dispenser, and its contents are vortexed and sonicated. The tube is placed in the capping station; the cap is removed and parked. The arm parks the multifunctional hand and retrieves the syringe had. An aliquot is taken from the sample and injected into a previously capped GC vial in the GC vial rack. After parking the syringe hand, the multifunctional hand will replace the cap on the second sample tube and return it to its original position. In this fashion, samples are accountably saved for further treatment if necessary.

This details the preparaton of one suppository for assay. We are also preparing standards along with each set of samples and have programmed the arm to do a suppository composite assay. There are minor differences in these preparations. For the standards, the analyst accurately weighs reference material into a 50 mL centrifuge tube and places it in the prescribed position in the rack. The arm picks up the standard tube and takes it to the capping station where the program picks up from the reconstitute sample operation; here, 20 mL of methanol is added instead of 40 mL. The rest of the operations remain the same for the standards and the samples.

In the composite assay, four suppositories are poured individually into one 50 mL centrifuge tube and the weight is taken. This composite sample is carried through the sample preparation adding more time for the initial dispersion of the suppositories and two more repeats of the excipient extraction steps. At the pipet sample aliquot to second tube operation, a subdilution is made by pipetting 4 mL to the second tube, adding 10.5 mL methanol to the second tube and then taking the 2 mL aliquot to a third tube to dry down and continue. As can be seen, all efforts were taken to keep each perparation as similar as possible.

VALIDATION

In validating the robotic system for this assay, two areas of concern were addressed. First, is the system accurate and precise in its measurements - weighing, volume dispensing and volume pipetting? Secondly, how do the results obtained for samples preapred on the robot compare to results obtained for these same samples prepared manually?

The weighing validaton was performed by placing standard weights into the sample test tubes in the sample rack and having the arm pour the weight into a previously tared 50 mL centrifuge tube in the balance. This mimics the way a sample is actually weighed. Each weighing was repeated twenty times and the results are listed in Table 2.

TABLE 2. Robot Weight Validation.

WEIGHT USED (mg)	$\bar{x}_{20}$ (mg)	C.V. (%)
100	100.1	0.277
500	500.1	0.042
1000	1000.0	0.023
2000	2000.0	0.010

The solvent delivery validation was perfomed by having the arm retrieve a capped 50 mL centrifuge tube from a rack, tare the tube in the balance, place it in the capping station, remove the cap, dispense the liquid into the tube replace the cap and return the tube to the balance for weighing. The volume delivered was calculated by dividing the weight delivered by the density of that liquid at the temrperature measured. All three solvents used in this assay were validated at the respective volumes used. The pentane and methanol were delivered from the fixed cannulae attached to the rotary actuator, while the chloroform-methanol mixture was delivered from the movable remote dispenser once the hand parked the tube cap. The results of the solvent delivery validation of listed in Table 3.

TABLE 3. Robot Solvent Delivery Validation.

SOLVENT	NOMINAL VOLUME DISPENSED (mL)	ACTUAL VOLUME DISPENSED (mL) ($\bar{x}_{20}$)	C.V. (%)
Pentane	20.0	19.9	0.247
	30.0	29.9	0.227
Methanol	10.0	10.0	0.018
	15.0	15.0	0.015
	20.0	20.0	0.018
	40.0	40.0	0.024
$CHCl_3$:MeOH (1:1)	2.0	2.0	0.278
	4.0	4.0	0.248

The differences in the coefficients of variation between the pentane, chloroform-methanol mix and the methanol are due, in the case of pentane, to the difference in volatility and compressibility of pentane versus methanol and, in the case of the chloroform-methanol mix, to the smaller volumes being dispensed.

The third measurement we validated was the pipetting function where the pipet tip adapter picks up a pipet tip, withdraws some sample from a tube and dispenses an accurate amount into a second tube. Here, the arm weighs an empty capped tube, takes it to the capping station, removes and parks the cap, parks the gripper hand, picks up the liquid dispersion hand, retrieves a pipet tip, withdraws some methanol from a nearby beaker and dispenses the amount needed into the tube. The arm then changes hands after shucking the pipet tip, recaps the tube and brings it to the balance for a weight determination. The operation of pipetting is performed as it is in the assay where an initial amount of liquid is brought up into the tip, expelled and then 0.5 mL more than is needed is drawn into the tip. A 1 mL pipet tip was used for the 1.0 mL pipetting while a 5 mL pipet tip was utilized for the other two volumes. The results of this validation are listed in Table 4.

TABLE 4. Robot Pipetting Validation.

SOLVENT	NOMINAL VOLUME OF ALIQUOT (mL)	ACTUAL VOLUME OF ALIQUOT (mL) ($\bar{x}_{20}$)	% RECOVERY	C.V. (%)
Methanol	1.0	0.996	99.6	0.329
	2.0	1.990	99.5	0.091
	4.0	3.985	99.6	0.121

The method validation consisted of preparing three lots of suppositories for content uniformity assay and two lots for composite assay. All five lots had previously been assayed by analysts in our laboratory. The results of the content uniformity comparison are listed in Table 5.

TABLE 5. MONISTAT Suppository Assay-Robot vs. Analyst.

	ROBOT		ANALYST
		% LABEL CLAIM	
Content	101.7		100.6
Uniformity	105.2		105.1
	99.1		99.6

The validations run for the suppository assay show the robotic arm to be accurate and precise in its measurements and demonstrated equivalency between robotic and analyst sample preparation.

DISCUSSION

The greatest asset of the robotic arm system is its inherent flexibility. The equipment and programming were designed before purchase to allow a variety of products to be assayed on one table. In this section, some of these design features will be discussed.

As can be seen from the overview diagram (Figure 1), the table is crowded. In order to have so many racks and have the arm access every position in every rack, most of our racks are dense packed in one or both directions. For example, the racks which hold the 50 mL centrifuge tubes are dense packed in one direction. The LC and GC vial racks are dense packed in both directions. The 1 mL pipet tip rack has one hundred positions and can be rack indexed as two 50 position racks. In this fashion, for example, half of the tips can be used for subdilution while the other half can have filter tips attached and be used to filter samples before chromatography.

Two small transfer racks were added to the table to provide an intermediate position for tubes when necessary. One two-position rack, next to the capping station, is used to hold tubes during a dilution operation. The other small rack has four positions and is next to the centrifuge. This rack holds tubes which will be used for balancing the centrifuge against a sample containing tube.

The sample test tube rack can be changed to hold varying sizes of test tubes. The third 50 mL centrifuge rack will only be used in those assays requiring a third clean tube for a subdilution.

The types of modules and the configuration of the racks on the table allows us to perform any combination of sample weighing, dispersion (vortex, shaker and/or sonicate), extraction, dilution, reconstitution and filtration. The final samples can be saved in capped vials or injected into an HPLC.

Of course, the other element that makes the table so flexible is the programming. We employ many small subroutines that direct unit operations. By stringing many of these small routines together any combination of the above mentioned functions can be performed.

We are using three levels of serial programming for this system--simple, semi-complex and complex. The validation for the suppository assay used simple serial programs. We are planning to use mainly complex serial programming for our sample preparations. In complex serialization, samples are being continuously introduced to the system as the previous sample completes a unit operation. In this assay, one sample repeats the extract excipient matrix operation several times using a number of modules. Therefore, a second sample cannot directly follow the first in a complex serial fashion since those modules would also be needed at the same time by the second (or third) sample. For the routine suppository assay we will use semi-complex serialization. Thus, we are preparing the suppositories in groups of two where the second sample will be one step behind the first. To expeditite the preparation, operations will be performed on the second sample while the first is in a time-consuming operation (such as dispersion, centrifugation or evaporation). Also, since most of our assays utilize liquid chromatography, future samples will be injected into an LC at the end of the sample preparation. The residual sample can be saved in an LC vial if desired. MONISTAT suppositories are assayed with gas chromatography and the samples are therefore injected into precapped GC vials for later transfer to an autosampler.

We would like to comment on the productivity gains we will make by automating the suppository preparation on the robotic system. An analyst can prepare one batch of suppositories for content uniformity assay (10 samples) in one day. The robotic system can prepare three batches in the same twenty-four hour period. Not only is the robotic system freeing the analyst to perform other duties, but considering we receive over one hundred batches of suppositories in our laboratory for assay a year, the productivity gains and time savings are evident.

Our immediate plans for this system include validating a second dosage strength of suppositories and programming for a composite capsule assay. The capsule assay is a "dilute, filter and shoot" assay and will involve powder pouring and LC injection. Long term plans include programming the sample preparaton for at least five other products, including dosage forms such as large tablets, capsule and tablet granulations. With this system, we believe we are well on our way to automating the sample preparations of all our solid dosage forms and, in the process, realizing the productivity gains needed to compete in today's pharmaceutical industry.

REFERENCES

1. Habarta, J. G., Hatfield, C. and Romano, S., Am. Lab., 17(10), 42 (Ocotber 1985).

THE EXTENSION OF
PHARMACEUTICAL ANALYSIS AUTOMATION
USING ROBOTICS

Brian P. Hatton, Peter Abley and Timothy J. Lux,
Beecham Pharmaceuticals, U.K. Division,
Worthing, West Sussex, England

ABSTRACT

Analytical automation in pharmaceutical laboratories is well advanced - for example, continous flow analysis, autosamplers and programmable data-handling. The remaining barrier to full automation of such analyses is the preparation of solutions from solid samples. This is still a labour intensive, costly operation and is subject to human fallability.

This paper will describe a flexible robot system for the preparation of β-lactum antibiotic solutions to a pre-programmed concentration and their subsequent analysis.

The sample type - included in a bar coded label - dictates all quantitative events. Serially diluted solutions are presented to the appropriate enalytical system. When the data are complete, a final report is generated automatically.

The system nucleus is a two-armed Zymate System. configured so that the combined working area forms a figure of eight. By scheduling tasks between the two arms, a highly efficient solid and solution handling capability is obtained.

INTRODUCTION

In this paper, I intend to review the applicability of robotics to quality assurance within the pharmaceutical industry.

After a brief introduction to our view of the benefits to be obtained from the use of robotics in quality assurance, I shall discuss some of the barriers to full automation in the laboratory and our solution to these problems. I will then discuss in detail our novel use of a two-armed robotic system to fully automate a routine laboratory procedure. Finally, I shall summarize the current status of our system, some suggested improvements to the Zymate System and future advances into other innovative areas.

The high level of quality assurance required within a pharmaceutically based company can represent a very significant percentage of product cost. The field of laboratory automation has already been fruitful for such companies in controlling a major part of Q.A. costs - that of routine testing of product samples. Indeed all modern quality assurance laboratories have sophisticated automatic anlaytical systems at their disposal.

All operations required in analysis of a solution, including reporting the data and printing the resulting certificate, can now be carried out without manual intervention. Unfortunately, most pharmaceutical products, actives and ingredients or excipients are not solutions. They are mainly solids. All of the antibiotics manufactured in my part of

Beecham are solids. The major single labour intensive (and therefore costly) step in the analysis of these antibiotics is the initial preparation of those all - important solutions. The successful extension of laboratory automation to include this early stage of solid sample analysis has therefore become our primary goal.

The Quality Control laboratory is unique in that each day a predictable, relatively small number of samples of each type is presented for analysis. These samples may be from a range of production areas and therefore contain a wide variety of different active ingredients. The various analytical techniques for this range of products require differing concentrations of solution for optimum sensitivity. In addition, many β-lactam products require special handling owing to their inherent instability. Nevertheless, analytical data must be generated rapidly and efficiently to maintain a healthy low stock level in a cost-effective manner and to provide quality feedback to manufacturing units.

From these factors it becomes clear that our goal is not dedicated automation, but a flexible system for automated solution preparation and handling. The necessary flexibility for this task can only be obtained using a robotic system.

The potential benefits of robotic assistance in the routine testing of pharmaceuticals are:

Reduced Costs: An unattended robotic system which is able to carry out a task at a speed equivalent to that of a human operator will result in considerable savings when used on a 24-hour basis. Although only the initial capital outlay is required, this must be carefully justified to maintain the credibility of robotics in general.

Reduced Analysis Time: With faster sample turnabround, stock levels in quarantine awaiting release can be reduced, improving cash flow. Further reductions in overall sample analysis time can be obtained by making relatively rapid serial dilutions of a single concentrate rather than multiple weighings of the solid.

Accuracy and Precision: The rigidly controlled chemistry resulting from robotusage will naturally afford improved precision of results. In addition, improved analytical precision resulting from the removal of human unpredictability will reduce the need for analysis in multi-plicate. This will then contribute to increased sample throughput. Accuracy may also be improved, depending on the mode of operation of a particular system and analysis.

Routine Tasks: Qualified technicians are currently required to spend part of their working lives carrying out routine tasks. Transfer of those tasks to a robot will release them for more challenging investigations suited to their skills and experience.

Hazard Avoidance: Some analyses involve the use of hazardous solvents, regants or conditions. Unnecessary contact with chemical and pharmaceutical products may need to be controlled. Pharmaceuticals, of necessity and by design, have pronounced physiological effects. Both this situation and the reverse - where a product is required in an especially pure form with no possibility of handling contamination - are amenable to a robotic solution.

At the time our initial assessment was taking place, only one general purpose-built laboratory robot was available - the Zymate System. The alternative, purchase of a small industrial robot and construction of a system using modules from individual manufacturers, was not viable within the time and resources constraints of the project and would have had a higher probability of failure.

Since the Zymate System also would need the addition of innovative modules to meet our requirements, and our robot would be the first such system in the U.K., a full Government grant of one third of the overall project cost was obtained.

DISCUSSION

The automation project selected took advantage of all the items listed earlier and simultaneously gave the installation a high likelihood of becoming financially viable, conforming to Good Laboratory Practices (GLP) and therefore being acceptable to the regulatory authorities. Almost all pharmaceutical actives (including the Beecham antibiotics) require analyses on every batch to include measurement of pH, purity and less frequently specific rotation. The pH is usually measured using a relatively concentrated solution (up to 10%), whereas purity and specific rotation require a lower content (0.2 to 0.01%) for optimum results.

The purity of β-lactam antibiotics is routinely measured by continuous flow analysis using an Autoanalyzer. Antibiotic solutions are poured into small cups and sampled at regular intervals by a dip tube as a carousel revolves. Intermixing of the different samples is eliminated by air segmentation and in liquid zones a sequence of chemical reactions takes place. After the air bubbles have been removed, the liquid stream passes into a spectrophotometer. Sample response may either be plotted on a chart or transmitted direct to a microcomputer for data manipulation. This type of system can reliably analyse about ten solutions per hour with little manual involvement.

The scope of the project, therefore, was to weigh and dissolve a quantity of the antibiotic to a pre-programmed concentration, then measure its pH. Serial dilution of this concentrate would provide a solution of the correct concentration for measurement of purity, andpossibly specific rotation, by interfacing with existing laboratory technology. Any available sample preparation time could be used for determination of additional parameters.

Several major barriers to a successful project outcome could be forseen. These, if not surmounted, would severely limit the usefulness of the system.

Sample Identity

Good Laboratory Practice demands that sample identity is rigorously maintained throughout analysis. All robot movements - and therefore sample movements - are essentially totally predictable within a validated program. However, the potential for loss of identity is relatively high when large sample racks are being loaded. This problem is rarely encountered in manual analysis of individual samples.

We have eliminated these difficulties, removing the need for technician data entry fo any kind, by use of bar coding, together with confirmation of sample presence by use of the bottle itslef to close a simple micro-switch. This system allows mixing of sample types in any sequence and even addition or insertion of extra samples at any time during analysis.

Immediately before sample preparation, the sample type bar code is read using a laser device as the bottle rotates on a turntable. This bar code dictates all future quantitive events. A second bar code identifying the sample is then read; this batch number is sent direct to the data output computer where it is matched with analytical results; and a printed record is produced.

We view this facility for positive identification of samples as essential for regulatory acceptance of a robotic system of this type. Consequently, bar coded labels are now added to all samples at source, reducing errors at all points of data transfer.

Weighing System

Various strategies are in use for automated weighing of powders. They include use of disposable weighing boats and funnel arrangements. Both of these utilise a volume approximation to the correct weight. However, we have elected to take the simple efficient route of direct tipping from the sample bottle into a tared container using a vibratory hand and balance feedback. Output from the balance controls the approach to a target weight by increasing or decreasing the angle of tip of the sample bottle. The prototype weighing hand produced in our own workshops has been superceded by a high performance product of Zymark's parallel development.

Liquid Dispensing

Under controlled conditions, a syringe pump can deliver a volume of solvent with great accuracy. However, when working for long periods unattended and even more so when carrying out serial dilutions, the system is potentially unreliable. To obtain consistent, convincing results we use weight/weight procedures throughout, checkweighing at each stage. The minimal errors associated with weight measurements afford excellent reproducibility.

Standardisation

Both the pH meter and Autoanalyzer require standardisation. The pH meter has been modified for direct program control. In the calibration mode, standard solutions are pumped through the pH flow cell and the set points adjust automatically. Closure of a relay takes the meter into a measurement cycle and results are sent direct to the data computer. Autoanalyzer calibration is achieved by robotic preparation of solutions from barcoded standards using their own distinct preparation routines.

Speed of Operation

The operation sequence described earlier, if performed in a single cycle by a robot arm surrounded by workstations is reliably estimated to require a time of greater than 15 minutes. This is considerably longer than the manual preparation time, and even in conjunction with the envisaged 24-hour operation would not have achieved the project objectives. A secondary consideration is that the number of distinct workstations required for 24-hour operation could not easily be sited within reach of one robot arm.

These limitations have been surmounted by inclusion of a second arm in the system, configured so that the working areas of the two arms overlap to form a figure of eight. By scheduling tasks between the two arms, using concurrent software, a highly effective solid and solution handling capability is obtained, giving a mean time for sample preparation of around 6 minutes. This introduces a further difficulty. It is now possible for both arms to move simultaneously into the overlap between their two areas of activity. Collisions of this type can be prevented either by hardware or software means. The hardware solution of incorporating detector beams is clumsy since robot movement is already under program control. The software method is more elegant, involving setting of a flag when either arm is in the overlap zone. This causes negligible operational delays since a flag check is required only when specific movements which would enter the overlap are carried out.

The sample preparation and analysis procedure will now be outlined to illustrate the mode of operation of this novel system.

Robot Arm 1 (Figure 1) places an empty bottle on the balance and this is tared automatically. For the initial weighing step, a four place balance is required to economise on usage of standards. Weighings of around 50 milligrams are then possible with acceptable precision.

The first sample is taken and the micro-switch check carried out to ensure that it has been located. Such checks are interspersed throughout the routine.

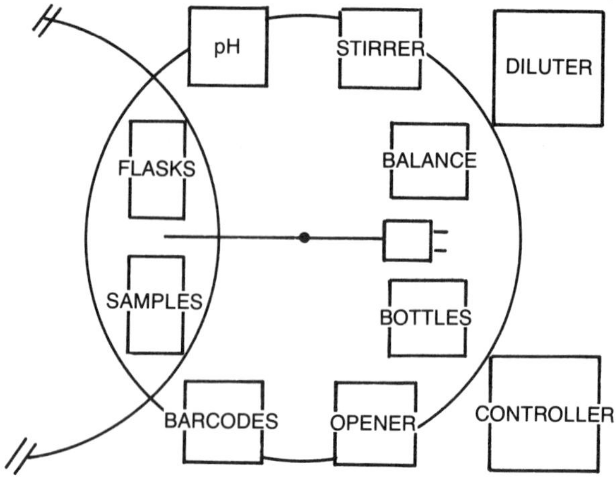

FIGURE 1. Robot arm 1 for sample weighing and pH measurement.

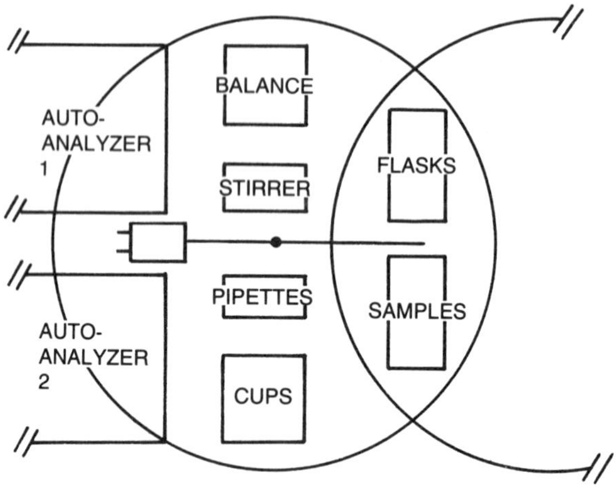

FIGURE 2. Robot arm 2 for sample dilution and introduction into Autosampler.

The sample is placed on the bar code reader turntable and from the bar code the appropriate preparation subroutine is selected. The bottle is opened and direct vibratory weighing carried out as described earlier.

The bottle is then resealed and replaced in the rack, giving the possibility of repeat analysis if necessary.

The volume of liquid required to reach the pre-programmed correct concentration is calculated automatically and dispensed. The total weight is then measured to ensure quantitive analysis and the suspension stirred to effect dissolution. The time required for this stage varies with sample type. For those which are slow dissolving, the preparation of sample 1. Most penicillins, however, are fairly rapidly water soluble and the procedure continues.

To avoid risk of contamination from the pH sampling probe,, part of the concentrate is first pipetted quantitively, again by weight, into a flask. Arm 1 then completes the cycle by measuring the pH of the remaining concentrate while Arm 2 (Figure 2) dilutes the solution in the flask on a three place top loading balance. After stirring to ensure distribution of the active ingredient, the dilute solution is transferred by pipette into Autoanlyzer cups and supplied to the correct analysis system - again dependent on the bar coded sample type. To ensure synchronisation with the availability of diluted solutions, the autosampler is under direct Zymate System control. As you see here we currently have two distinct Autoanalysis systems ready to accept samples of different types.

This last step involving filling of Autoanalyzer cups, although occupying a large part of the operating time of Arm 2, has been retained temporarily. The cups and Autosampler will be replaced by a three position sampler, dipping directly into the sample and standard flasks, as the next phase of development. In addition to providing free time for Arm 2, this will also clear the necessary space around that Arm for further techniques of analysis to be installed - for example measurement of specific rotation.

As a final step of the procedure, Arm 2 carries out a flask wash ready for re-use. By this means a very large flask storage rack is eleminated.

Current Status

Validation studies to demonstrate equivalence of pH and purity values from manual and robotic methods have been successful. These studies also confirm that the programmed sequence of events gives reliable positioning and recovery of equipment at each stage of the operation.

A direct comparison of manual pH and robotic pH is shown in Table 1. In all cases the agreement is very good, giving a correlation of greater than 0.999 and a standard deviation of residuals of 0.03.

TABLE 1. pH Data.

Sample	Manual	Robotic	Residual
1	3.78	3.81	-0.03
2	4.26	4.23	+0.03
3	4.61	4.59	+0.02
4	5.33	5.36	-0.03
5	5.82	5.82	0.00
6	6.11	6.10	+0.01
7	6.95	6.99	-0.04

Results comparing manual and robotic purity determinations of Flucloxacillin of a range of purities artificially achieved by sampling the product at various stages of drying are shown in Table 2. Again in all cases there is good agreement between manual and robotic assays, giving a correlation coeficient of 0.983 and a standard deviation of residuals of 0.3.

Full validation data, involving many more samples, has been generated by use of the Zymate in parallel to the existing system of manual routine analysis over a period of several weeks.

TABLE 2. Purity Data.

Sample	Manual	Robotic	Residual
1	91.5	91.7	-0.2
2	92.8	93.1	-0.3
3	93.8	93.6	+0.2
4	94.5	94.4	+0.1
5	95.1	95.6	-0.5
6	95.6	95.5	+0.1
7	95.6	96.1	-0.5
8	95.9	95.4	+0.5
9	96.3	96.4	-0.1
10	97.5	97.3	+0.2

Speed of Operation

I illustrated earlier some of the measures we have found necessary to reach an acceptable speed of operation. The introduction of variable speed would dramatically improve the viability of Zymate Systems for all applications and ease the problem of financial justification.

Disc Operating System

As a result of the complex programming and safeguards required for our pharmaceutical application, the current program occupies 58K of memory. Our original intention had been to call a program from disc for each preparation rather than jumping to a subroutine as necessary with the Zymate System. There are many more products and analyses we would wish to handle without manually re-loading programs from disc. A more sophisticated disk operating system is required.

Bar Code Interface

Since there is no bar code facility in the Zymate System any system requiring these must have a microcomputer to control the device and send data to the controller. I have illustrated earlier that we consider bar coding to be essential. The necessary interface should be available as standard.

At this point, I must add that we are very pleased with our Zymate System. These points would improve the system, and encourage a broadening of the natural application development. Pereliminary contacts with several regulatory bodies have not any serious flaw in our thinking and system design.

FUTURE APPLICATIONS

In conclusion, I should like to mention a few possible future applications for laboratory robotics. Some of these we shall ourselves be progressing.

Complex Analyses: Where an analytical technique is not only labour intensive but also requires concentration and a highly qualified operator, robotisation should be seriously considered. This may apply to some of the multi-active pharmaceuticals or natural product analyses. **Optimisation:** The repetitive operations required in optimisation, assessment and development of chemical analyses are an obvious candidate as a robotic task. This may be covered more fully elsewhere in the symposium.

Sterility Testing: Testing of pharmaceutical products for low level micro-organism contamination is made more complex by use of human operatives. False positive results are a pitfall. This aspect of robotics is under active investigation within Beecham.

Zymark and Zymate are registered trademarks of Zymark Corporation.

AUTOMATED SAMPLE PREPARATION OF PHARMACEUTICAL PARENTERALS FOR ANALYSIS USING ROBOTICS

John H. Johnson and Ragu Srinivas
American Critical Care
1600 Waukegan Road
McGaw Park, IL 60085

and

Thomas J. Kinzelman
Zymark Corporation
Hopkinton, MA 01748

ABSTRACT

An automated system method for preparation of liquid samples for subsequent analysis has been developed and validated. Samples were processed with a commercially available Zymate System. All software routines and many hardware components were designed at American Critical Care. These included special test tube holders, sampling holders, pipet holders, cannulas and an autosampler vial capping system. Computer routines were designed specifically to accommodate dilutions ranging form 1:10 through 1:10,000. Samples are removed from ampules or vials and diluted in accordance with established procedures using an internal standard solution. The resulting diluted solutions are then put into autosampler vials for subsequent HPLC analysis.

The efficiency of the system lies in its ability to run unattended, including overnight and weekends. The operator/setup time is typically

0.5-2 hours compared with manual sample preparation time of 4-16 hrs for a set of 50 samples. This eight-fold increase in efficiency is accomplished with accuracy and precision comparable to manual dilution methods. The application described here will accommodate all liquid products requiring dilution and/or addition of internal standard.

INTRODUCTION

The Analytical Chemist has been able to support a substantial increase in sample load during recent years by automating two aspects of chromatographic methods. First, autosamplers permit expansion of analysis time to a 24-hour day. Second, the integration feature of the laboratory computer has automated data acquisition and calculation previously performed manually.

Analytical chemists who perform large numbers of determinations for parenteral pharmaceuticals are still left with the problem of performing serial dilutions, addition of internal standards and transfer to autosampler vials. This seemingly endless task is tedious and time-consuming. In 1984 alone this laborartory performed the repetitive task of opening, diluting, internal standard addition, and filling/capping autosampler vials for over 7,000 samples requiring 80 analyst hours per week. Thus, the focus of improving productivity in this laboratory has been automating sample dilution of small volume parenterals for subsequent high performance liquid chromatographic (HPLC) analysis.

This report describes a robotic system adapted for dilution of samples from ampuls and vials. The robotic system at American Critical Care has been programmed for dilution applications ranging from 1:10 through

1:10,000 to include addition of an internal standard and subsequent transfer to various HPLC automatic sampler trays for subsequent HPLC analysis.

EXPERIMENTAL

Robot

The robotic system consists of a Zymate Laboratory Automation System equipped with two Model 510 master laboratory stations, a Model 830 power and event control station, a Model 900 general purpose hand, a Model 920 blank hand, a Model 620 vortex station and a Model 520 fraction collector (Zymark, Hopkinton, MA).

Work Stations

Figure 1 shows how the devices described are arranged about the robot within its work envelope. Since the robot cannot see or otherwise perceive its environment, objects within the work envelope must be located at predefined positions.

Work stations include a master lab station (MLS_1) which draws up internal standard from a reservoir followed by the sample via the dilution/dispenser hand (DH) fabricated from the blank hand. A second master lab station (MLS_2) is used to pipet diluted samples into the autosampler tray. Sample vials and ampuls are placed in Racks R_1, R_2, R_3 and R_4. Racks R_1 and R_2 hold 5 mL and 10 mL ampuls, respectively. Disposable Autosampler vial rack (ASR) and pipet rack (PR) hold 5 mL and 10 mL vials, respectively. Racks R_5 and R_6 hold small autosampler vials and disposable pipet tips, respectively. A test tube dispenser (TTD)

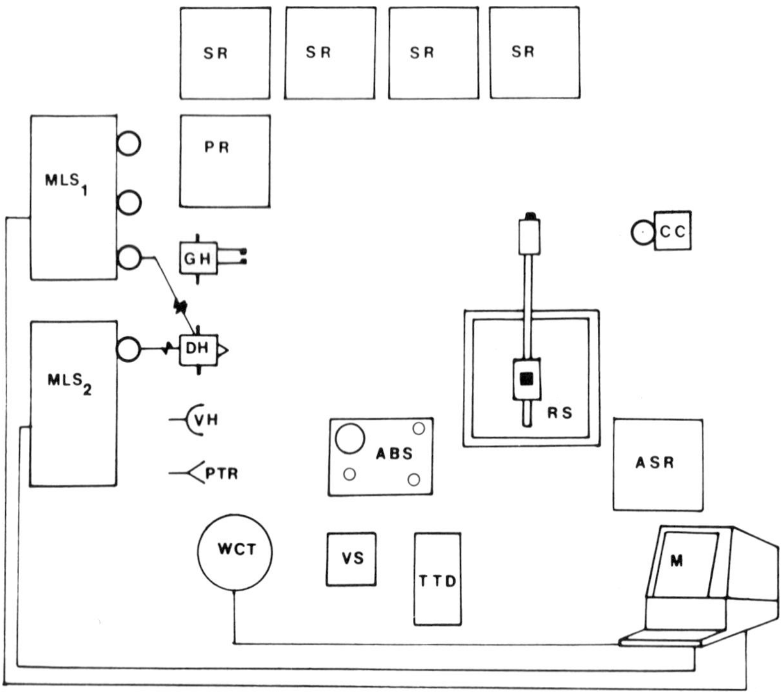

FIGURE 1. Robotic work station (top view) for dilutions. (SR) sample racks, (PR) disposable pipet rack, (MLS) master laboratory stations, (GH) general purpose hand, (DH) dilution dispenser hand, (VH) vial hanger, (PTR) disposable pipet tip remover, (WCT) Wisp autosampler carousel tray station, (ABS) ampul breaker station, (VS) vortex station, (TTD) test tube dispenser, (RS) robot arm, (ASR) autosampler vial rack, (M) microprocessor controller, (CC) crimp capper.

was designed to hold 500 16 x 100 mm (12 mL) tubes. Stations to hold vials (VH) and ampuls (ABS) during sample withdrawal were designed at ACC. Diluted samples are mixed via the Vortex Station. The Wisp station (WCT) is composed of the fraction collector fitted with an aluminum plate designed to support a 48-position autosampler tray (Model 72700-48, Waters Associates, Milford, MA).

Dilution - Dispenser Hand

A master lab station (MLS_1) was arranged whereby 1 mL, 5 mL and 10 mL gas-tight syringes were coupled via 6 inch lengths of 1/16 in Teflon tubing to a 3-port Teflon manifold. This was then connected via an 8 ft length of 1/16 in Teflon tubing to a modified blank hand equipped with a machined end to hold a 1-1/2 in, 18 gauge disposable needle as shown in Figure 2(d). The hand sits in a holder needle side down when not in use. A 3 mL plastic disposable syringe barrel is mounted as a funnel under the needle to collect diluent during the purge sequence. The barrel is connected to a waste container via Teflon tubing. A second tip holder is mounted at 180 to accommodate 1 mL disposable pipet tips. The tip is connected via a 8 ft length of 1/16 in Teflon tubing to a second master lab station (MLS_2) having a 1 mL gas-tight syringe.

Ampul Breaker Station

A polycarbonate block was used to hold ampuls during sampling. This block also holds test tubes for a serial dilution routine. Holes were drilled to hold 10.7 cm x 19 mm (10 mL) and 90 cm x 6 mm (5 mL) ampules. Additional holes were drilled for 16 x 100 mm test tubes and a nominal 1.8 mL autosampler vial number 2701-B9 (Wheaton Tubing Products,

FIGURE 2. Ampul processing operation. (a) ampul procurement, (b) ampul opening operation, (c) obtaining a test tube, (d) ampul sampling operation, (e) sample dilution operation, (f) dispensing sample into a Wisp vial.

Milleville, NJ). Ampuls are opened by the general purpose hand (Figure 3) which has a section of 2 in x 2 in angle iron mounted at a right angle to the fingers. The angle iron is pushed through the ampul top in a karate-type motion.

Vial Hanger

Vials were held in place for sampling via the vial hanger assembly shown in Figure 4. The holder was machined to accommodate the 11 mm neck and recessed for the 13 mm finish of 5 and 10 mL crimp cap vials. An additional metal arm is mounted to the common post. This steadies the dilution/dispenser hand to consistently enter the 6 mm inner diameter of the vial opening. The dilution/dispenser hand enters from the side and is guided by this hand stop as the dilution/dispenser hand moves down into the vial.

Test Tube Dispenser

Test tubes used for serial dilution are held by the dispenser shown in Figure 2(c). It consists of a 500-tube capacity plexiglass holder, a motorized belt drive with a cam attached to the two forward gears. When a test tube is removed, the motor is activated by a switch at the front of the holder and turns till a new test tube is in place. The holder design, belt drive and cams attached to the forward gears ensure that test tubes are jostled until a new tube is presented at the front of the holder. The tube then rests against the switch which deactivates the motor. After the robot picks up the tube it presses the tube against a switch for verification. The program continues only after this verification step. If there is no tube, the switch is not activated and

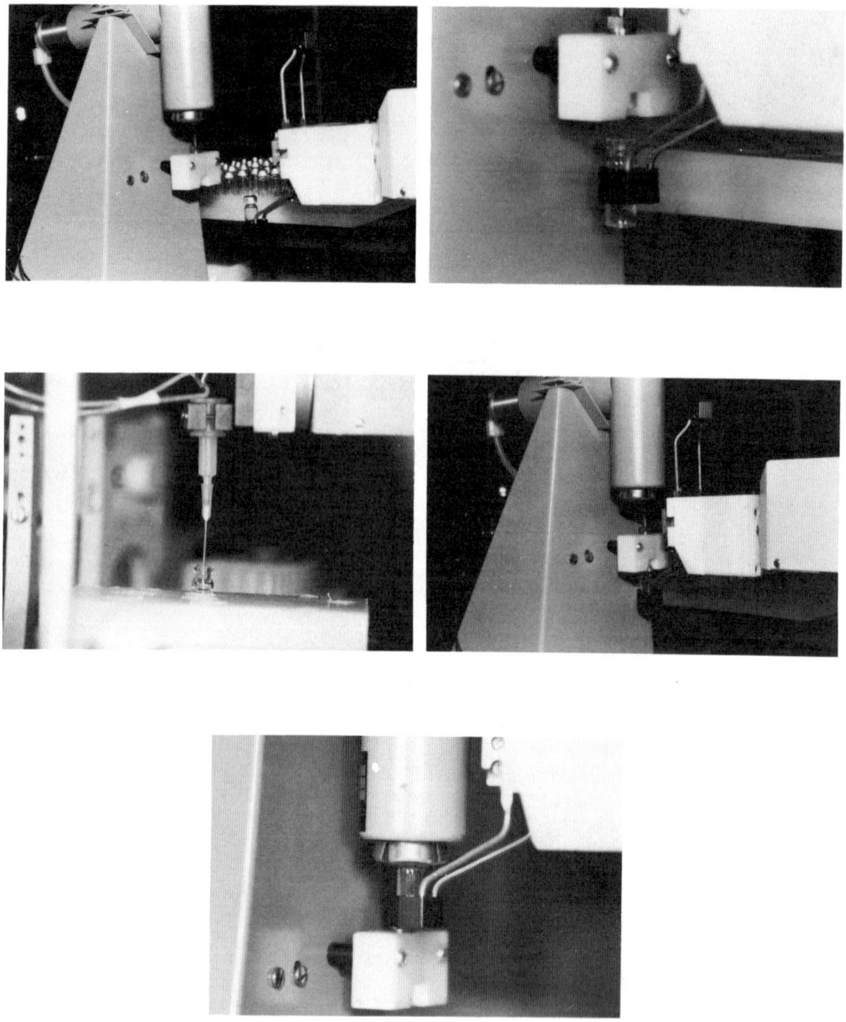

FIGURE 3. Crimp capping operation. (a) autosampler vial procurement, (b) cap placement into holder, (c) dispense sample into vial, (d) cap withdrawal, (e) crimp cap.

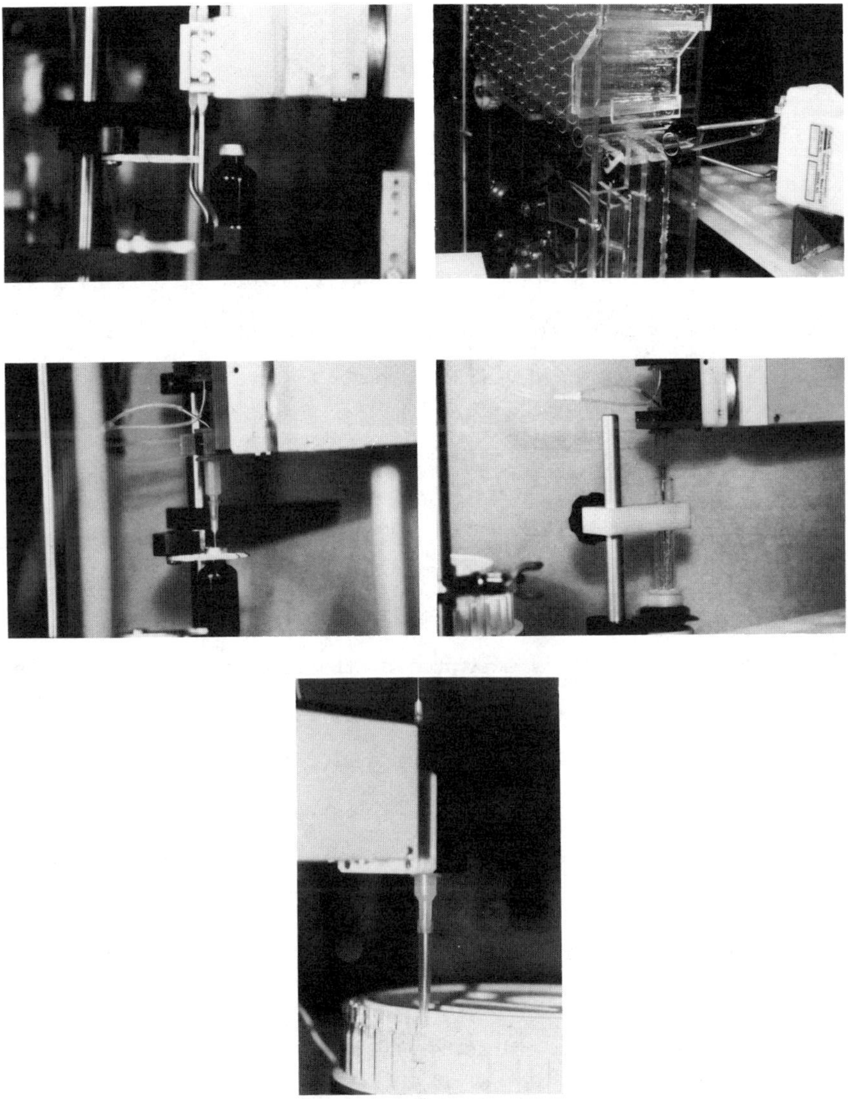

FIGURE 4. Vial processing operation. (a) vial placement into hanger, (b) obtaining a test tube, (c) vial sampling operation, (d) sample dilution operation, (e) dispensing sample into a Wisp vial.

the general purpose hand will go back to the test tube dispenser to get a tube.

WISP Vial Carousel

The final diluted sample is placed into a vial compatible with the autosampler of choice. Samples to be injected into an HPLC system equipped with a WISP autosampler are dispensed into the corresponding 48 vial capacity tray with vials in the appropriate slots. The tray is mounted to the fraction collector station shown in Figure 2(f). The stepping motor is programmed via the computer to advance one autosampler vial at a time by rotating 1/48th revolution.

Small Autosampler Vial Capper

The capper consists of a pneumatically activated crimper (Model 0201, Wheaton Corp, Milleville, NJ). A pneumatic valve (Model 903-00, Rainin Instruments, Woburn, MA) is activated by the robotic computer via the power and event control station. An assembly was fabricated to hold the caps while the vial is filled. It consists of a teflon holder shown in Figure 3. The vacuum is supplied by a pump (Model 13144, Gelman Instruments, Ann Arbor, MI) which is activated by the power and event control station. The vial with cap is placed in the holder where the vacuum holds the cap while the vial is moved to another station where the sample is added. After filling, the vial is returned to the holder. The vacuum is turned off, releasing the cap onto the vial. The vial with cap is then placed into the crimper. The crimper is activated by the computer, sealing the vial. The vial is then returned to its original position in the rack.

Disposable Pipet Apparatus

Disposable pipet tips are arranged in a rack and are attached to the dilution/dispenser hand as described above. A pipet tip remover (Zymark cat. no. 37152) provides the means for tip removal. The hand with tip is put in the station then lifted. The tip is pulled off and drops into a waste container.

Sample and Standard Preparation

Nitroglycerin: The sample solution of nitroglycerin was prepared by dissolving 10 mL of nitroglycerin (10% w/w) in ethanol (ICI Americas, Wilmington, DE) in 2 L of a solution containing 50% alcohol, USP and 50% distilled water yielding a 0.5% or 5 mg/mL solution. A solution of the internal standard (isosorbide dinitrate) was prepared by dissolving 4 g of isosorbide dinitrate (25% w/w) on lactose, USP (Henley Co., New York, NY) in 4 L of 50% alcohol, USP, 50% water yeilding a 250 mg/L solution.

Bretylium Tosylate: The sample solution of bretylium tosylate was prepared by dissolving 100 g bretylium tosylate (Delmar Chemicals, Montreal, Canada) in 2 L distilled water yielding a 50 mg/mL solution. A solution of internal standard was made by dissolving 400 mg pyrilamine maleate (Reisman Co., New York, NY) into 4 L distilled water.

Esmolol Hydrochloride: Sample solutions of esmolol hydrochloride at a concentration of 100 mg/mL were prepared by dissolving 200 g esmolol hydrochloride (American Critical Care, McGaw Park, IL) in 2 L distilled water. An internal standard solution was prepared by dissolving 80 mg 2-chlorobenzyl alcohol (Aldrich Chemical Co., Milwaukee, WI) into 4 L distilled water to yield 20 mg/L.

Chromatography

A series 2/2 liquid chromatograph (Perkin-Elmer, Norwlk, CT) equipped with WISP autosampler (Waters Associates, Milford, MA), a Model 160 selectable wavelength detector (Beckman Instruments, Berkley, CA), an on-line data system Model 3356 (Hewlett-Packard, Avondale, PA) and a 10 mV Model 555 recorder (Linear Instruments, Irvine, CA) were used throughout the studies.

The chromatographic procdure used for analysis of the Tridil (nitroglycerin) samples was previously described.[1] In this study, the samples were monitored at 214 nm.

The chromatographic procedure used for analysis of Bretylol (bretylium tosylate) was previously described.[2] Sampler were monitored at 229 nm. Esmolol hydrochloride samples were analyze using a procedrue described earlier.[3] Samples were monitored at 214 nm.

Software

The software routines which drive the system are written in EasyLab language. These routines consist of robotic positions, switches and valve operation entered into a system dictionary. Table I outlines the routines performed by the system to accomplish the sample withdrawal, sample dilution, vortexing, standard addition and dilution into the autosampler vial of choice from ampuls and/or vials.

System Operation

The sequence begins with a dialogue in which the user inputs the dilution factor, the autosampler type, the number, size and type of sample containers. The system assumes that the user has placed the samples in their proper places along with pipet tips and test tubes as shown in Figure 1. The inlet tubes from one master lab station are placed into the internal standard diluent. Samples are processed in the following order: 5 mL ampuls, 10 mL ampuls, 5 mL vials then 10 mL vials. Ampuls are accessed and diluted as shown in Figure 2. The robot attaches the general purpose hand and removes the first ampul from the corresponding rack. The ampul is then placed in the breaker station. The general purpose hand is then rotated 90 exposing an angle iron. The general purpose hand then pushes across the top of the ampul breaking the neck which then falls into a waste container. A test tube is obtained from the test tube dispenser and placed in the vortex station.

The sample is now ready to be diluted and the general purpose hand is replaced with the dilution/dispenser hand. The appropriate dilution is performed in the following manner. Sample is drawn from the ampul into the needle of the dilution/dispenser hand subsequently into the Teflon tubing attached to a 1 mL syringe of the master lab station. Simultaneously, diluent containing internal standard is drawn with a 5 or 10 mL syringe using the same master lab station. The vortex mixer is actuated. The sample followed by the appropriate amount of internal standard is then forced back through the needle of the dilution/ dispenser hand into the vortexing test tube. Carryover is prevented by drawing up excess internal standard. This excess is used to flush the needle after the sample and internal standard are dispensed. The vortex mixing is continued while the dilution/dispenser hand is parked and the general purpose hand is reattached. For dilutions greater than 1:25, a second tube is needed (see Table 2). The vortex mixer is turned off, the test tube is removed from the vortex mixer and placed into a holder. This then becomes the sample for the second dilution. This process is repeated a third time for dilutions greater than 1:500. The fully-diluted sample is now ready for placement into the appropriate autosampler vial.

For Wisp vials, the carousel is rotated to the proper autosampler vial. The dilution/dispenser hand rotates 180 and picks up a disposable pipet tip. The tip is then placed into the test tube containing the fully-diluted sample located in the Vortex Station. A 1 mL syringe on master lab station #2 draws a maximum of 0.9 mL of the diluted sample. The disposable pipet tip in the dilution/dispenser hand is placed over the autosampler vial and the sampler is discharged into the autosampelr vial. This process is repeated three times, delivering a total of 2.7 mL to the Wisp autosampler vials. The disposable tip is then discarded. Each time the sample is dispensed into an autosampler vial, the printer lists the sampler number.

For the smaller 1.6 mL crimp-cap autosamples vials, the general purpose hand picks up a vial with a loose cap attached earlier by the operator. The vacuum pump is turned on. The vial/cap is placed under the cap holder (Figure 3). As the vial is removed, the cap is held in place by the vacuum. The vial is then placed in a hole in the breaker station block where the fully-diluted sample is dispensed from the needle side of the diluter dilution/dispenser hand. The needle is used as it is easier to "hit" the smaller target (3 mm) of this autosampler vial. To eliminate contamination, 3.2 mL are drawn; 1.6 mL are added to the vial. The dilution/dispenser hand is parked and the excess 1.6 mL is then ejected to waste followed by internal standard to prevent carryover. The autosampler vial is then returned to the cap holder. The vacuum is turned off releasing the cap onto the autosampler vial. The capped vial is then placed in the crimper. The crimper is actuated using the power and event controller, sealing the autosampler vial. The sealed vial is then replaced in its original place in the rack.

Test tubes are then emptied into a funnel fitted with a Tygon tube which leads to a waste container. Each tube is slowly rotated over the funnel and the contents emptied. The hand then moves over a waste can then releases the tube.

The ampul is picked from the breaker station and returned to its original place in the rack. The counter is increased by one. The count is compared with the number of 5 mL ampuls entered by the operator. Unless the count exceeds this value, the above procedure is repeated for the next ampul sample.

When the count exceeds the number of 5 mL ampuls, the system checks to see if the operator entered 10 mL ampuls or vials. The larger ampul samples are sampled and diluted in the same manner as the smaller ampuls. The only difference is the position in the breaker station for breaking open the large ampul and sampling its contents. The next sub-routine processes 5 mL and 10 mL vials much the same as ampuls as shown in Figure 4. The vials are placed in the vial hanger for sampling. The routine for dilution and placement into autosampler vials are similar to those used for ampuls. When all samples have been processed the robot parks its hand and returns to a rest position.

RESULTS AND DISCUSSION

Dilutions using the robotic system are performed using master lab station programmable syringe pumps and test tubes in contrast with volumetric flasks and pipets with manual methods. The robot performs a series of tasks: (1) it opens vials and ampuls to draw samples, (2) transfers specific volumes to a test tube, (3) dilutes the sample with a specified volume of internal standard solution and (4) transfers the fully-diluted sample to one of two autosampler vials for subsequent HPLC analysis. Appendix A delineates the steps required to perform dilutions of magnitudes ranging form 1:10 through 1:10,000. Table 1 shows the schemes for individual dilutions.

The most precise dilutions were achieved when the syringes of the master lab stations were filled to at least 20% of their capacity. Precision was also improved by over-filling each syringe then expelling excess into the sample and internal standards reservoir, respectively. This

TABLE 1. Dilution Schemes.

Dilution Factor	First Dilution(mL)		Second Dilution(mL)		Third Dilution(mL)	
	Sample	Diluent	Sample	Diluent	Sample	Diluent
1:10	0.9	8.1				
1:20	0.5	9.5				
1:25	0.4	9.6				
1:50	0.9	3.6	0.9	8.1		
1:100	0.9	8.1	0.9	8.1		
1:200	0.9	8.1	0.5	9.5		
1:250	0.9	8.1	0.4	9.6		
1:500	0.5	9.5	0.4	9.6		
1:1000	0.9	8.1	0.9	8.1	0.9	8.1
1:2000	0.9	8.1	0.9	8.1	0.5	9.5
1:2500	0.9	8.1	0.9	8.1	0.4	9.6
1:5000	0.9	8.1	0.5	9.5	0.4	9.6
1:10000	0.5	9.5	0.5	9.5	0.4	9.6

eliminated any error associated with the syringe stepping motors when reversing direction from the fill to dispense modes. Because the largest syringe the master lab station could accommodate was 10 mL, all dilutions were made in 10 mL or less for each dilution factor. A maximum volume of sample and internal standard favors better accuracy and precision. For example, two steps for a 1:50 dilution were found to be more precise using a 0.9 mL:3.6 mL dilution followed by 0.9 mL:8.1 mL (Relative Standard Deveiation = 0.80) than a one-step 0.2 mL:9.8 mL dilution scheme (Relative Standard Deviation = 1.4).

Carryover between samples was eliminated by placing an 8 ft x 1/16" I.D. tubing between the dilution/dispenser hand and the syringes in the master lab station. The sample is drawn through the needle into this tube. This tube is connected to a manifold then to the master lab station. The volume of the tubing between the dilution/dispenser hand and the manifold exceeds 5 mL. Thus, the sample never reaches the syringes in the master lab station. Both the 5 and 10 mL syringes were also connected to the same manifold. During the dilution step, the 1 mL syringe draws the sample through the needle into the line while the appropriate volume of internal standard is drawn into the 5 or 10 mL syringe in the master lab station as delineated in Table 1. Hence the tubing and the needle are flushed with internal standard solution after each dilution. Blank samples resulted in no peaks at all in the HPLC system. Experiments with blank samples showed the carryover between samples was less than 0.01%.

The same solution of internal standard was used for all dilutions of a compound. This eliminated the need for additional valving to accommodate a series of internal standard concentrations for various samples. Although greater volumes of internal standard were used than in most manual methods where internal standard is added after or as a part of the last dilution, large amounts of the compound were not needed. For nitroglycerin experiments, one half gram of isosorbate dinitrate were used for each 50 sample set, 200 mg of pyrilamine maleate for bretylium tosylate and 40 mg of 2-chlorobenzyl alcohol for esmolol hydrochloride.

The vortex mixer station was activated before addition of the sample and internal standard to provide a continuous mixing of a more viscous sample with the internal standard. Vortexing continued an additional 20 seconds while robotic hands were changed during serial dilution or as the disposable pipet tip was mounted on the robotic hand for subsequent transfer to the Wisp vial. Separate disposable pipet tips were used for each sample with Wisp autosampler vials to minimize carryover. The excess sample drawn into the dilution/dispenser hand in the case of the smaller 1.6 mL autosampler vials described above accomplished the same purpose.

Fail-safe features were built into the system to prevent jamming of ampuls, test tubes and disposable pipet tips. A switch was actuated with the particular item in the general purpose hand. This insured the robotic hand actually retrieved the article before an other was placed into a common station. This is particularly useful when ampuls with defective scoring shatter during the "break ampul" subroutine. Hence, they cannot be sample. Shattered ampuls which are not retrieved from the breaker station prevent mounting of subsequent ampuls. In the case of a shattered ampule, the gripper hand cannot retrieve it. The contact closure is not made and the program is interrupted until the operator intervenes.

Early experiments were designed to sample sealed vials directly. The dilution/dispenser hand was used to charge sealed vials with an amount of air equivalent to the sample volume which is subsequently withdrawn. A % Relative Standard Deviation (RSD) of 2.8 was observed for a 1:10 dilution. Removing stoppers from vials resulted in precision equivalent to those for ampuls (Table 2). However, leaving samples open to the air can result in evaporation and thus a change in concentration between the first and last sample. This was remedied by removing the stopper from each vial and recrimp-capping vials with a thin teflon cover, same size as that used with Wisp vials. This affords a rigid seal to allow air to

TABLE 2. Automated Dilutions.

Sample Solution	Dilution Factor	Average of 15 Determinations[a] x Dilution Factor	% Relative Standard Deviation (% RSD)	Theoretical % Relative Error from Reference 5
Nitroglycerin	1:10	100.00[b]	0.16	0.13
	1:20	97.37	0.47	0.13
	1:25	93.90	0.48	0.16
	1:50	92.44	0.80	0.15
	1:100	89.61	2.17	0.18
	1:100[c]	89.60	0.93	0.18
Bretylium Tosylate	1:100	100.00[a]	0.29	0.18
	1:200	99.40	0.14	0.18
	1:250	98.49	0.20	0.20
	1:500	98.87	0.33	0.20
	1:1000	99.65	0.78	0.20
	1:100[c]	102.84	0.27	0.20
Esmolol Hydrochloride	1:1000	100.00[b]	0.64	0.20
	1:1000[d]	97.37	0.20	0.20
	1:2000	99.74	0.43	0.20
	1:2500	97.47	1.47	0.22
	1:5000	97.58	0.86	0.24
	1:10000	95.08	0.40	0.24
	1:10000[c]	96.38	1.44	0.24
	1:10000[d]	96.32	0.34	0.24

[a]These are based on 1:10 for nitroglycerin, 1:100 for bretylium tosylate and 1:1000 for esmolol hydrochloride, respectively.

[b]Normalized to 100%

[c]Vials

[d]One sample injected 15 times on the HPLC system.

flow through the hole pierced by the 18 gauge needle on the dilution/dispenser hand. Removal of the seal results in extra effort by the operator but which is also necessary when manually pipetting samples from vials.

Table 2 shows the results of fifteen separate automatic dilutions into autosampler vials from dilution factors 1:10 through 1:10,000. Within each sample set, e.g. nitroglycerine, the deviation form 100% increases with increasing dilution factor. However, there is little difference in deviation form 100% or difference in % standard deviation between 1:100 for nitroglycerin or 1:10,000 for esmolol hydrochloride. An increase in RSD follows the same pattern with the exception of 1:100 (ampul) and 1:10,000 (vials). Note that the % RSD for 1:100 is 2.17 for nitroglycerin but only 0.29 for 1:100 using bretylium tosylate. This indicates the error or bias is a function of the stepping motors in the master lab station. These errors are well within % RSDs reported for dilutions made by hand; 1.8% for Tridil (nitroglycerin),[1] 1.3% for Bretylol (bretylium tosylate)[2] and 1.6% for esmolol hydrochloride.[3] The % RSD values found with the robotics system also compare favorably with theoretical values generated with optimum pipet and volumetric flask sizes for any given dilution.[4]

A comparison of vials and ampuls shows the % RSD is less for vials in each case. This indicates no significant difference between ampuls which are samples open to the air and vials which have a Teflon seal.

In order to separate the error associated with automated dilution from the HPLC system, one sample (1:1,000) was injected 15 times into the HPLC system. A % RSD of 0.34 is similiar to the reported theoretical value.[4] Thus, when the error from HPLC analysis is subtracted, the resulting errors associated with the robotic system then become proportionately smaller.

CONCLUSIONS

The efficiency of the robotic system described here lies in its ability to prepare samples unattended. The operator need only place samples, autosampler vials and standards in the appropriate racks. The dilution parameter and autosampler choices described earlier are entered into the system and the reservoir tubes from the two diluter syringes are placed in the appropriate internal standard/diluent. A set of six one-point standards is used to calibrate the robotic/HPLC systems as is done in this laboratory when samples are prepared by hand. The total setup time is typically less than one hour for 50 samples compared with 16 hours for a manual three-step 1:5000 volumetric dilution. This robotic system is currently used on a routine basis in this laboratory. The system has proven to be an effective way to free analytical personnel for more original investigative work.

ACKNOWLEDGEMENTS

The authors wish to thank Mr. Jack Van Eck for fabrication of the test tube dispenser, Ms. Valerie Bergman for her laboratory assistance and Ms. Chris Maier for assistance in preparation of this manuscript.

REFERENCES

1. D. M. Baaske, J. E. Carter and A. H. Amann, J. Pharm Sci, 68, 481 (1979).

2. Y. Lee et al., Am. J. Hosp. Pharm., 38, 183 (1981).

3. Y. Lee, D. M. Baaske and A. S. Alam, J. Pharm. Sci, 73, 1660 (1984).

4. R. B. Lam and T. L. Isenhour, Anal. Chem., 52, 1158 (1980).

Appendix A

General Dilution Routine

Step	Robotics Action	Actor[a]
1	Enter experimental parameters to include dilution factor, number size and type of samples and autosampler	User
2	Set counters.	Robotic controller
3	Purge each dilution syringe.	Master lab station
4	Grasp first ampul and place in breaker station.	Robot/general purpose hand
5	Remove sealed tip from ampul.	Robot/general purpose hand
6	Grasp test tube, check for tube in hand then place in vortex.	Robot test tube dispenser/power and event controller/vortexer
7	Draw up sample volume as per dilution factor selected. Simultaneously charge dilution/dispenser with volume of internal standard as per dilution factor selected.	Robot/master lab station #1/dilution dispenser hand/master lab station #1
8	Dispense sample then diluent into vortexing tube.	Robot/dilution dispenser hand/vortex/power and event controller
9	If dilution < 1:50 go to 15.	
10	Remove test tube from vortex and place in block for dilutions >1:50. Obtain second test tube and place in vortex. For dilutions <50 go to step 15.	Robot/general purpose hand
11	Place dilution/dispenser syringe into test tube containing first dilution. Draw up appropriate sample volume. Go to 8.	Robot/master lab station/ dilution dispenser hand
12	Repeat step 9.	

13	Remove test tube from block. Empty and dump tube. For dilutions > 1:500, remove tube from vortex and place in block. Obtain third test tube and place in vortex station. For dilutions <1:500 to to step 15.	Robot/general purpose hand
14	Repeat steps 11 and 12.	
15	Attach disposable tip to dilution dispenser hand. Check to confirm. Draw up fully-diluted sample from test tube in vortex station.	Robot/dilution dispenser hand/master lab station #2/power event controller
16	For small autosampler vials, obtain samll vial. Fill vial, crimp cap.	Robot/gripper hand/small vial capper/dilution dispenser hand/power event controller/master lab station #1
17	For Wisp autosampler vials, dispense diluted sample into vial on carousel.	Robot/dilution dispenser hand/master lab station #2/wisp vial carousel
18	Empty and dump test tubes. Return ampul to position in rack. Add 1 to counter.	Robot/gripper hand
19	Check counter. If count is less than number of small ampuls in count then repeat steps 4-19 with additional small ampuls.	Robot
20	Grasp first large ampul and remove sealed tip.	Robot/general purpose hand
21	Repeat steps 7-18 with large ampuls	Robot
22	Check counter. If count is less than number of large ampuls then repeat step 21.	Robot
23	Grasp first small vial and put into vial hanger.	Robot/general purpose hand
24	Place dilution/dispenser syringe into vial. Draw up sample volume as per dilution factor selected.	Robot/dilutin dispenser hand/vial hanger station/ master lab station #1
25	Repeat steps 9-18 with small vials.	
26	Check counter. If count is less than number of small vials, repeat steps 23-25 with additional small vials.	Robot

27	Grasp first large vial and put into hanger.	Robot/general purpose hand
28	Repeat steps 9-18 with large vials.	Robot
29	Repeat step 25.	Robot/dilution disperse hand/vial hanger station/ master lab station #1
30	Check counter. If count is less than number of large vials, repeat steps 27-29 with additional large vials.	

[a]The microprocessor participates in each step.

Zymark, Zymate and EasyLab are registered trademarks of Zymark Corporation.
Wisp is a registered trademerk of Waters Associates.
Tygon is a registered trademark of the Norton Co.
Teflon is a registered trademark of E.I. duPont de Nemous and Co.

APPLICATION OF ROBOTICS FOR THE ROUTINE PRODUCTION OF FLUORINE-18-LABELED RADIOPHARMACEUTICALS

James W. Brodack, Michael R. Kilbourn,
Michael J. Welch and John A. Katzenellenbogen
Mallinckrodt Institute of Radiology
Washington University School of Medicine
St. Louis, MO 63110

ABSTRACT

The first synthesis of a positron-emitting radiopharmaceutical, 16α-[^{18}F]fluoroestradiol-17β, using a laboratory robotic system has been accomplished. All operations in the preparation of theradiopharmaceutical, which involves three synthetic steps followed by HPLC purification, are controlled by the robot system (Zymate Laboaratory Automation System). The product is obtained in a shorter time period than when prepared manually and without any radiation dose to personnel. This represents a new and exciting approach to the automation of the routine synthesis of short-lived, positron-emitting radiopharmaceuticals.

INTRODUCTION

Positron emission tomography[1] (PET) has emerged as a powerful tool for the in vivo study of the biochemistry and function of the human body. Previously unobtainable information about hemodynamics,[2] metabolism,[3] and receptors,[4] under both normal and diseased conditions, can now be routinely studied. For example, Figure 1 shows a PET study of cerebral blood flow, blood volume, and oxygen metabolism of a patient with stroke. Such biological and medical studies rely heavily on the routine preparation of appropriate radiopharmaceuticals labeled with positron-emitting radionuclides (a positron is a positiviely charged particle with the mass of an electron; upon collision with an electron it annihilates to produce two orthoganol 511 KeV gamma rays).

More than 300 compounds labeled with oxygen-15, nitrogen-13, carbon-11, gallium-68 or fluorine-18 have been reported. A number of these are in routine use in institutions around the world. The radionuclides have very short half-lives (from 2 min for O-15 to 110 min for F-18) and are produced by in-house cyclotron bombardment of solid, liquid or gas targets, to give high specific activity (100 to $1x10^8$ Ci/mmol) precursors (e.g., $^{11}CO_2$, $H^{11}CN$, $F^{18}F$, $H^{18}F$) suitable for use in organic syntheses. In the synthesis of radiopharmaceuticals labeled with these radionuclides, the short half-lives dictate the need for high levels (100 to 2000 mCi) of radioactivity at the start of the synthesis as the overall preparation of the radiopharmaceutical may take 2-3 half-lives to complete. Thus, the routine preparation by a manual procedure inevitably leads to unacceptably high cumulated radiation doses for the personnel involved. For this reason remote, heavily shielded apparatus are utilized. We and others have described remote, manually operated apparatus for several important radiopharmaceuticals (^{11}C-gluose,[5] ^{11}C-palmitate,[6,7] ^{15}O-water,[8] ^{18}F-2-fluoro-2-deoxyglucose,[9,10] ^{11}C-2-deoxyglucose,[11] ^{11}C-butanol,[12] ^{11}C-acetate[13]). Completely

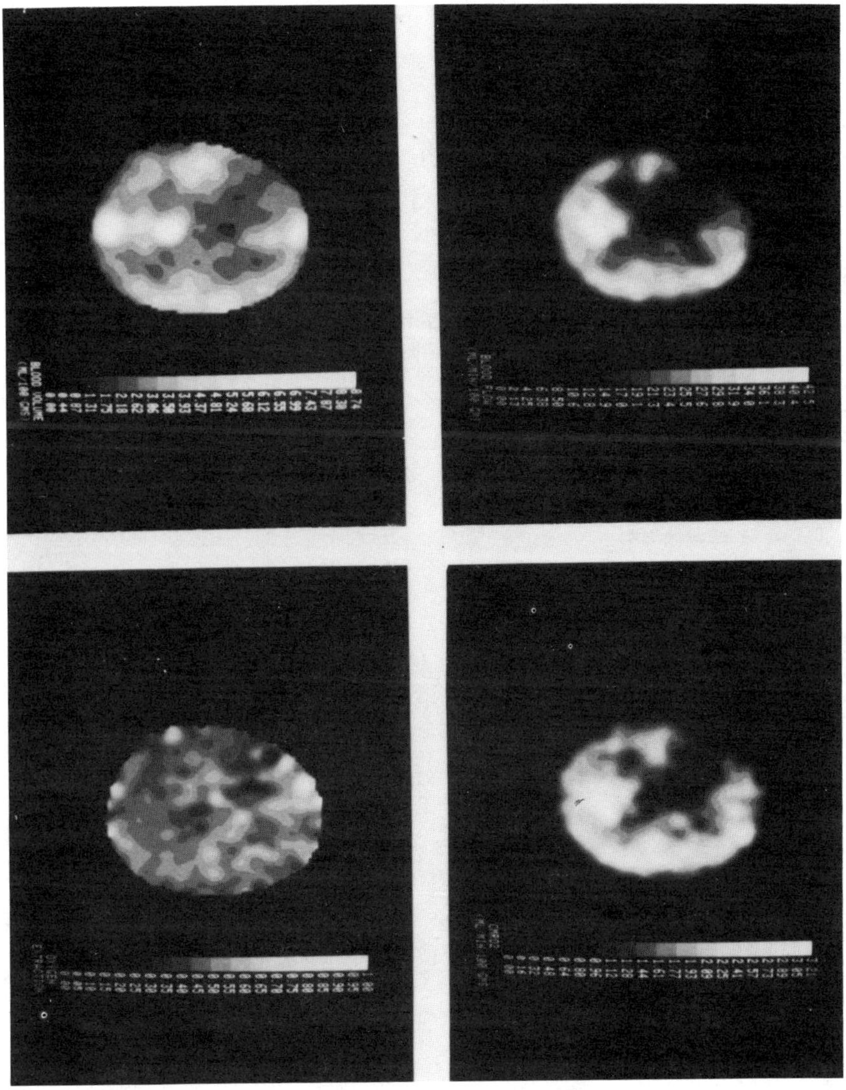

FIGURE 1. PET scans of patient with left frontal infarction. Within the region of the infarct, cerebral blood flow (rCBF, upper left) and oxygen metabolism ($rCMRO_2$, upper right) are markedly decreased. Cerebral blood volume (rCBF, lower left) and oxygen extraction (rOEF, lower right) also decreased, but to a lesser degree.

automated apparatus have been described for a few radiopharmaceuticals such as $[^{18}F]$2-fluoro-2-deoxy-D-glucose,[14] ^{11}C-methionine,[15] ^{13}N-glutamate,[16] and ^{11}C-antipyrine.[17]

Both of the current approaches to routine radiopharmaceutical preparation (remote manual and remote automated) suffer from limitations. Such apparatus are dedicated to the preparation of a single radiopharmaceutical (e.g., ^{11}C-glucose or ^{18}F-2-FDG) or a group of closely related radiopharmaceuticals (e.g., C-11 fatty acids). For a PET program employing a variety of diverse radiopharmaceuticals, separate apparatus have to be designed, constructed, and maintained. These apparatus are constructed with valves, tubing, and vessels in a fixed arrangement. Changes (due to improvements in, or new ideas for, the chemistry) cannot be rapidly and simply incorporated.

Laboratory robotic systems are an attractive alternative to the current approaches to radiopharmaceutical automation. Such systems are designed to work repeatedly (and dependably) and can operate under hostile conditions (such as a high radiation area). Progrmamable, microprocessor controlled robotic systems offer a versatility unmatched in any specifically designed and constructed apparatus for radiopharmaceutical preparation. We describe here the first application of laboratory robotics to the preparation of a complex, positron-emitting radiopharmaceutical, 16α-$[^{18}F]$fluoroestradiol-17β,[18] by the reaction sequence shown in Figure 2.

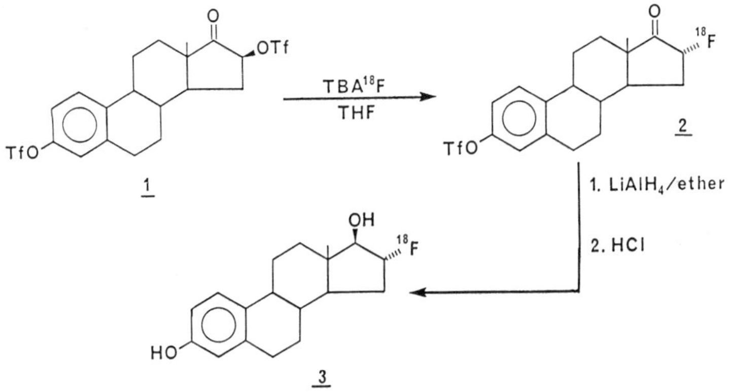

FIGURE 2. Steps in the robotic synthesis of 16α-[^{18}F]fluoroestradiol-17β.

EXPERIMENTAL

The components of the Zymate Laboratory Automation System used in this application were arranged in a 180°C arc (Figure 3). The bench top was trimmed to 28 in x 58 in, the dimensions of the shielded hood available. The controller, HPLC pumps, and detectors were placed either below the bench or outside of the hood.

Several specialized accessories were constructed for this applicaton. A free-standing nitrogen line was made by welding a length of stainless steel tubing to the bottom of a small lead pig. Glass pipette tips, which could be handled by the Z960 Pipet Kit, were made from short lengths of disposable glass pasteur pipets with a polypropylene collar cut from a polypropylene pipet tip.

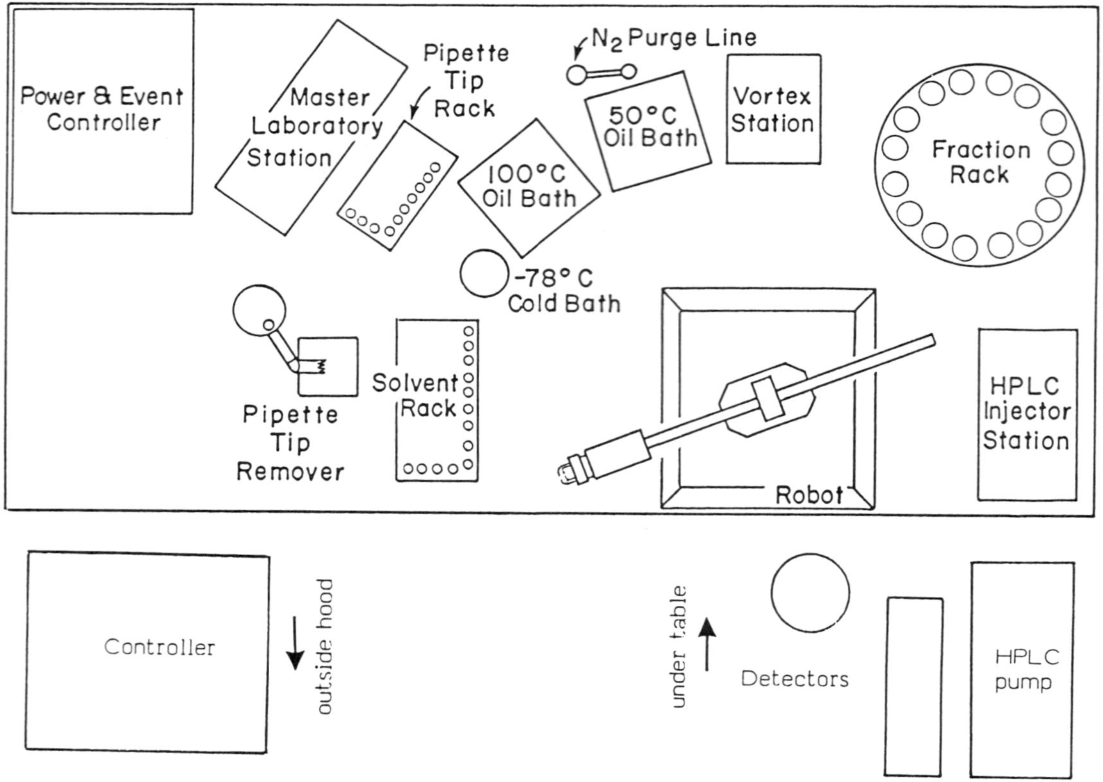

FIGURE 3. Layout of components of Zymate Laboratory Automation System with shielded hood.

Vacutainer tubes (#6436, 5 mL, Becton-Dickinson) were used both as the reaction vessel and for storage of solvents and solutions. A 1 mL Reacti-vial (Pierce Chemical) with teflon-faced septum was used to hold the lithium aluminum hydride (LAH).

Fluorine-18 was produced by proton bombardemnt of an oxygen-18 water target, as previously described.[19] 3,16β-bis(trifluoromethane-sulfonyloxy)-estra-1,3,5,(10-triene-17-one) was prepared as previously described.[18]

Chromatography was done using a Spectra-Physics SP8700 HPLC system and a Whatman Partisil M9 10/50 silica gel column. The "run gradient" and "pump stop" terminals on the HPLC pump were connected to the robotic system via closure switches on the HPLC injector station. Mass detection was done by UV at 254 mm (Schoeffel Instruments). Radioactivity detection was done using a NaI(Tl) crystal, high voltage supply, amplifier, single channel analyzer and rate meter. The analog output form the rate meter was split and provided a signal to both a chart recorder nd the HPLC injector station. A software program was written to utilize the integration parameters built into the injector station memory, allowing for integration of the radiometric signal during fraction collections.

Methods

The entire synthetic procedure was broken down into 5 levels of subroutine. The major steps in the systhesis are:

1. Resolubilization of fluorine-18-tetrabutylammonium fluoride (^{18}F-TBAF)

2. Reaction of ^{18}F-TABF with the triflate[1]

3. Reduction of the intermediate fluoroketone with LAH

4. Quenching of the reduction reaction and liquid-liquid. extraction

5. Injection into the HPLC

These program steps are briefly summarized below.

1. **Resolubilization of ^{18}F-TBAF:** Fluorine-18 fluoride ion in water (~200 uL) was placed in a Vacutainer. The vessel was placed in an oil bath, the nitrogen line moved to direct the flow into the tube, and the water evaporated. Aliquots (0.5 mL) of acetronitrile are added and evaporated at timed intervals. Finally, the nitrogen line is removed, the THF added, and the resolubolized ^{18}F-TABF solution in the THF pipetted to a second Vacutainer.

2. **Reaction with triflate:** An aliquot of a solution of triflate in THF was added to the resolubilized TBAF adn the solution allowed to set for 5 minutes.

3. **Reduction:** The THF solution was placed in the 50°C oil bath, the nitrogen line moved into position, and the THF evaporated. The nitrogen line was removed; the tube was removed from the oil bath, deithyl ether added, and placed into the dry ice-actone bath. Using a needle attached ot the remote dispenser, the septum on the Reacti-vial was punctured, and the LAH solution withdrawn and added to the ether solution. The reaction vessel was then removed from the dry ice-acetone bath, placed in the test tube rack, and allowed to warm to room temperature.

4. Quench and extract: A solution of dilute hydrochloric acid was added, followed by an aliquot of diethyl ether, and the mixture vortexed. The tube was replaced in the rack, and the ether layer drawn off using the pipette and transferred to a small drying column of Na_2SO_4. The extraction step was repeated twice. The combined dried ether extracts were collected in a Vacutainer tube.

5. HPLC injection: The ether solution was placed in the 50°C oil bath, the nitrogen line set in place, and the ether evaporated. The residue was redissolved in 0.5 uL of methylene chloride. The tube was then placed next to the HPLC injector, and the solution pipetted from the tube into the stainless steel cone on top of the injector. The solution was then drawn into the injector loop using a robot-controlled syringe attached to the vent port on the injector.

RESULTS

The final product, 16α-[^{18}F]fluoroextradiol-17β, was obtained in 5-6% radiochemical yield (decay corrected) in a total synthesis time of 80 minutes. Radiochemical purity was >99%.

DISCUSSION

We have successfully applied a laboratory robotic system to the synthesis of a complex, positron-emitting radiopharmaceutical. The product is obtained in suitable quantities and radiochemical purities for use in a clinical PET study. There is no operator involvement in the radiopharmaceutical synthesis and thus no absorbed radiation dose problem.

In the development of this robot-controlled procedure, several problems were met and overocme. The free-standing nitrogen line, used in the evaporation steps, allows for free movement of the hand during this step. In this application the hand is then utilized to add solvents at the appropriate time. The use of glass pipetts, modified so they can be handled by the remote dispenser on the robot handle, was crucial. It was found that orgnaic solvents leach a substance (unknown) from the polypropylene pipette tips, which interfered with the reaction of the resolubolized ^{18}F-fluoride with the triflate. Such glass pipettes may be of general interest in applications where solvents or solutions are incompatible with polypropylene. The use of a small, teflon-septum capped Reacti-vial to contain only the needed amount of LAH proved an excellent solution to the nettlesome problem of handling an air and moisture sensitive reagent.

The overall radiochemical yield (5-6%) is less than we have obtained in manual syntheses (22%). We have found three remaining problem areas. First, the resolubilization of the F-18 TBAF is accomplished by the robot in only 50% yield, compard to 90% when done by hand. Second, the reduction of the ^{18}F-fluoroketone to the ^{18}F-fluoroestradiol appears incomplete. Finally, the robot-controlled injection into the HPLC is not as efficient as a careful manual filling of the injection loop. We are presently examining each of these problem areas in closer detail. It should be noted, however, that with the combination of the F-18 targetry currently in use and robotic control of the preparation, up to 1 Ci of fluorine-18 could be used at the start of the synthesis, to yield >30 mCi of the final radiopharmaceutical.

The use of a robot system which combines the movements of a hand and arm with the use of valves and switches for liquid and gas handling - all under simple-to-program microprocessor control - represents a radically new approach to the synthesis of positron emitting radiopharmaceuticals. Fixed layout, dedicated apparatus will no longer be necessary. Changing from one radiopharmaceutical preparation to another may simply require entering a new program and, perhaps, alterations (additons, deletions) to the layout of the equipment (see Figure 3).

The flexibility of the EasyLab programming, where robot movements are broken down into increasingly simpler subroutines, allows for:

1. simple and rapid changes in synthetic procedures (e.g., optimization of yields or preparation time)
2. easy programming of additional syntheses

The RESOLUBILIZE.F-18.IN.THF subroutine of the ^{18}F-fluoroestradiol program can be applied to any synthesis utilizing nucleophilic ^{18}F-fluoride, such as [^{18}F]2-fluoro-2-deoxy-D-glucose via cyclic sulphate opening[20], or syntheses of [^{18}F]spiroperidol[21] or [^{18}F]N-methylspiroperidol[22] by aromatic nucleophilic substitution. We have used this subroutine to prepare ^{18}F-TABF in a variety fo solvents (CH_2Cl_2, $CHCl_3$, benzene, dichlorobenzene) for a study of solvent effects on F-18 halofluorination of olefins[23]. Many of the subroutines of the ^{18}F-fluoroestradiol synthesis represent basic laboratory operations (e.g., liquid-liquid extraction) and thus can be used, with little or no modification, in syntheses of other positron-emitting radiopharmaceuticals.

Our ^{18}F-fluoroestradiol program utilizes about 40% of the available computer memory. Thus, a second synthesis program could be stored directly in memory or, alternatively, a separate disk could be used for each synthesis. Changing from one radiopharmaceutical to another may involve nothing more than entering a new program, placing reaction vessels/solutions/solvents in the correct places on the board, and starting the program. Repeat syntheses should also be very simple, requiring merely a second set of vessels/solutions/solvents. Repeat preparations could be done without operator intervention, and with no need to enter a (possibly) radioactive hood. Such repetitive preparations are not important with F-18 radiopharmaceuticals such as ^{18}F-fluoroestradiol, but are very important with radiopharmaceuticals such as carbon-11 labeled 2-deoxyglucose[24] or carbon-11 labeled palmitic acid[25], where repeat studies at short time intervals are part of the clinical protocol.

CONCLUSIONS

Radiopharmaceuticals labeled with short-lived, positron emitting radionuclides can be successfully prepared using a laboratory robotic system. The use of robotics provides a versatility unmatched by previous methods of radiopharmaceutical automation. The application of robotics should widen the clinical availability of radiopharmaceutials for positron emmission tomography.

ACKNOWLEDGEMENTS

The work described herein was supported by a Departemnt of Energy Contract No. DE-FG02-84ERG0218 and National Institutes of Health Grant HL 13851. We thank Drs. William J. Powers and Marcus E. Raichle for providing Figure 1.

REFERENCES

1. Ter-Pogossian, M. M., Raichle, M. E., and Sobel, B. E., Sci. Am., 243, 139 (1980).

2. Raichle, M. E., Ann. Rev. Neurosci., 6, 249 (1983).

3. Grietz, T., Ingvar, D. H., and Widen. L., eds., The Metabolism of the Human Brain Studied with Positron Emission Tomography, Raven Press, New York, 1985.

4. Kilbourn, M. R., and Zalutsky, M. R., J. Jucl. Med., 26, 655 (1985).

5. Dence, C. S., Lechner, K. A., Welch, M. J., and Kilbourn, M. R., J. Labeled Compd. Radiopharm., 21, 743 (1984).

6. Welch, M. J., Dence, C. S., Marshall, D. R., and Kilbourn, M. R., J. Labeled Compd. Radiopharm., 20, 1087 (1983).

7. Padgett, H. C., Robinson, G. D., and Barrio, J. R., Int. J. Appl. Radiat. Isot., 33, 1471 (1982).

8. Welch, M. J., and Kilbourn, M. R., J. Labeled Cmpd. Radiopharm., in press.

9. Fowler, J. S., MacGregor, R. R., Wolf, A. P., Farrell, A. A., Karlstrom, K. I., and Ruth, T. J., J. Nucl. Med., 22, 376 (1981).

10. Barrio, J. R., MacDonald, N. S., Robinson, G. D., Najafi, A., Cook, J. S., and Kuhl, D. E., J. Nucl. Med., 22, 372 (1981).

11. Van Haver, D., Rabi, N. A., Vandewalle, M., Goethals, P., and Vandecasteele, C., J. Labeled Cmpd. Radiopharm., 22, 657 (1985).

12. Dischino, D. D., Welch, J. J., Kilbourn, M. R., and Raichle, M. E., J. Nucl. Med., 24, 1030 (1983).

13. Pike, V. N., Horlock, P. L., Brown, C., and Clark, J. C., Int. J. Appl. Radiat. Isot., 35, 623 (1984).

14. Iwata, R., Ido, T., Takahashi, T., and Monma, M., Int. J. Appl. Radiat. Isot., 35, 445 (1984).

15. Davis, J., Yano, Y., Cahoon, J., and Budinger, T. F., Int. J. Appl. Radiat. Isot., 33, 363 (1982).

16. Suzuki, K., and Tamate, K., Int. J. Appl. Radiat. Isot., 35, 771 (1984).

17. Vandewalle, T., Vandecasteele, C., De Guchteneire, F., Meulewaeter, L., Van Haver, D., Denutte, H., Goethals, P., and Slegers, G., Int. J. Appl. Radiat. Isot., 36, 469 (1985).

18. Kiesewetter, D. O., Kilbourn, M. R., Landvatter, S. W., Heiman, D. R., Katzellenbogen, J. A., and Welch, M. J., J. Nucl. Med., 25, 1212 (1984).

19. Kilbourn, M. R., Jerabek, P. A., and Welch, M. J., Int. J. Appl. Radiat. Isot., 36, 327, (1985).

20. Tewson, T. J., J. Nucl. Med., 24, 718 (1983).

21. Kilbourn, M. R., Welch, M. J., Dence, C. S., Tewson, T. J., Saji, H., and Maeda, M., Int. J. Apopl. Radiat. Isot., 35, 591 (1984).

22. Arnett, C. D., Fowler, J. S., Wolf, A. P., Shive, C-Y., and McPherson, D. W., Life Sci., 36, 1359 (1985).

23. Chi. D. Y., Katzenellenbogen, J. A., Kilbourn, M. R., and Welch. M. J., J. Nucl. Med., 26, P37 (1985).

24. Russell, J. A., and Wolf, A. P., J. Nucl. Med., 25, P33 (1984).

25. Bergmann, S. R., Fox, K. A., Ter-Pogossian, M. M., Sobel B. E., and Collen, D., Science, 220, 1181 (1983).

Zymark, Zyamte and EasyLab are registered trademarks of Zyamrk Corporation.

USE OF THE ZYMATE ROBOT FOR MICROBIOLOGICAL INOCULATION AND MIXING OF COSMETIC PRESERVATION TESTING SAMPLES

J. L. Smith
Chesebrough-Pond's Incorporated
40 Merritt Boulevard
Trumbull Industrial Park
Trumbull, Connecticut 06611

ABSTRACT

The Zymate Laboratory Automation System was adapted to interface with the sample inoculation and mixing phase of cosmetic preservation testing.

This test is usually performed as described in ASTM Method E 640-78 or the CTFA Guideline for Preservation Testing of Aqueous Liquid and Semi-Liquid Eye Cosmetics. This involves dividing a cosmetic sample into several equal portions, each of which is inoculated with test microorganisms including gram-negative and gram-positive bacteria, yeasts and fungi. The samples are then mixed. Each time a determination is performed it is followed by inoculation and thorough sample mixing. This mixing step is usually performed manually using a wooden coffee stirrer or similar disposable tool.

The manual mixing step is tedious and may be non-uniform; therefore, a source of variation in test results. In our laboratory, a standard mixing of fifty rotational stirs is commonly used. There is also concern of variation between operators. With over one hundred individual formulas simultaneously on test, the sample mixing becomes rather onerous and time-consuming.

We determined the Zymate System could be used to intervene in the mixing portion of this test to allow staff time for more productive and ineresting work.

The robot is located some distance from the laboratory. A buzzer is used to indicate progress after every fifth sample and signals via a special program when the last sample has been processed. A separate alarm buzzer is also located in the laboratory and is activated by either microswitch or photocell feedback at several points throughout the procedure.

INTRODUCTION

Description Of The Method

Mechanically, the operational aspect of cosmetic preservation testing is not a very complicated affair, particularly with creams, lotions, shampoos and other fluid products to which this discussion is limited. The basis of the technique is similar to that described in the USP for testing preservative activity of ophthalmic products. Challenge inocula (bacteria, yeasts, and fungi) are introduced into test portions of the product. At various time intervals after introduction of the challenge inocula, microbiological tests are performed to determine if the preservatives have reduced the level of these microorganisms to acceptable limits. The amount of reduction required and other such microbiological considerations will not be discussed here, because they are not particularly pertinent to the automation of the test as treated in this application.

We chose to initially deal with the most onerous portion of the test, that is, the inoculation and mixing of the samples. This is normally carried out manually because of the relatively high viscosity of the cosmetic samples and the varying viscosity of these different cosmetics which makes mechanical mixing (use of magnetic stirrers or propeller mixers, for instance) inadequate and/or difficult. Mixing is usually performed using wooden coffee stirrers or other disposable devices which are sterilized; however, strict aseptic procedures are not required for carrying out this test.

The operator will usually inoculate all samples to be tested and then mix each sample for fifty stirs. Because numerous stability samples as well as experimental back-up test formulas are required to assure a good chance of making the ship-to-trade dates for the new product, the testing is extensive and time-consuming.

Why Automate This Test?

Since viscous cosmetics adhere to the jar, samples must be mixed carefully and thoroughly to assure reproducibility. One can easily imagine the boredom and variability involved in performing this test manually. With thirty or so samples per operator, a major portion of the test is taken up with inoculation and stirring, which may take the major part of an afternoon. With some versions of the test, samples must be inoculated and mixed at least three days per week. We found in our practice the maximum number of samples one technician could practically handle was fifteen formulas at one time (this means a total of 75 test samples). in times of heavy sample testing, we found everyone in the laboratory became an operator, using up time better spent on data analysis, report preparation and other more creative pursuits.

EXPERIMENTAL DESIGN

Description of the Automated Mechanical System

The system (Figure 1) consists of the following components: Zymate controller, printer, robot arm module, modified general purpose hand with two sets of finger (one for jar lids and one for stirring sticks), two master laboratory stations (MLS) with two remote nozzles, capping station, and power and event controller (PEC). In addition, the following custom designed equipment is included: three jar racks holding ten jars apiece (five per column), four stick racks holding eight sticks apiece, and a lid station (for now, fashioned from wood and used to hold jar lids while other operations are functioning). A manually activated magnetic stirrer is used to hold the bottles of microbial suspensions which are dispensed by the master laboratory stations.

Operation of the Mechanical System

The robot arm lifts the first sample jar, places it in the capper and uncaps the lid which is then placed in an inverted position in the lid station. The arm then proceeds to the remote nozzle, lifts it from its station and lowers it to a position above the jar. One millimeter of microbial suspension is dispensed into the cosmetic sample and the remote nozzle is returned to its station. The arm then selects a stick from the appropriate stick rack and brings it into position over the sample jar in the capper. The stick is lowered into the sample and the capper instructed to achieve maximum torque. On a "NO MODULE WAIT" command both the capper and the arm function to stir the sample. Both sides of the stick are then wiped on the jar rim to remove excess sample

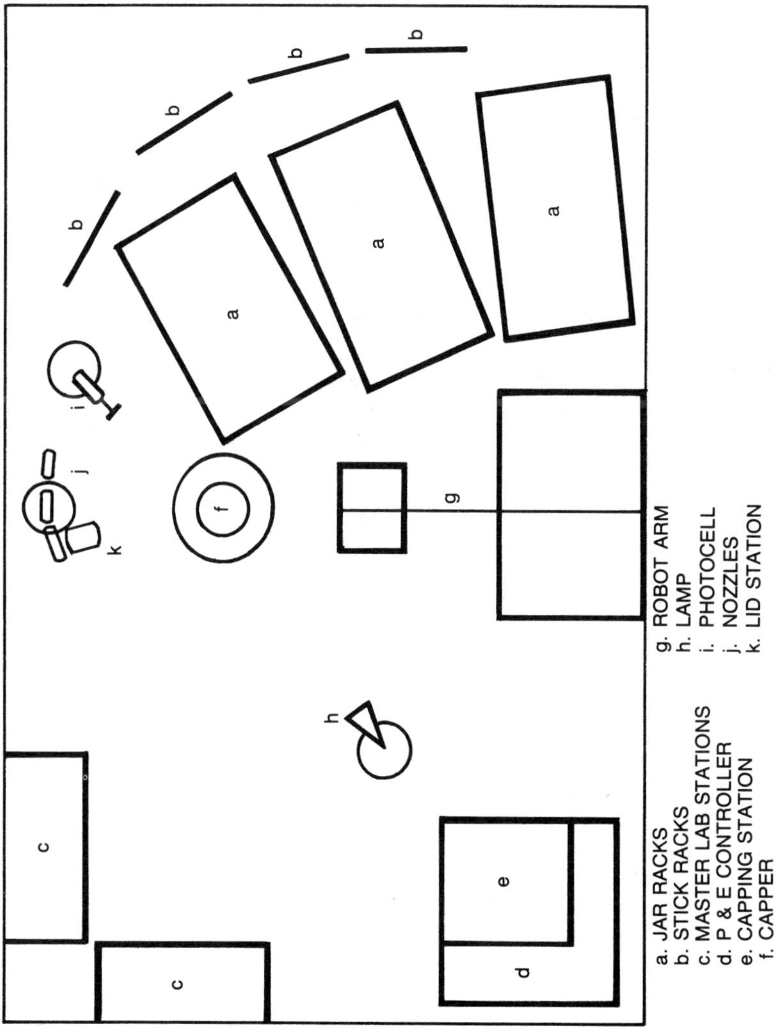

Figure 1. Diagram of bench layout for cosmetic preservation testing.

and the stick is then dropped into a beaker of disinfectant. The arm then lifts the lid from the lid station and caps the jar which is then returned to the jar rack.

Description of the Automated Electrical System

The feedback electrical system consists of the following items: A microswitch (Archer Subminiature SPDT Lever Switch, Radio Shack catalog #275-016) which is placed in a position on the lid station to sense whether the lid is correctly situated in the lid station; a photocell (Archer Photo Cell, Cadmium Sulfide, Radio Shack catalog #276-116A) set opposite a Duracell flashlight (modified to hold an Archer #47 Lamp, 6.3 volts at 150 milli-amps Radio Shack catalog #272-1110). The beam of the lamp crosses over the capping station to activate the photocell placed approximately 83 cm from the light source. This setup provides additional monitoring of events occurring over the capper. Two buzzers are used. One buzzer (Archer Mini-Buzzer, Radio Shack catalog #273-053) is used as a Remote Alert and the other (Archer Piezo Alert, Radio Shack catalog #273-066), as a Progress Signal. Both remote buzzers are used to notify the laboratory which is located across a hallway in another room and at least thirty feet away. A red light signal (Archer Sub Mini Screw Lamp, 6v, Radio Shack catalog #272-11142 with a bayonet assembly) is used in conjunction with the Remote Alert Buzzer to visually signal alert problems in the proximity of the machine.

Operation of the Alarm System

Since the robot is not equipped with any sensing mechanisms and especially since the entire operation is located at some distance from the microbioloby laboratory, it is very important to install the feedback devices just mentioned to monitor critical activities during the operation. Fortunately, most of these activities occur over the capper and are monitored by detecting the status of the light beam from the flashlight with the photocell. The photocell is connectd to an input lead on the power and event controller. The monitoring points in the program for the photocell system are:

1. **clear capper:** prior to pickup of a jar from the rack the status of the beam is checked to determine if the capper is clear; i.e., the beam must be uninterrupted. This monitoring point also has the bonus of indicating whether the light bulb has burnt out or if the voltage on the light source needs to be boosted in order to properly activate the input from the photocell to the power and event controller.

2. **Jar over capper:** interruption of the beam by the sample jar to verify that a jar has indeed been picked up by the hand.

3. **Nozzle over jar:** interuption of the beam by the remote nozzle to verify the nozzle has been picked up.

4. **Stick check:** interruption of the beam by the stick to verify the stick has been picked up.

5. **Jar over capper:** interruption of the beam by the jar on its way back to the rack. The microswitch on the lid station is set to trip when the lid is in the station. It is connected to an input lead on the power nad event controller and its status is checked when the lid is placed in the station to verify its presence and again when the lid has been removed from the station to verify the station is again clear.

Description and Results of Validation Tests

Three types of validation tests were run:

Mixing: This was initially performed using dye solutions instead of microbial suspensions. Various food dyes were prepared and placed in the MLS reservoirs and the program run using various creams and lotions. Adequacy of mixing was determined by visual observation of the thoroughness of dye dispersion in each test sample. Adequate dispersion was achieved using a stir program of 2 minutes and 15 seconds duration. A total of 90 samples were tested in this manner. Many more dye tests were actually run during the stir program development phase. The inital stirring phase of the program was reduced from five minutes to two and one quarter minutes. A more efficient stir program is being developed to reduce this time further.

Uniform Inocula Delivery: The second validation test set was performed to evaluate uniformity of the delivery of microbial suspensions over the two and one-half hour time period of the thirty sample mixing program. Microbial suspensions, including bacteria, yeasts, and molds were placed in the reservoirs (bottles containing stirring bars) and agitated with the magnetic mixers. Levels of microorganisms delivered to test jars containing sterile saline were determined at the beginning and at the end of the program run. Results indicated that the delivery at the end of the run was the same as at the beginning. There was some tendency noted for the molds to settle out in the delivery syringe but these were evidently heavier particles which did not significantly affect the total count delivered.

Method Validation: The third validation involved comparison of the manual test to the mechanized test using duplicate samples. One set of thirty samples was tested by both methods and additional set of shampoo samples from a competitive market survey were also run in the same manner. Results from the manual method matched the mechanized method closely. Since this mechanical interface at this point in the method involves only the inoculation and stirring portion of the test and additional work is planned to further automate the test, additional validation was not performed.

General Programming Features

A general description of the program as follows: Both the jar racks and the stick racks are indexed and positions are determined by the controllr via the internal triangulation method. The NO MODULE WAIT command is employed at several points in the program (1) whenever the capping or uncapping operation is used, (2) during the stir operation, and (3) to fill syringes while the sample jar lid is being placed in the lid station. This command allows the robot hand to be used in synchrony with the capping station and in the last instance for the MLS syringes to fill while the robot arm is preparing the jar for inoculation. All the alarm programs are written using timers and switches to activate the buzzers via the PEC. The alarm programs are very brief and DO LOOPS are used to achieve a repetitious sound. The stirring program is also on a DO LOOP.

During the stir operation the capper is, as mentioned earlier, instruced to achieve full torque. This command is given at several points in the stir program while the robot arm is instructed to move the stick to various positions in the jar to achieve mixing. In the stir program, the DO LOOP is used to run the program through two cycles. This allows operator discretion to be used should some difficulty be encountered with thoroughness of mixing with especially viscous samples the stir program may be increased simply by instructing the controller to execute the stir.

DISCUSSION AND SUMMARY

In summary, we have been able to show the utility of the Zymate System in performing a routine portion of the cosmetic preservation test as carried out in our laboratory. This was initiated as an exploratory venture, but it is apparent from our experience that the system can be expanded to handle more than the thirty samples initially used in the qualifying tests. This application did not generate highly technical data which was subjected to statistical analysis. The functions of the test are primarily mechanical in nature and a moderate degree of variability is acceptable compared to methods where exact quantitative assays are carried out. This can be clarified more to indicate that the data are handled largely on a judgemental basis, and interest is primarily in the antimicrobial effect of the preservatives expressed as orders of magnitude. Initial challenge levels to the individual samples are on the order of log 5.0 - 7.0. In addition, several data sets on each formula are examined in order to make a final decision. The usual criteria which must be applied in making a pass/fail judgement on a particular sample are not covered here, but they include ratings of desired presrvative activity based not only on the reduction achieved during the test but also, on the configuration of the container, the manner in which the product will be used or applied and the degree of contact the user will have with the product in question. In short, the usefulness of the robot in this instance does not require a high degree of accuracy, only dependability and reasonable reproducibility within the manipulative parameters of the method. The importance of feedback controls cannot be overstressed. The operation was found to be very reliable but at this point not at a stage where the robot could be left totally unattended in the evening for instance. This is primarily due to the non-standardized nature of the jar lids used to hold the samples. This will be explored further in the future but the jars are standard items described in the ASTM Method for testing cosmetic preservatives and they have been especially selected as acceptable, not containing any antimicrobial adhesives or other such materials in the jar lid. We currently have experienced approximately one failure in a process involving 90 samples, usually associated with an operation involving the lid or the stirring sticks.

The other non-standardized items used are the coffee-stirrer type wooden sticks. These must be selected to be free of warping, a time-consuming process. We are currently exploring use of more uniform plastic sticks which can be loaded into a stick dispensing machine. We anticipate this should improve performance considerably.

FUTURE PLANS

Complete automation of the method is planned as follows:

1. Integration of a stick dispensing machine which will allow expansion of the number of samples which may be processed during a single program. The goal is to expand to a capacity of 120 samples.
2. Replacement of the wooden stick stirrers with plastic sticks of a more uniform nature to reduce failures due to warped wooden sticks.
3. Interface the system with a modified spiral dilutor using prepoured plates.
4. Replacement of the sample jar lids with lids of a more unifrom construction to facilitate manipulation by the robot fingers.

ACKNOWLEDGEMENTS

The help of the Zymark staff, in particular Nancy Robertson, is gratefully acknowledged.

BIBLIOGRAPHY

American Society for Testing and Materials, Standard Test Method for Preservatives in Water-Containing Cosmetics, ANSI/ASTM E-640-78, May 1978.

C. Haynes and N. Estrin, eds., A Guideline for Preservation Testing of Aqueous Liquid and Semi-Liquid Eye Cosemtics, 2nd ed., CTFA Technical Guideline, The Cosmetic, Toiletry and Fragrance Association, Washington, D.C., 1982.

Smith, J. L., Cosmetics and Toiletries, 96 (March):39-41 (1981).

Smith J. L., in The Cosmetic Industry, Scientific and Regulatory Foundations, Norman F. Estrin, ed., Marcel Dekker (1984) New York, pp. 316-318.

USPXXI, Antimicrobial Preservatives - Effectiveness., p. 1151, (1985).

GENERAL PURPOSE ROBOTIC PREPARATION OF COMPOSITE TABLET SAMPLES FOR HPLC ANALYSIS

Guy W. Inman Jr. and David D. Elks
Analytical Development Laboratories
Burroughs Wellcome Company
P. O. Box 1887
Greenville, NC 27834

ABSTRACT

An automated system for preparing composite tablet samples for HPLC analysis has been developed. A typical batch of ten samples consisting of 5 to 32 tablets per sample can be serially processed in approximately three hours. Analyst intervention is required only at the beginning and end of the preparation sequence, which consists of the following steps: the robot tares a sample, adds diluent, homogenizes the sample to disintegrate the tablets, dilutes the sample when necessary, removes and filters replicate aliquots, dispenses each aliquot into a Micromeritics HPLC vial, and caps each vial. After this sequence is repeated for each sample, the analyst removes the vials and places them on an autoinjector tray for HPLC analysis. The robot can select, via a series of 3-way teflon valves and a power and event control station, one of three diluent reservoirs. Diluent is first added rapidly with a magnetic gear-drive pump and then more slowly with a master laboratory station. All diluent volumes, including those associated with the dilution step, are determined on a weight basis with an electronic balance. Tablets are disintegrated with a Brinkmann Polytron homogenizer, which is connected to the system through a mechanical relay and a power and event control station. hplc vials are capped with a vacuum-actuated accessory, designed and fabricated in-house, which attaches to a syringe hand. vials and caps reside in a special rack that allows the capper to pick up a cap by vacuum and force the cap into a vial.

methods were validated for lanoxin 0.125 and 0.250 mg tablets and sudafed 30 and 60 mg sugar-coated tablets. the system is currently being used by quality assurance to prepare production samples and by analytical development to develop new procedures. this flexibility is possible because samples can be prepared during the night, the system's design does not limit its application to a single product, and the system is not dedicated to a particular HPLC system or analytical technique.

For typical batches of ten samples, the occupied analyst times related to sample preparation (4.5 hours for lanoxin and 2.5 hours for Sudafed) were reduced to 45 minutes.

INTRODUCTION

A Zymate System was purchased by Burroughs Wellcome Co. to help reduce the amount of analyst time required for the preparation of composite tablet samples for HPLC analysis. Two major problems had to be solved before the system could be considered successful. The first was to find a method for rapidly and reliably disintegrating tablets and extracting all of the active ingredients from excipient materials. The second was to make the system flexible. No product tested by our Quality Assurance Laboratories had a large enough sample load to justify a dedicated Zymate System. Varying production schedules and timetables meant that the system would need to switch from one application to another with minimal setup time between applications. A similar situation existed in the Analytical Development Laboratories where there was an even greater sample variety. The uncertainty of final product approval at this stage of assay development was an added impetus for flexibility. Our laboratory environment needed a system, not dedicated to a single HPLC station, that could prepare a group of samples, store those samples, and then be ready to prepare a different group of samples.

A computer-controlled Technicon SOLID Prep II, which is normally used as part of an Auto-Analyzer train for content uniformity assays, was used in our early attempts at automating composite (5 to 20+ tablets) sample preparations. This device had serious limitations for composite sample preparations:

1. Its maximum volume was limited to 125 mL.
2. In-line sonication was not possible.
3. In-line filtration was difficult if not impossible for composite samples
4. The amount of added diluent could not be verified.

The Zymate System allowed us to adapt a more suitable device, a Brinkmann Polytron tissue homogenizer, for tablet disintegration. Equipped with an appropriate generator tip, the Polytron can handle diluent volumes up to 2.5 liters and will sonicate when operated at high speeds.

Flexibility was achieved through system design. The addition of a capping device to the Zymate System to prepare samples for HPLC injections circumvented the problematical nature of HPLC - allowing sample preparation to proceed independent of the operating condition of the HPLC system. The vacuum-actuated device, designed and fabricated in-house, was attached to a syringe hand and used to cap Micromeritics HPLC vials, which are widely used in our Development and Quality Assurance Laboratories.

DEVELOPMENT

Our primary goal was to ensure that the Polytron tissue homogenizer could successfully disintegrate tablets and remove the active ingredient from excipient materials as efficiently as conventional laboratory methods - such as shaking followed by extended treatment in a sonic bath. Other areas of development included optimizing the speed and accuracy of diluent delivery by the addition of a pump and electronic balance, determining the extent of sample cross-contamination and diluent evaporation, and developing a capping device for Micromeritics HPLC vials.

Homogenization (Tablet Disintegration and Active Recovery)

A Brinkmann PT 10-35 Polytron homogenizer, equipped with a PT20ST generator probe and interfaced to the system through a power and event control station and a mechanical relay, was used in all sample preparations. Composite samples of either 16 (0.250 mg) or 32 (0.125 mg) Lanoxin tablets could be disintegrated and the active ingredient, digoxin, dissolved in two minutes with the Polytron set at speed 10. As shown by Table 1, disintegration was almost complete for Sudafed tablets after about one minute. For these products, two minutes of homogenization with the Polytron was equivalent to 30 minutes on a sonic bath with intermittent swirling. This efficiency was partially attributed to special 250 mL Erlenmeyer flasks, modified by a local glassblower, with four rectangular baffles. These baffles helped disrupt the swirling action and forced tablet pieces back into the probe tip. A disadvantage of the Polytron is the obnoxious and painful high-frequency sound emitted by the generator and probe assembly. The sound intensity was reduced by installing the Polytron in a foam-lined wooden enclosure and by placing the Zymate System in its own room.

TABLE 1. Effect of Homogenization Time on HPLC Assay Results for Sudafed S/C Tablets.

Time (Seconds)	Mean % I.s. for 5 Samples Based on Peak Area	Std. Dev.	Range
15	94.3	7.9	20.6
30	100.4	7.1	19.6
60	97.6	1.2	2.9
150	98.3	0.7	2.3

Diluent Addition and Dilution

Although adequate in the early stages of development, the master lab station could not deliver relatively large volumes (100 mL or greater) of diluent with the speed and accuracy we desired. A rapid diluent delivery system was developed using the following components: a magnetically-coupled gear-drive pump (Model 405 MicroPump), a series of Teflon three-way valves (Neptune Research Model 360T), a Mettler PE 1600 top-loading balance, a portable dispenser, and a master lab station. Diluent was first added rapidly (about 85 mL in 10 seconds) with the MicroPump to within 10% of a target weight and then topped off with the master lab station. The speed of the MicroPump was controlled through a variable AC outlet on a power and event control station. Dispensed volumes were determined by the software from the known diluent density and the diluent weight read from the balance. In addition to being more accurate (+0.01 mL) than either the master lab station alone or manual delivery, the system recorded and printed the actual dispensed volumes for each sample.

Dilutions, necessary only in the Sudafed preparation, were also made on a weight basis using a pipet hand, 10 mL pipet tips, the master lab station, and a vortexing station.

Capping Device for HPLC Vials

A simple capping device, designed in-house, was fabricated from Teflon and mounted on a syringe hand. A special rack was built to hold Micromeritics HPLC vials and their polyethylene caps, which were placed on small posts behind each vial. The robot capped the vials by rotating the syringe hand 180 degrees, moving the capper over a cap, turning on the vacuum pump, switching on the vacuum/pressure valve to the vacuum position, and withdrawing the cap into a tapered cylindrical shaft inside the capper. The capper then moved to a clear position over a vial and moved down over the vial while simultaneously switching the vacuum/pressure valve to pressure. The cap was forced partially into the vial forming an airtight seal. Prior to placing the vials on the Autoinjector tray, the analyst pushed the caps the remaining distance into the vials.

Sample Cross-contamination

The sample cross-contamination due to sample solution remaining on the homogenizer probe tip was approximately 0.9% for a blank solutions following sample solutions. Since the amount of carryover for similar samples would be insignificant (about 0.1% for samples that differed by 10%), the tip was not cleaned between each sample.

Diluent Evaporation

Since the Lanoxin procedure used a 50:50 alcohol:water diluent, we were concerned that diluent evaporation might be significant. The average sample loss during homogenization per 100 mL of sample was 0.48 g. The average amount of sample which adhered to the generator probe tip was 0.42 g. The difference, about 0.06 g or 0.06%, attributed to evaporation was not significant.

Reproducibility (Manual Versus Robotic Sample Preparation)

Tables 2, 3 and 4 compare the HPLC assay results for sample and pseudoephedrine hydrochloride standard solutions prepared manually and by the robot. Good agreement between the methods was observed. Statistical F- and T-tests showed no significant differences in the variances was found for the pseudoephedrine standard solutions. However, this was a result of the exceptionally low standard deviation (<0.2%) for the manually prepared standards. Standard solutions prepared by the robot gave very good HPLC assay results with a standard deviation of 0.6%. The sample preparation sequence for composite tablet samples is given in Table 5.

TABLE 2. HPLC Assay Reproducibility for Lanoxin Tablets Automated and Manual Preparation Percent label Strength by Peak Area.

	0.125 mg		0.250 mg	
Sample	Manual	Automated[1]	Manual	Automated[1]
1	99.1	97.3	98.6	98.3
2	98.0	96.8	100.4	98.8
3	97.9	98.0	99.8	98.9
4	97.5	99.0	100.2	99.6
5	98.1	98.0	100.8	98.7
6	98.0	99.0	99.3	99.0
7	96.3	97.7	98.2	97.9
8	97.9	98.6	99.8	100.0
9	98.1	98.7	100.4	100.3
mean	97.9	98.1	99.7	99.0
Std. Dev.	0.729	0.769	0.871	0.780
% RSD	0.745	0.784	0.874	0.787
Range	2.8	2.2	2.6	2.0

[1]Values corrected for diluent displaced by sample.

TABLE 3. HPLC Reproducibility for Sudafed S/C Tablets Automated and Manual Preparaton Percent Strength by Peak Area.

	30 mg		60 mg	
Sample	Manual	Automated[1]	Manual	Automated[1]
1	95.8	93.9	99.7	101.0
2	97.2	97.4	99.7	99.9
3	95.8	96.7	101.0	100.6
4	96.4	97.5	101.3	99.6
5	96.9	98.5	102.2	100.4
6	95.9	98.1	99.8	102.5
7	97.2	96.8	98.9	102.2
8	96.9	95.2	99.2	99.9
9	98.3	97.0	100.3	99.8
10	96.8	95.9	99.7	101.2
11	96.1	97.4	99.6	102.6
12	97.4	94.8	99.7	99.9
13	95.4	98.9	100.3	99.2
14	95.7	97.8	99.4	99.2
15	98.3	97.9	99.2	102.0
mean	96.7	96.9	100.0	100.7
Std. Dev.	0.910	1.41	0.891	1.17
% RSD	0.941	1.46	0.891	1.16
Range	2.5	5.0	3.4	2.8

[1]Values corrected for diluent displaced by sample.

TABLE 4. HPLC Assay Reproducibility for Psuedoephedrine HCL Reference Standard Automated and Manual Preparation Percent Label Strength by Peak Area.

Sample	Manual	Automated[1]
1	100.3	100.8
2	100.5	99.8
3	100.5	99.4
4	100.4	100.0
5	100.6	100.4
6	100.3	100.9
7	100.2	100.3
8	100.5	100.7
9	100.2	99.9
Mean	100.4	100.2
Std. Dev.	0.145	0.508
% RSD	0.144	0.507
Range	0.3	1.5

[1]Values corrected for diluent displaced by reference standard.

TABLE 5. Sample Preparation Sequence

System Setup

1. Check and, if necessary, add diluent to the reservoir.
2. Attach filters to 1 mL pipet tips and ensure that the correct number of pipet tips with filters, 10 mL pipet tips, test tubes, and sample flasks are present and located properly.
3. Ensure that the robot arm is positioned properly.
4. Add the correct number of tablets to the sample flasks.
5. Enter the number of samples, standards, and standard weights in to the controller.
6. Initiate the procedure.

Robotic Sample Preparation

1. Transport a sample flask with tablets to the balance and tare the sample.
2. Add diluent to the sample.
3. Homogenize the sample.
4. Transfer a sample aliquot to a test tube.*
5. Dilute the sample aliquot and vortex.*
6. Attach a pipet tip with filter and withdraw an aliquot of the prepared sample solution.
7. Shuck the filter and transfer the filtered aliquot to an HPLC vial.
8. Cap the vial and shuck the used pipet tip.
9. Repeat steps 6 through 8 for the same prepared sample solution.
10. Repeat steps 1 through 9 for each sample and standard.+

* Dilution was not necessary for Lanoxin Tablets.
\+ Digoxin standard solutions were prepared manually.

CONCLUSIONS

The robotic sample preparation system prepared composite Lanoxin and Sudafed Tablet samples that were equivalent to those prepared manually by established procedures. For typical batches of ten samples, the occcupied analyst times related to sample preparation (4.5 hours for Lanoxin and 2.5 hours for Sudafed) were reduced to 45 minutes. The total assay times were also reduced, because the analyst could perform a number of tasks (mobile phase preparation, interaction with the laboratory computer system and system suitability tests) in parallel with the robot. Additional advantages of the Sudafed system were a 90% reduction in solvent consumption and the elimination of duplicate sample preparations. No hardware changes were required to shift from a Lanoxin to a Sudafed procedure. Since samples were stored in capped vials, completion of a preparation sequence was not dependent on the operating state of the HPLC system. Although manual intervention was required to transport samples to an instrument, the most labor intensive portions of the assays were automated. Robotic diluent delivery on a weight basis was more accurate and precise than manual delivery and the actual weights or volumes delivered were printed for each sample. The system's flexibility has allowed the development of new procedures to proceed in parallel with production-related sample preparation.

Future plans for the system include expansion of its dilution range, the addition of liquid-liquid and solid-phase extraction capabilities, and sample preparations for quantitative FT-IR, FT-NMR, and UV analyses.

Zymark and Zymate are registered trademarks of Zymark Corporation.
Lanoxin and Sudafed are registered trademarks of Burroughs Wellcome Co.

FULLY AUTOMATED DISSOLUTION TESTING

L. J. Kostek, B. A. Brown
and J. E. Curley
Pfizer Inc.
Groton, CT

ABSTRACT

Tablet dissoultion rate testing has become one of the more common pharmaceutical laboratory analyses. More than 400 tablet and capsule monographs in USP XXI have dissoultion testing requirements. A robotic method has been developed that runs up to 12 sets of six tablets without manual intervention. The robot adds dissolution test medium to a standard testing apparatus, introduces tablets, samples at pre-set time intervals, assays aliquots and calculates results. The robot then prepares the test vessels for the next assay by emptying, washing and refilling them with test medium. Each vessel is checked for cleanliness before the robot proceeds. The Zymate System is interfaced with a DEC Pro 350 computer for preparation of assay reports.

The development of this system and its validation using USP Prednisone calibrating tablets are described. Benefits of this system include dramatic cost savings and compression of the time required to gather data. By running overnight the robot can perform in a 24-hour period what a person would accomplish in the usual 40-hour work week.

INTRODUCTION

Dissolution rate testing is performed on tablets and capsules to evaluate the rate of drug release under uniform conditions. The preamble to USP XXI points out that in the past decade further development of commercial dissolution equipment and more extensive experience of analysts have made the dissolution test a more mature technology. Almost 400 drug monographs now have dissolution test requirements.[1] Dissolution rate data are used both as a quality control test during manufacture and as an indicator of bioavailability. The dissolution rate test is perhaps the fastest growing demand on any pharmaceutical testing laboratory.[2]

The apparatus for the test is described in the United States Pharmacopeia.[3] The approach is to drop a capsule or tablet into each of six kettles that are agitated by paddles. At predetermined intervals, say every 15 minutes for an hour, a sample is withdrawn and assayed for dissolved drug content. Figure 1 depicts two hypothetical dissolution rate profiles. The left hand curve depicts the normal or expected behavior. The right hand curve indicates a cause for concern, because the rate of dissolution is slower than expected.[4]

Because of the usefulness of this test, a significant demand has arisen for it to be run on samples associated with formulation development, clinical supply release and stability evaluations.

For just one product under development at Pfizer, 1100 dissolution tests were performed in a 10 month period.

The steps of a dissolution test are shown in Table 1. The manual method

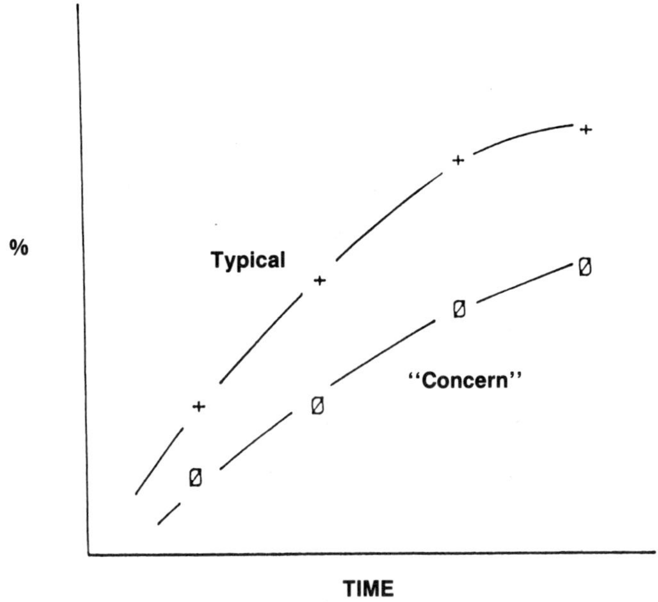

FIGURE 1. Hypothetical dissolution rate profiles.

Table 1. Time Analysis of Steps in Dissolution Rate Tests.

Step	Personal Time Required, Minutes Manual	Semi-Automated Robot
Preheat Bath and Test Fluids, clean testing vessels	20	20
Prepare glassware and filters	30	—
Drop tablets at timed intervals Withdraw aliquots at timed intervals Filter and dilute aliquots	60	5
Measure drug concentration spectrophotometrically	20	--
Calculate Results	15	--
Record Results in legal record	15	15
TOTALS	160	40

requires nearly three hours for a single determination.

In the fall of 1984,[4] we reported how the Zymate System had been developed to perform the dissolution rate test on six tablets according to USP Method II. The robot was capable of taking each tablet in turn from a rack, introducing each to a dissolution vessel, sampling at pre-set time intervals, assaying, calculating and printing results. Although the technician was spared the tedium of doing the assay, a person was still required to add testing medium to each vessel at the beginning of each run and to clean the vessels afterwards.

Table 1 shows that this semi-automated approach reduced the person-time required from 160 minutes to 40 minutes. Although this is a significant reduction, our goal was complete automation.

This paper describes a fully automated system that achieves the steps of cleaning vessels and fillng them with test fluid before each run. We have now developed a system that can test up to 12 sets of six tablets without manual intervention.[5] The key to developing this fully automated procedure was designing robot hands capable of filling and washing vessels.

EQUIPMENT

Automatically filling the test vessels with dissolution medium became a rather simple task using a commercially available dispenser. The Hanson Media Mate has an 80 liter tank for heating and storing test fluid. The unit also degasses and dispenses a measured amount of medium for all six

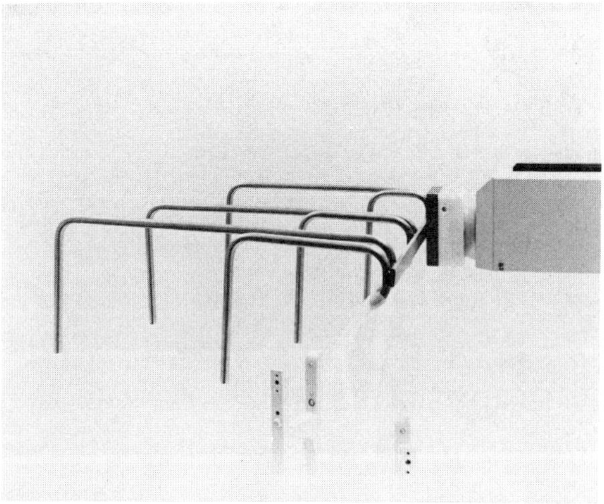

FIGURE 2. Dissolution vessel filler.

vessels simultaneously. The trick was to design a custom robot handcapable of delivering the medium to the test vessels. This hand consists of six different lengths of stainless steel tubing (Figure 2). The tubes are welded onto a stainless steel bar. These tubes are spaced in a manner which allows access to all six dissolution test vessels simultaneously.

The automation of the filling routine involved interfacing the medium dispensing unit through the power and event controller. Software was written that placed the fill hand over the vessels and then dispensed a measured amount of degassed, pre-heated fluid into the test vessels. Test medium is held at 39°C to allow for cooling during dispensing.

Washing each vessel between runs was the more difficult task. Several concepts were considered including removing each vessel and inverting them over a whirling brush or having a robot hand contain a scrub brush. The final wash hand design is shown in Figure 3. The hand we fabricated is made from two pieces of stainless steel tubing. One tube is used for draining the vessels and the other tube is used for spraying rinse water into the vessels. The drain tube is equipped with a small piece of flexible tubing. At the end of this flexible tubing is a small piece of beveled and notched stainless steel tubing (Figure 4). The small piece of stainless steel tubing helps the drain tube bend around the dissolution paddles. It also keeps the drain tube from sticking to the walls of the test vessels. The second tube is equipped with a ten-hole nozzle (Figure 5) for spraying a cone-shaped stream of rinse water into

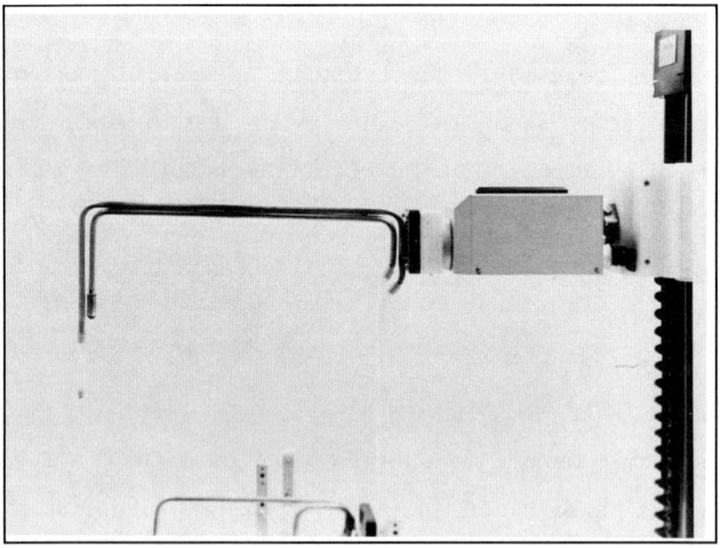

FIGURE 3. Dissolution vessel washer.

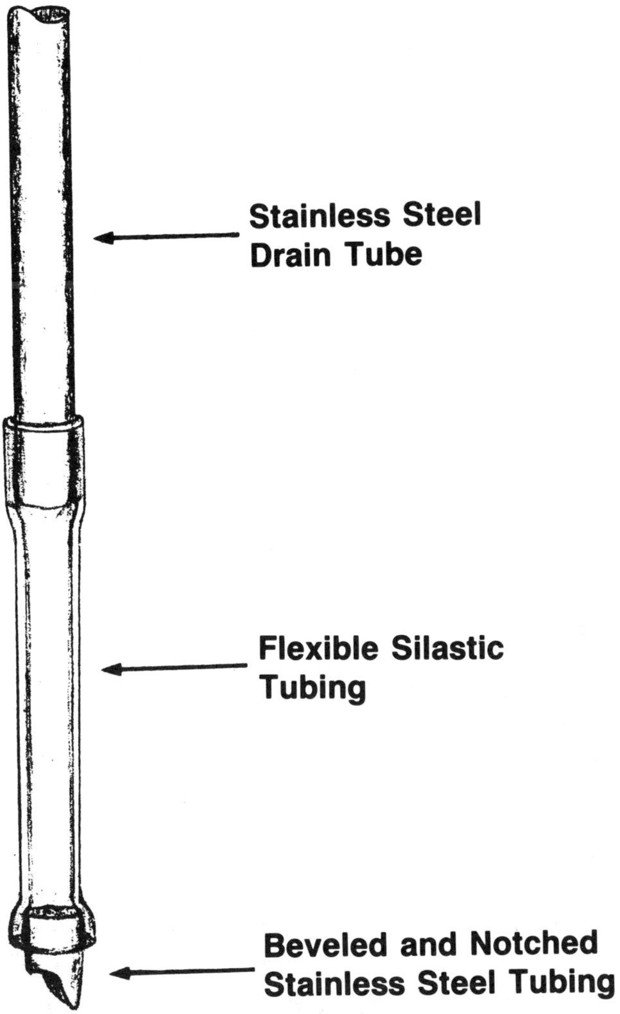

FIGURE 4. Wash hand drain tube.

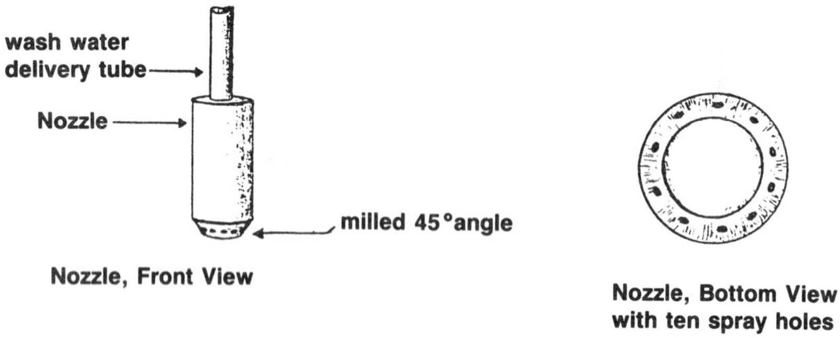

FIGURE 5. Spray nozzle.

the dissolution vessels. The rinse water is domestic hot water filtered through a cartridge type filter to remove particulates. An electronically activated solenoid switching valve was connected to a solid state relay and placed in line after the filter. The relay was interfaced with the power and event controller. Programming was written to move this hand from vessel to vessel for draining and cleaning.

Several final touches were added to guarantee trouble free operations. Verification switches were added to check that fill and dispense hands were in place before fluids were dispensed. This prevents flooding. A thermistor measures fluid temperature before samples are added to test vessels, thus ensuring the USP requisite of testing at $37^{\circ}C \pm 0.5^{\circ}C$ is met. The system was interfaced with an IBM model 5150 computer for report generation. The final bench layout is shown in Figure 6.

The setup for the dissolution system includes a Zymark robot with controller, power and event controller, printer, sample rack, filter rack and filter shucker. Five individual custom robotic hands are required:

1) Sample Handler (Figure 7)
2) Solution Sipper (Figure 8)
3) Dissolution Vessel Filler (Figure 2)
4) Dissolution Vessel Washer (Figure 3)
5) Temperature Recorder

The Zymark components of the system are interfaced with a heated dissolution bath, a spectrophotometer, a supply of heated dissolution medium, domestic hot water and a IBM model 5150 Computer. A list of equipment used is given in Table 2.

PROCEDURE

Figure 9 shows the sequence of events for automated tablet dissolution. Setting up the system for automated dissolution testing involves the following steps: fill the sample rack with appropriate tables; enter run numbers, tablet lot identities, standard concentration and tablet potency into the IBM computer; prepare a standard solution of drug for spectrophotometric calibration, add filters to the filter rack, set dissolution bath paddle speed and check temperatures of medium and dissolution baths. This preparation takes approximately one half hour.

At the start of the automated dissolution procedure, the robot attaches

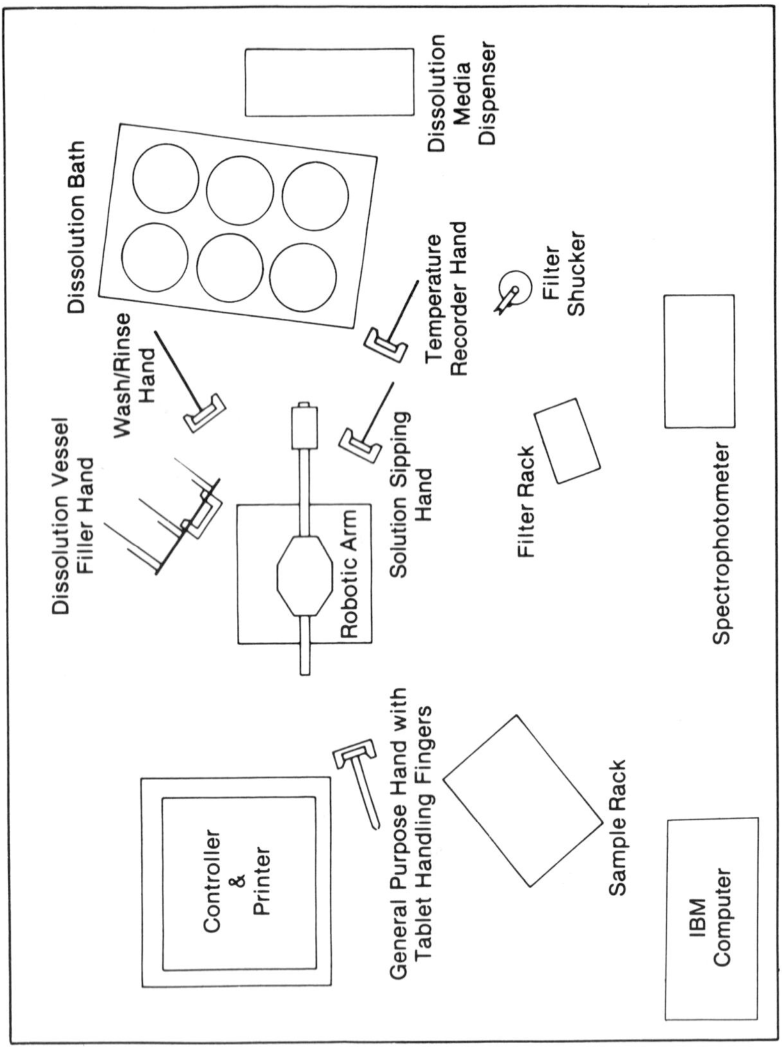

FIGURE 6. Diagram of robotic system for fully automated dissolution testing.

FIGURE 7. Sampler handler.

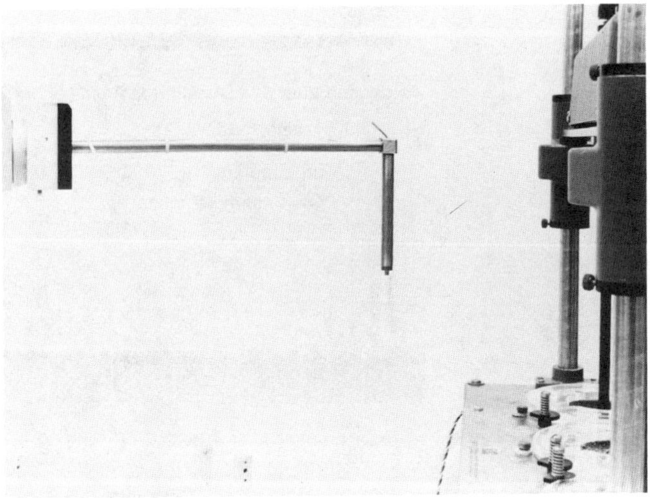

FIGURE 8. Solution sipper.

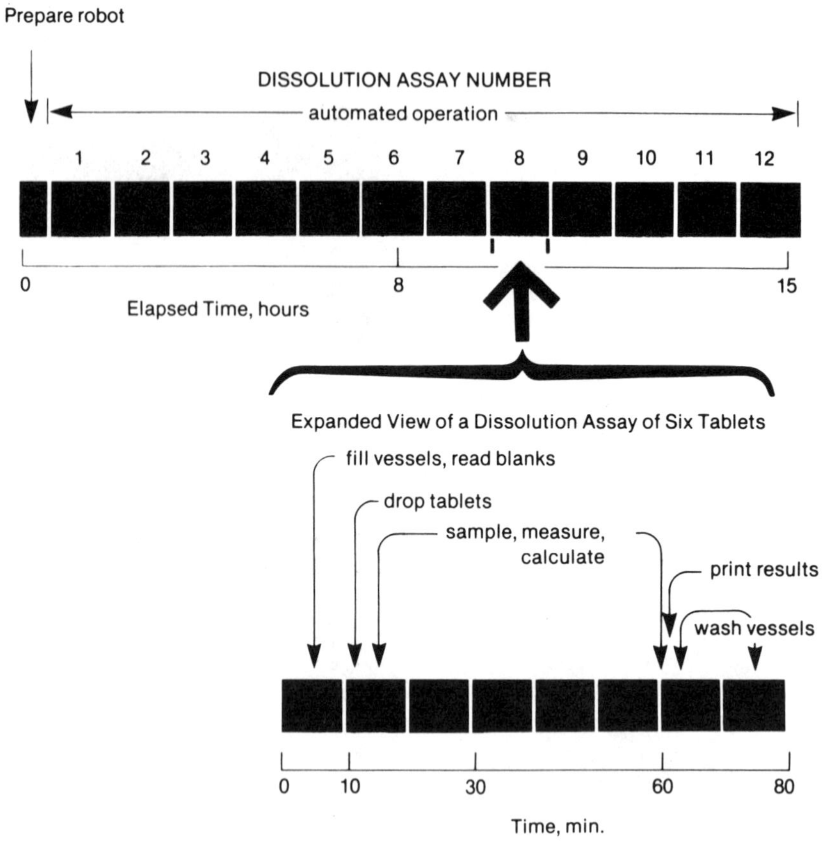

FIGURE 9. Sequence for automated tablet dissolution.

TABLE 2. Components of Automated Dissolution System.

Zymate Robotic System	
Zymate Controller	
Zyamte Printer	
Zymate Arm	
Zymate Power and Event Controller	
Tablet Handling Forcep	
Solution Sipping Finger*	
Dissolution Vessel Filler*	
Wash/Rinse Finger*	
Temperature Recorder*	
2 Check Switches	
Filter Schucker	
Filter Rack	
Sample Rack	*Pfizer design
Gilford 240 Spectrophotometer (substitutes acceptable)	
Hanson Dissolution Test Unit 72 RL	
Hanson Media Mate	
Sage Instruments Pump 371	
Masterflex Pump 7019	
Automatic Switch Company Solenoid	
Cartridge Type Water Filter	
IBM Computer 5150	

the dissolution filler hand, verifies the hand pickup and proceeds to fill the dissolution vessels with 920 mL of test medium. Next, the robot checks dissolution medium temperature with the temperature recorder hand. If the temperature is outside the specified range of $37^{o}C \pm 0.5^{o}C$, the robot will wait and check again after 5 minutes. When a proper temperature is achieved, that temperature will be printed and the robot will proceed to attach the solution sipper hand. A filter is attached and its presence verified. The robot proceeds to check each vessel for cleanliness. A 20 mL aliquot is removed from each vessel, an absorbance blank is recorded and an average blank for all six vessels is calculated. If a blank absorbance reading from any one of the test vessels is higher or lower than a predetermined limit, the robot will rewash and refill all six vessels. The robot will again check the blank. If a second failure occurs, the system will shut down.

After obtaining the blank readings, the robot samples the standard solution of drug under test. A standard absorbance is determined and stored in the robot controller for future calculations. This standard is measured only once, prior to the start of the first run of a dissolution setup. The computer checks to see that this measurement is within $\pm 3\%$ of the expected value. If not, the system is shut down. Next, the robot attaches the sample handler hand and proceeds to drop one tablet into each of the six dissolution vessels. With the drop of the first tablet, five timers come into play. One timer is set and reset for 45 seconds delay between individual tablet drops. This sequence allows all six vessels to be sampled at exactly the same time intervals. Four other timers control the prescribed intervals for sampling. Typical times used are 5, 15, 30 and 45 minutes. After all six tablets are dropped, the robot will attach the solution sipper hand and prepare to sample the test vessels. At the appropriate times, filtered dissolution medium is pumped through a flow cell into a spectrophotometer. The absorbance readings are sent to the robot controller and the percent dissolved calculations are performed. The data are recorded on the printer plus stored in an array in the robot controller. When the sampling routines are completed, the robot controller prints the assay results. In addition, the controller passes the accumulated data to the IBM Computer for the preparation of a report. A typical report is shown in Figure 10.

When the assay is completed, the robot attaches the dissolution vessel washer hand, verifies the hand pickup and begins the vessel cleaning routine. A single vessel cleaning routine consists of a total drain of the test vessel followed by a 2 second spray of rinse water, a 7 second pause to drain the rinse water and another spray/drain cycle. The wash hand then moves to the top of the test vessel and emits a 2 second spray of rinse water to rinse the vessel walls. The hand moves back down, drains the vessel and performs two more spray/drain cycles at the bottom of the vessel. This is followed by another rinse at the top of the vessel and two more spray/drain cycles at the bottom. This whole routine of 8 spray/drain cycles is performed in approximately two and one half minutes. A total of 450 mL of rinse water is used per vessel. All six vessels are cleaned in approximately 15 minutes. After finishing the wash cycle for the six vessels, the robot proceeds to attach the dissolution filler hand. The whole dissolution cycle begins again and continues until the designated tests are completed.

RESULTS

The performance of this robotic system was evaluated using 50 mg USP prednisone tablets. These tablets are used universally to calibrate dissolution apparati. To evaluate the equivalency of the robotic system to the manual procedure, concurrent assays of the same set of tablets

Dissolution Rate Testing by Robotics Unit #1

Run Number = 1
Notebook Serial Number: 12345
Lot Number: ROBOT TEST
Dose: 50 mg Potency = .4
Dilution Factor = 900 Concentration = 207
S.T.P. Number: U.S.P. F.I.D. #: NONE
Stability Storage Conditions: ROOM TEMPERATURE
Sample Supplied by: L. KOSTEK Operator: RFC
Q Value: 80 Q Interval Test Performed at: 30 Min
Comments: 50 RPM
Date: 10-09-85

	TABLET NUMBER						
% Dissolved after	***1***	***2***	***3***	***4***	***5***	***6***	***Avg.***
15 min	**69**	**70**	**70**	**70**	**71**	**71**	**70**
30 min	**71***	**71***	**72***	**72***	**90**	**91**	**78**
45 min	**91**	**91**	**92**	**92**	**92**	**92**	**92**
60 min	**93**	**93**	**93**	**94**	**94**	**94**	**94**

*** U.S.P. Criterion Failed**
****

	TABLET NUMBER						
Absorbance after	***1***	***2***	***3***	***4***	***5***	***6***	***Avg***
15 min	**23.1**	**23.2**	**23.3**	**23.4**	**23.5**	**23.6**	**23.35**
30 min	**23.7**	**23.8**	**23.9**	**24**	**30.1**	**30.2**	**25.95**
45 min	**30.3**	**30.4**	**30.5**	**30.6**	**30.7**	**30.8**	**30.55**
60 min	**30.9**	**31**	**31.1**	**31.2**	**31.3**	**31.4**	**31.15**

RUN START TIME: 2 : 17 : 22
Run Temperature = 37.1
Standard Constant = 3

FIGURE 10. IBM report.

were performed. A technician sampled each vessel at the same time as the robot. The technicain then filtered, diluted and assayed the samples. Table 3 describes the excellent agreement between results from the robotic system and the technician.

TABLE 3. Comparison of Manual and Robotic Methods.

Dissolution Rate Testing of Prednisone Tablets
USP Lot G
USP Method II Dissolution Media: Water
Paddle Speed: 100 rpm

		% Dissolved	
	Tablet #	Manual	Robot
5 Min.	1	24	20
	2	23	21
	3	24	22
	4	24	23
	5	23	22
	6	24	23
	Average	24	22
15 Min.	1	42	41
	2	42	41
	3	42	41
	4	43	42
	5	41	41
	6	43	42
	Average	42	41
30 Min.*	1	57	57
	2	58	57
	3	58	57
	4	58	58
	5	59	58
	6	59	59
	Average	58	58
45 Min.	1	67	68
	2	68	68
	3	68	68
	4	68	68
	5	68	68
	6	70	70
	Average	69	69

*USP specification for lot G: 48-67% dissolved after 30 minutes.

TABLE 4. Summary of Results.

Dissolution Rate Testing of Prednisone Tablets
USP Lot #G
USP Method II
Dissolution Media: Water

Tablet Set #	Robot Run	Average % Dissolved After 5 Minutes	15 Minutes	30 Minutes*	45 Minutes
1	Day 1	22	41	58	69
2		22	41	58	68
3		22	41	57	68
4		22	42	58	68
5		21	41	57	68
6		21	41	58	69
7	Day 2	22	42	59	69
8		22	42	58	69
9		22	41	57	68
10		22	42	58	69
11		23	45	62	73
12		22	44	61	72
13		22	42	59	70
14		22	41	57	68
15		23	43	58	69
16		22	42	58	68
17		22	41	58	68
18		22	41	58	68

*USP specification for Lot G: 48-67% after 30 minutes

The reproducibility of the method was demonstrated by gathering data on two separate days for USP 50 mg prednisone calibrators. These results are summarized in Table 4. Samples were taken at 5, 15, 30 and 45 minutes to demonstrate the utility of the system for determining dissolution rate profiles. Thus, the experiment on Day 1 for six sets of six tablets actually entailed 144 analytical measurements. This work was completed by the robot in 7.5 hours. The second day's experiment for 12 sets of tablets or 288 measurements required 15 hours of robot time. The 18 dissolution rate profiles in Table 4 were completed in just under 24 hours of robot time. We estimate that one person would require 6 working days to perform the same set of 18 analyses.

CONCLUSION

There are several advantages of a robotic approach. The obvious is labor savings. A half hour of setup time invested by a technician saves 32 hours of testing and calculating! The person is now freed up to do other, hopefully more interesting, work.

Another advantage of a robot is that because the rate of data output increases, the waiting time for data decreases. In formulation development, this allows for quicker decision making. Another aspect that makes robots well suited for dissolution testing is their tendency to produce highly accurate and reproducible data. The robot samples each vessel in the same place after an exact time interval. Humans just cannot repetitively work with this same precision.

Because we are encouraged by the success of this application, our next goal for dissolution testing is automating the USP method for capsule dissolution.

REFERENCES

1. The United States Pharmacopeia XXI, United States Pharmacopeial Convention, Inc., Rochville, MD. 1985, p. xlv.

2. L. M. Sattler, T. J. Saboe, and T. F. Dolan, "Automated Dissolution Testing", Zymark Application Note AP305, Zymark Corporation, Hopkinton, MA 01748, November 1984.

3. The United States Pharmacopeia XXI, in loc cit p. 1244.

4. J. Curley, in Advances in Laboratory Automation - Robotics 1984, G. L. Hawk and J. Strimaitis, eds., p. 299, Zymark Corporation, Hopkinton, MA, 1984.

5. L. J. Kostek, B. A. Brown and J. E. Curley "Fully Automated Tablet Dissolution Testing", Zymark Application Note AP307, Zymark Corporation, Hopkinton, MA 01748, April 1985.

Zymark and Zymate are registered trademarks of Zymark Corporation.
Media Mate is a registered trademark of Hanson Reseach Corporation.

A TOTALLY AUTOMATED ROBOTIC PROCEDURE FOR ASSAYING COMPOSITE SAMPLES WHICH NORMALLY REQUIRE LARGE VOLUME DILUTIONS

Allan Greenberg and Richard Young
Research Laboratories
Ortho Pharmaceutical Corp.
Raritan, NJ 08869

ABSTRACT

This paper presents the approach used by our laboratory to transfer an existing manual composite assay to a totally automated robotic method. The robotic method eliminates the large consumption of solvents and multiple dilutions previously required in the composite assay. Data will be presented to demonstrate the equivalence of the manual and robotic versions of the assay.

INTRODUCTION

Typically, the first day a robotic system is put in place it has to be demonstrated to everyone's satisfaction that the robot can indeed pick up glassware, place it where you want it and not smash anything in the process. Hopefully, the times when the robotic arm crashes into anything are few and far between. Usually, the first assignment given to a robot will be one in which a pre-existing method is to be transferred to the robot, verbatim. At this point in time, very little robotic method development time is allotted. There is some skepticism that the robotic system can perform the assays as well as or better than the human analyst. The steps and actions of the robot must mimic the manual method as much as possible. Several months later after congratulating themselves on what an excellent choice of equipment the robot was and how well it is performing, credibility in the quality, quantity, and reliability of the robotic performance begins to emerge. After several more months, when the junior chemists are swamped with the paper work resulting from the robotic assays and cannot keep up with its output, a second phase of development begins. This second phase begins with a lack of robot time for method development due to continuous routine analytical use and lends serious thoughts to the purchase of a second robotic system.

This second phase also encompasses an attitude toward the robot that will allow robotic method developemnt to be performed - not just simply the verbatim transfer of pre-existing methodology. It is this robotic method development of pharmaceutical assays that I am addressing which truly starts to use the capabilities of the robot. It is the next logical stage of development.

METHOD DEVELOPMENT CONSIDERATIONS

The manual version of the stability indicating assay on this particular pharmaceutical dosage form involved:

- making a sample composite by crushing 20 tablets
- weighing and transferring a specified sample weight of the powder

- extracting the analyte with solvent using shaking and ultrasonic agitation
- diluting the sample and adding internal standard
- followed by injection onto an HPLC system for quantitation versus a standard.

The sample weight was chosen to be large enough for an analyst to weigh accurately and to require minimal dilution in order to be within the standard linear response range of the detector. In this method, multiple dilution of the sample, was necessary to bring the sample size injected into the HPLC into the working linear range of both the HPLC detector and the standard response curve. The manual procedure required about 150 mL of extraction solvent per sample for a powdered sample assay and about 250 mL per content uniformity assay.

When first thoughts were put on paper about a transfer of the manual method to the robot, the following items were listed:

- gallons of solvent needed for the assay
- peristaltic pumps to deliver the extraction solvent
- gallons of waste solvent generated
- where to place the large volumetric flasks so that the robotic arm may access them
- the small capacity of the system because of the use of the large size volumetric flasks, etc.

Several constraints also complicate any new method developed for the robot:

1. Stability indicating properties to be kept intact
2. Accuracy and reproducibility to be similar or superior
3. Solubilities of analyte and internal standard were required to be large

4. Linear range of standard response curve amenable to change
5. The assay had to operate with minimal human interaction

Keeping the stability indicating properties of the original manual method intact would alleviate the need to recreate and retest stressed drug substance. By not changing the chromatographic system (i.e. the mobile phase or the HPLC column), revalidation would not be required. Not changing the stability indicating properties of the assay would also prevent any questions as to the relevance of the old or new assay data.

The analytical accuracy and reproducibility of the robotic method had to be similar or superior to the manual composite assay. A robotic method whose accuracy and reproducibility was significantly less than that of the manual version would not be an acceptable trade off for increased throughput.

The solubility of both the analyte and the internal standard have to be sufficiently large to allow the extraction of the analyte in a limited volume of solvent. Typically a method is validated for an assay in the range of 50 to 150% of the nominal dosage content. We chose to use a 5 tablet sample size as a representatibe composite sample. For a hypothetical dosage form containing 50% drug substance and a nominal dosage size of 400 mg, a five tablet composite sample would contain 2000 mg of drug substance. The requires validating the 1000 to 3000 mg range of analyte in a small extraction volume. The same type of solubility problems also existed with respect to the internal standard that was used.

The linear range of the standard response curve must be amenable to extension to either larger or smaller concentrations. On the high end of the concentration range, care must be taken not to exceed the linear response range of the detector or overload the column. There is also a possible problem as well on the low end of the concentration range. A certain minimum value for the area counts of the analyte is required. For accurate and reliable computer integration of our chromatographic peaks we normally like to have a minimum of 20 to 50,000 area counts.

The assay must run with minimal human assistance. Our goal was to run the assay as much as possible under total robotic control. There are limitations - someone has to "feed" the robot (i.e. extraction solvent, samples, etc. must be supplied by a human), but disintegrating and extracting hard tablets, weighing, diluting, filtering and injecting into an HPLC system should be accomplished by the robot.

The original manual stability indicating composite assay is illustrated in Figure 1. What was desired of the robotic method is shown in Figure 2.

We decided that the chromatography, a reversed phase ion-pair system, would be kept intact. A typical chromatogram of analyte and internal standard under assay conditions is shown in Figure 3.

The largest sample extraction or dilution volume that we could accomodate in our system was 30 mL in a 50 mL centrifuge tube. We were easily able to dissolve 3000 mg of analyte in the extraction solvent.

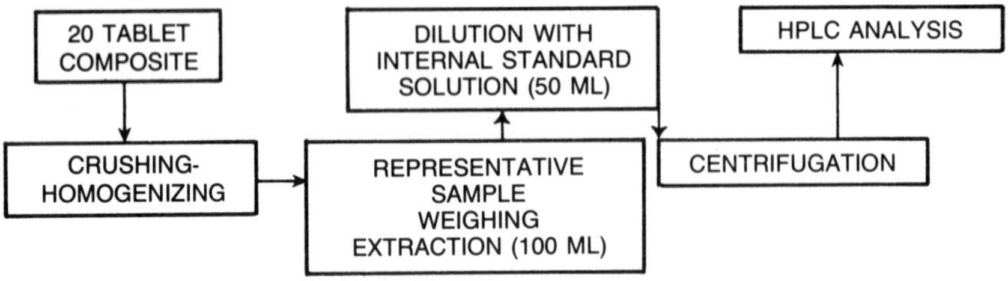

FIGURE 1. Flow diagram of a manual stability indicating composite assay.

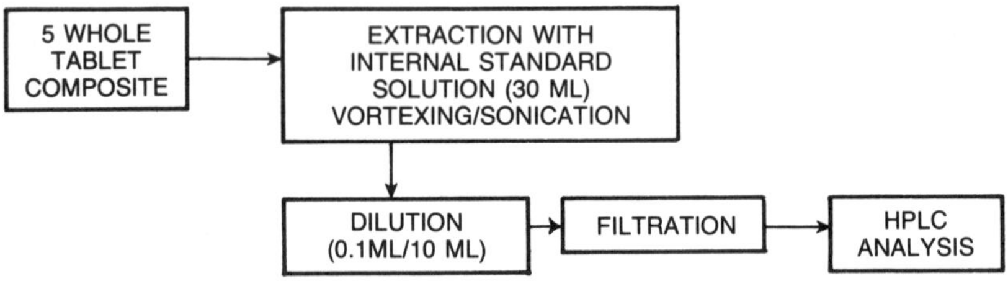

FIGURE 2. Diagram of an automated stability indicating composite assay.

However, the internal standard was not sufficiently soluble. A nearly saturated internal standard solution was employed for a linearity study of 75 to 125% of the composite sample range. A linear regression analysis of the data yielded an acceptable straight line with zero intercept. However, the area count ratio of analyte to internal standard peak was 16:1. This caused concern about the accuracy of the measurements of the area ratios.

A new internal standard was sought, one whose solubility and/or response at the analytical wavelength was greater than that of the previous internal standard. It was also required, in order not to affect the stability indicating properties of the method, that the new internal standard elute at a retention time similar to that of the old internal

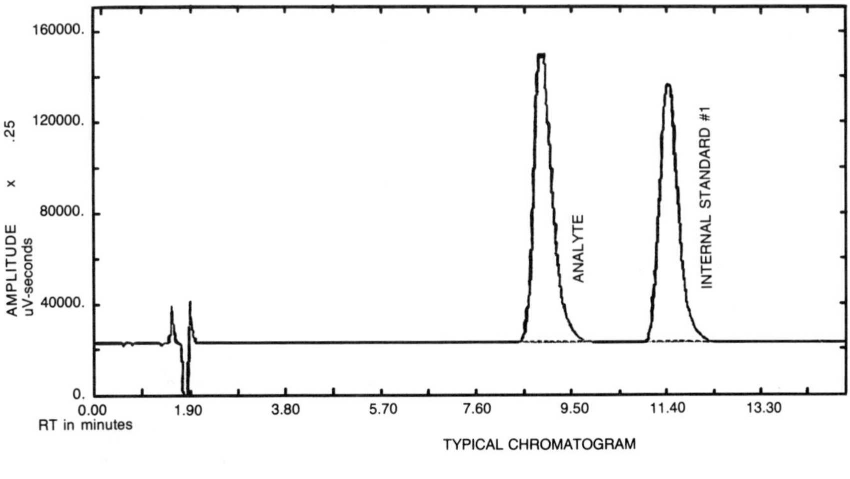

FIGURE 3. Chromatogram of an analyte and internal standard for composite assay.

standard. If this was not possible, then the new internal standard should elute in a region of the chromatogram that would not create an interference with any other peaks in the chromatogram, such as the analyte or potential degredation peaks. Fortunately, a new internal standard was found that eluted within 0.2 minutes of the old internal standard. At a typical dosage concentration level, the area count ratios of analyte to internal standard were 3:1. A typical chromatogram is shown below (Figure 4).

It should be noted here that the difference in chromatographic peak shapes between this chromatogram and the one previously shown is due to a change in the manufacturing procedure that produced the HPLC columns and is not related to the work being discussed here.

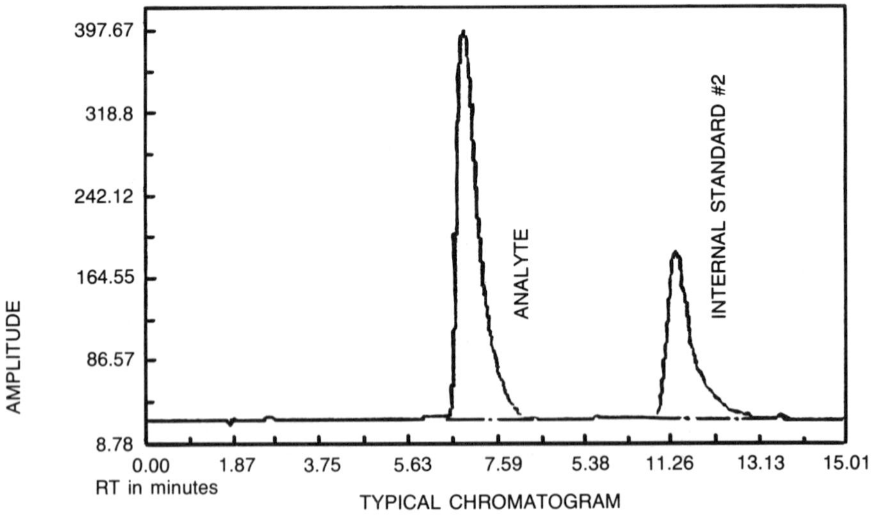

FIGURE 4. Chromatogram of new internal standard for composite assay.

The bench lay-out of our robotic system is shown in Figure 5. The master laboratory station, as well as the HPLC pumps and detector, are beneath the table top.

Recovery studies were repeated with the new internal standard. A linear regression analysis of the new linearity data yielded an acceptable straight line with zero intercept.

The recovery data for the maual version of the method is shown in Table 1. The average recovery was found to be 101% with a RSD of $\pm$1.19%. Synthetic samples were used in both the manual and robotic recovery studies, and peak areas were employed for quantitation.

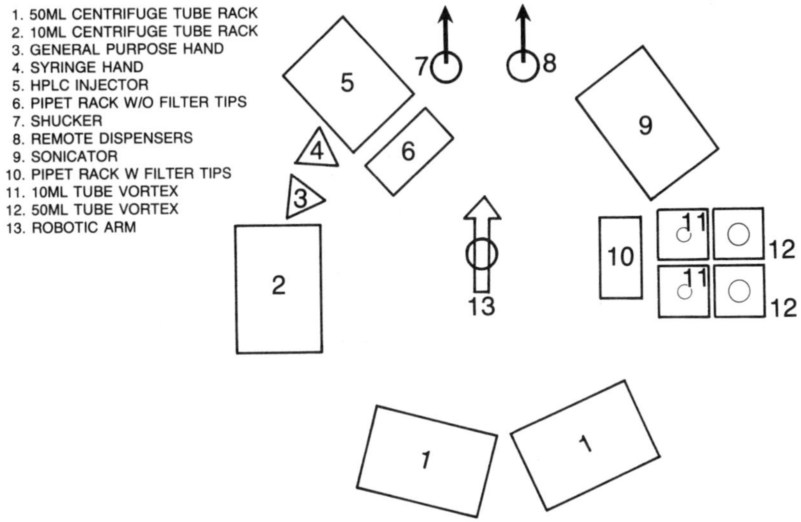

FIGURE 5. Bench layout for assaying composite samples by HPLC.

Table 1. Manual Recovery Studies for Composite Assay.

Nominal Composite Dosage (mg)	mg Added	mg Recovered	% Recovery
850	846.49	860	103
	846.92	862	102
	843.27	863	102
	857.70	875	102
	853.59	875	103
1130	1123.88	1140	101
	1183.14	1210	102
	1107.46	1130	102
	1126.61	1150	102
	1133.95	1150	101
1700	1728.70	1730	100
	1684.17	1700	101
	1709.10	1730	101
	1696.96	1710	101
	1704.31	1720	101
	1703.92	1720	101
2300	2330.02	2340	100
	2331.12	2330	100
	2309.83	2310	99.8
	2309.67	2310	100
2875	2882.18	2860	99.2
	2917.41	2910	99.7
	2857.43	2830	99.2
	2861.51	2830	99.0
	2859.28	2840	99.2
	2865.80	2850	99.4
			$\bar{x}_{26}$ = 101
			RSD = $\pm$ 1.19

The data from the recovery study employing a robotic version of the maual method is shown in Table 2. The average recovery was found to be 99.6% with a RSD of $\pm$1.50%.

Table 2. Robotic System Recovery Studies for Composite Assay.

Nominal Composite Dosage (mg)	mg Added	mg Recovered	% Recovery
850	853.1	858	101
	848.2	852	100
	853.3	857	100
	852.5	846	99.2
	850.6	847	99.6
1130	1131.8	1140	101
	1129.1	1120	99.2
	1130.0	1160	103
	1130.6	1110	98.2
	1134.0	1140	101
1700	1703.1	1700	99.8
	1702.7	1650	96.9
	1704.2	1680	98.6
	1699.6	1670	98.6
2300	2302.4	2276	98.9
	2301.8	2230	96.8
	2302.4	2254	97.9
	2303.3	2293	99.6
2875	2877.6	2890	100
	2876.7	2860	99.4
	2875.0	2880	100
	2872.2	2900	101
	2876.7	2940	102
			$\bar{x}_{23} = 99.6$
			RSD = $\pm$ 1.50

CONCLUSION

In conclusion, we have shown that:

1. The stability indicating properties of the original method remained intact
2. The extraction solvent and procedure were modified to allow the utilization of fairly concentrated solutions of analyte and internal standard
3. The standard response curve was extended to greater concentrations

4. The accuracy and precision of the manual and robotic methods were shown to be similar

5. The robotic composite method was run with minimal human assistance.

Zymark and Zymate are registered trademarks of Zymark Corporation.

DEVELOPMENT OF DATA-PROCESSING SYSTEM FOR STABILITY STUDIES (I)*

Reiji Shimizu, Masaaki Matsuo,
Takaaki Miyamoto, Yukio Shimaoka,
Hideyuki Mano, Kiyoshi Banno**,
Yoshiki Fujikawa***

**Analytical Chemistry Research Laboratory
Tanabe Seiyaku Co., Ltd.
16-89 3-chome, Kashima
Yodogawa-ku, Osaka

***Analysis & Computing Division
Tanabe Seiyaku Co., Ltd.
21, 3-chome, Doshomachi
Higashi-ku, Osaka

ABSTRACT

A laboratory automation system was developed for acquistion, analysis, compilation and recording of the data obtainable in the research on long-term stability tests of pharmaceutical raw materials. This system is composed of several sets of frequently used measuring instruments connected by RS-232C interface with personal computers. The measuring instruments are one micro-electronic balance, two semimicro-electronic balances, one spectrophotometer, one potentiometric titroprocessor and two integrators for high performance liquid chromatograph. This system has functions for, among others, reading of stability information on stored samples with a bar code method, automatic transmission of the value weighed of samples, on-line processing of the data obtained, recording with the floppy discs and print-out of analog data and

*Previously published in the Journal of the Society of Japanese Pharmacopia, 16 (5).

analytical results with the XY-plotter. Furthermore, to make this system more capable and then to make the experiment more efficient, we have established an integrated on-line system from the weighing of samples to the data processing and recording, by connecting the Zymate Laboratory Automation System with the personal computer, especially as to the high-performance liquid chromatographic method which is most widely applied in the ordinary stability tests.

INTRODUCTION

This is the first report, conceing long-term stability tests of pharmaceutical raw materials, on the successful development of a laboratory automation system for collection, transfer, analysis, editing, recording and filing of data with frequently used analytical instruments connected to personal computers. By connecting a host computer to this system in the near future, we plan to promote on-line processing for compilation of lists from recorded and stored data, and listing and retrieval of stability information.

COMPOSITION OF THE SYSTEM

The system configuration for measurement is shown in Figure 1. The specifications of the apparatus and interface used in this system are summarized in Table 1.

METHOD OF CONNECTION

Printers are connected according to the centronics specification, and the robot controller and semimicro-electronic balance are coupled by current loops, while all others are connected by RS-232C interface. Data of each instrument is processed by connecting hand-held personal computers PC-8201[1] for the ease of operation, and the PC-8201 and C-280[2] are connected to handle the data comprehensively in the C-280. The data transfer speed[3,4] is 9600 bps between spectrophotometer and personal computer, 4800 bps between micro- or semimicro-electronic balance[5] and personal computer, and 2400 bps between all others.

Besides, if the number of channels is insufficient in the RS-232C interface between the C-280 and PC-2801 personal computer, an RS-232C multiplexer[6] is connected as shown in Figure 1 in order to change the channels by the software where necessary, so that the operator may experiment without being conscious of the multiplexer.

DATA PROCESSING

Data Processing by C-280 (No. 1)

The C-280 (No. 1) personal computer connected to the balance,[5] spectrophotometer[7] and potentiometric titroprocessor[8] is designed to process the data in items (1) through (6) simultaneously by the multijob and multitask functions.

(1) The weight data from the balance is read in through the PC-8201.

(2) The weight data and measurement data of spectrophotometer (AD converted value capable of reproducing adsorption spectrum, maximum wavelength, minimum wavelength, absorbance) are collected as a set, which is read in, recorded, and stored through the PC-8201.

(3) The weight data and measurement data of potentiometric titroprocessor (AD converted value capable of reproducing titration curve, equivalent point) are collected as a set, which is read in, recorded, and stored through the PC-8201.

(4) A collection of information about analog chart, processing result, and measurements is printed out to the XY-plotter[9].

(5) The recorded data of items (2) and (3) is referred to.

(6) The recorded data of residue on ignition and loss on drying is referred to.

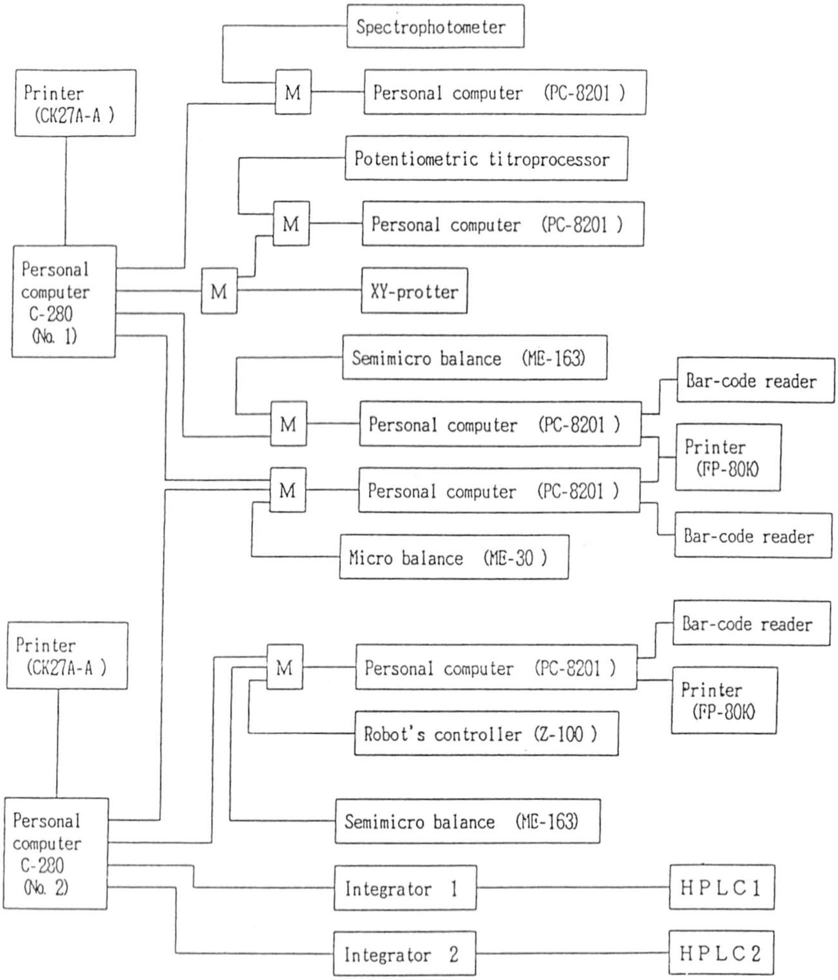

FIGURE 1. Block diagram of the data processing system for stability tests.

TABLE 1. List of Apparatus.

Apparatus	Maker	Model	Number	Remarks
Personal computer	Pana-Facom	C-280D	2	Main memory 384 KB Hard disk 10 MB Floppy disk 1 MB
	NEC	PC-8201	5	Main memory 64 KB
Bar code reader	NEC	PC-8246	3	
Printer	Pana-Facom	CK27A-A	2	24 dot X 136 columns
	Epson	FP-80K	2	9 dot X 80 columns
XY-plotter	National	UP-6802A	1	Color printing
RS-232C Multiplexer	Techno Park	M-100A	6	4 channels
Micro electronic balance	Mettler	ME-30	1	
Semimicro electronic balance	Mettler	ME-163	2	
Spectrophotometer	Shimadzu	UV-365	1	
Potentiometric-titroprocessor	Metrohm	636	1	Digital titrator
Integrator	Shimadzu	C-R2AX	2	User memory 20 KB RS-232C interface
High performance liquid chromatograph	Shimadzu	LC-5A	2	
Zymate (Robot)	Zymark	Z-100	1	Contorller, Robot RS-232C interface Injector

Data Processing by C-280 (No. 2)

Parallel processing[2] of items (1) through (6) is designed in order to process the output data from the integrator[10] connected to the high performance liquid chromatograph (HPLC).

(1) In the case of manual weighing, the weighing data is read in through the PC-8201; or in the case of weighing by robot, the data is read in through the robot's controller.[11]

(2) The peak information (AD converted value capable of reproducing chromatogram, number of peaks, status, retention time, area) is received respectively from two integrators which operate independently, and the peak is identified, while this information is combined with the weighing data of item (1) to calculate quantitatively. The data processing result is formatted, and is transmitted and printed out to the integrators.

(3) All data of item (2) are recorded and stored.

(4) The master table for qualitative and quantitative determination is referred to or altered.

(5) The recorded data of items (1) and (2) are referred to.

(6) Using the data of item (2), the resolution factor showing the column efficiency and tailing factor are determined, and are sent to the printer.[2]

EXPERIMENT

Preparation of Sample Bar Code Label[12,13]

In order to facilitate inputs of information about stability test samples into the computer and to avoid input errors when weighing samples, the labels to be attached to the sample bottles are prepared in bar code so as to be directly read by the bar code reader. The bar code is composed of ten digits, comprising the encoded sample name, storage condition, storage term, and lot number as shown in Table 2. This program is shown in the flow chart in Figure 2.

As the bar code, the JAN code[12] is used.

TABLE 2. Sample Code.

Code's name	Unit number
Sample	3
Storage condition	2
Term unit	1
Storage term	3
Lot No.	1

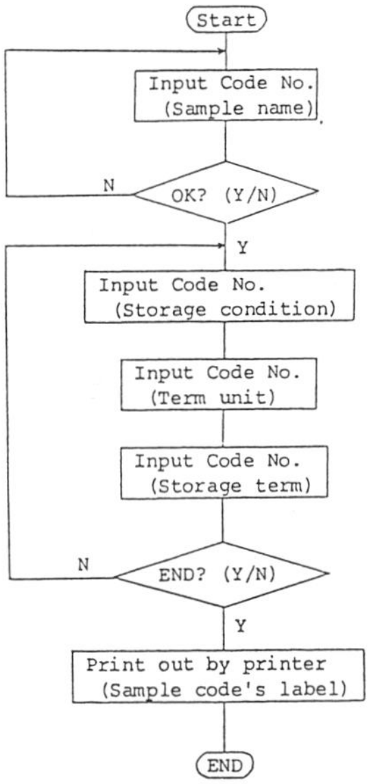

FIGURE 2. Printing system of sample code's label.

Specifically, the storage condition is specified in fourteen code numbers necessary for stability test as shown in Table 3, the storage term is designated in ten code numbers, and the term unit consists of the day, month and year.

TABLE 3. Relationship Between Code No. and Storage Conditons

Code No.	Storage Conditions
00	Initial sample
01	Refrigerator
02	Room temperature, Closed colorless bottle
03	Room temperature, Closed brown bottle
04	Room temperature, Opened colorless bottle
05	Room temperature, Opened brown bottle
06	25°C, 75% RH
07	40°C (Incubator)
08	40°C, 75% RH
09	60°C (Incubator)
10	Sunlight, Closed colorless bottle
11	Sunlight, Closed brown bottle
12	Xenon light, Closed colorless bottle
13	Xenon light, Closed brown bottle
14	40°C, 75% RH, Closed bottle

FIGURE 3. Example of sample code's label. (1) Sample Name; (2) Storage condition and storage term; (3) Lot No.; (4) Barcode; (5) Code No.

An example of output of sample bar code label is shown in Figure 3.

Weighing of Samples by Balance

By operating the weighing program of the PC-8201 connected to the micro- or semimicro-electronic balance,[5] the apparatus number, experiment number (SEQ#), opeartor's name, and experiment phase are sequentially fed. The information about the sample is read by the bar code reader, and the sample code adhered to the sample bottle is decoded, and displayed on the liquid crystal display of PC-8201, so that the content may be checked. To weigh the sample, the total weight of sample and tare is measured; first then the tare weight is determined. By pressing the end key, the input parameter, sample information, and weight measurements are printed out to the printer[14] connected to the PC-8201 and are also transferred to the C-280. Meanwhile, in the event of misoperation in this weighing system, it is designed to re-sample by specifying SEQ#.

Measurement of Loss on Drying and Residue on Ignition

After the inputs of operator's name, experiment phase, and container number (weighing bottle or crucible), the sample code label adhered to the bottle is read by the bar code reader. In succession, the tare, and the sum of tare and sample are weighed, and the loss on drying or residue on ignition is measured according to the respective measuring method. By feeding the container number and weighing it again, the weight is automatically calculated. The result of processing is printed out. All necessary items of measurement data are transferred to the C-280 to be recorded and stored in the on-line system.

Measurement by Spectrophotometer

The weighed sample is used as sample solution and is measured in the flow chart shown in Figure 4.

By operating the PC-8201 connected to the spectrophotometer and feeding SEQ#, the itemized set values relating to measurement registered in the C-280 (wavelength scanning speed, wavelength scale, start wavelength, and end wavelength) and the sample information in the sequence of measurement are sent into the PC-8201, then shown on the liquid crystal display. To collect data, the absorbance (0.2 mm/point) delivered from the spectrophotometer is read. After the measurement, the maximum and minimum absorption wavelengths are determined. The acquired data (SEQ#, operator's name, date of measurement, and spectral information) is transferred to the C-280.

Measurement by Potentiometric Titroprocessor

Titration is effected according to the flow chart shown in Figure 5.

By operating the PC-8201 connected to the potentiometric titroprocessor[8] and feeding the SEQ#, operator's name, and normality factor of standard solution for volumetric analysis, the sample information (blank test or sample) in the sequence of measurement registered in the C-280 is sent to the PC-8201 and is shown on the liquid crystal display. To collect data, all data delivered from the titroprocessor is taken in, and the equivalent point correction and quantitative calculation are performed after the titration. The acquired data (SEC#, operator's name, data of measurement, and titration infromation) is transferred to the C-280.

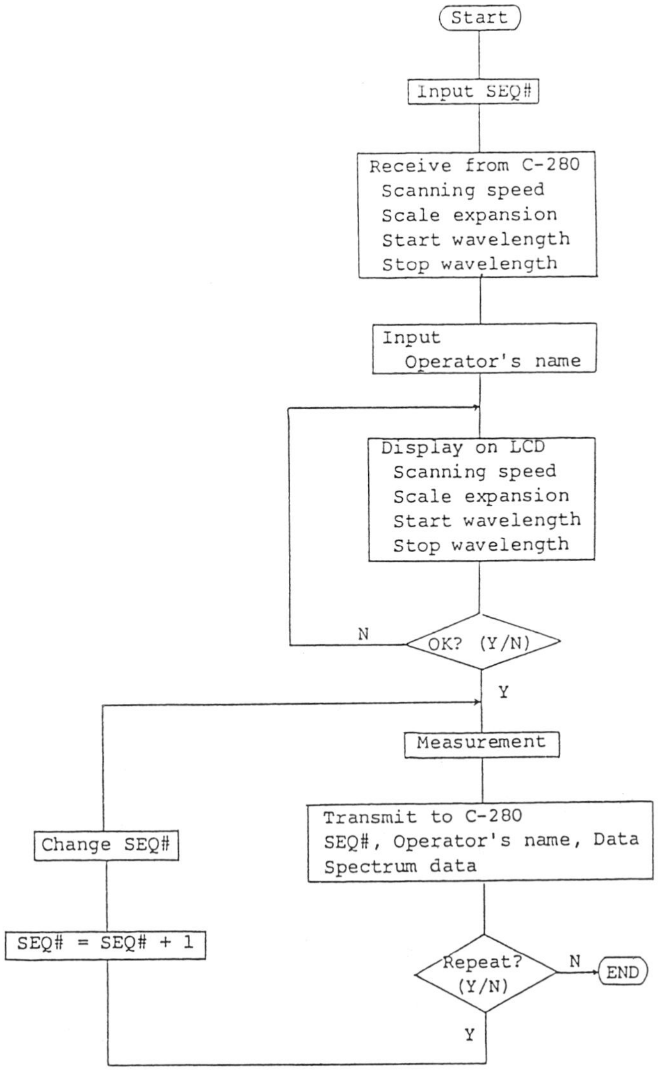

FIGURE 4. Flow chart of operation program for spectrophotometer.

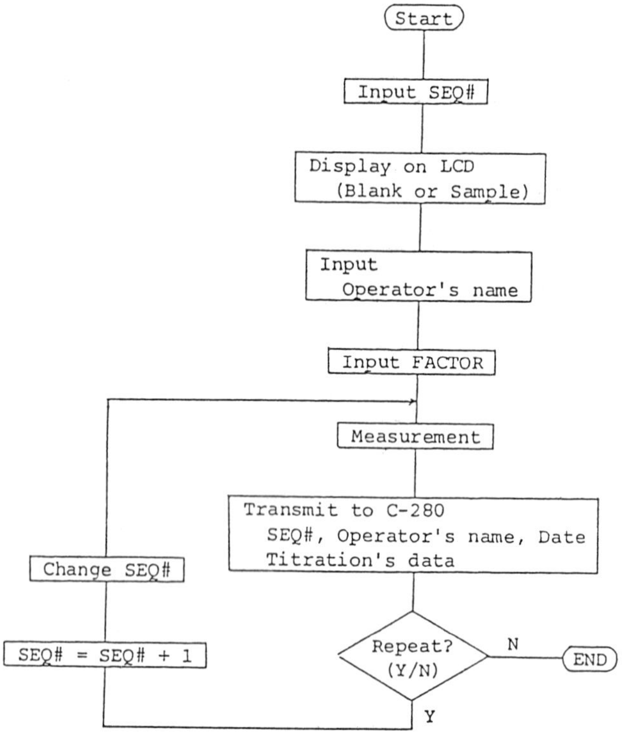

FIGURE 5. Flow chart of operation program for potentiometric titroprocessor.

Measurement by HPLC

The C-280 receives data (number of peaks, retention time, shape of peak, peak area, and AD converted value) from the integrator connected to the HPLC. Furthermore, using the sample weight value, the necessary data processing is done, and the result of processing is formatted depending on each assay, and is printed out again to the integrator in the on-line system. A data processing flow chart is shown in Figure 6.

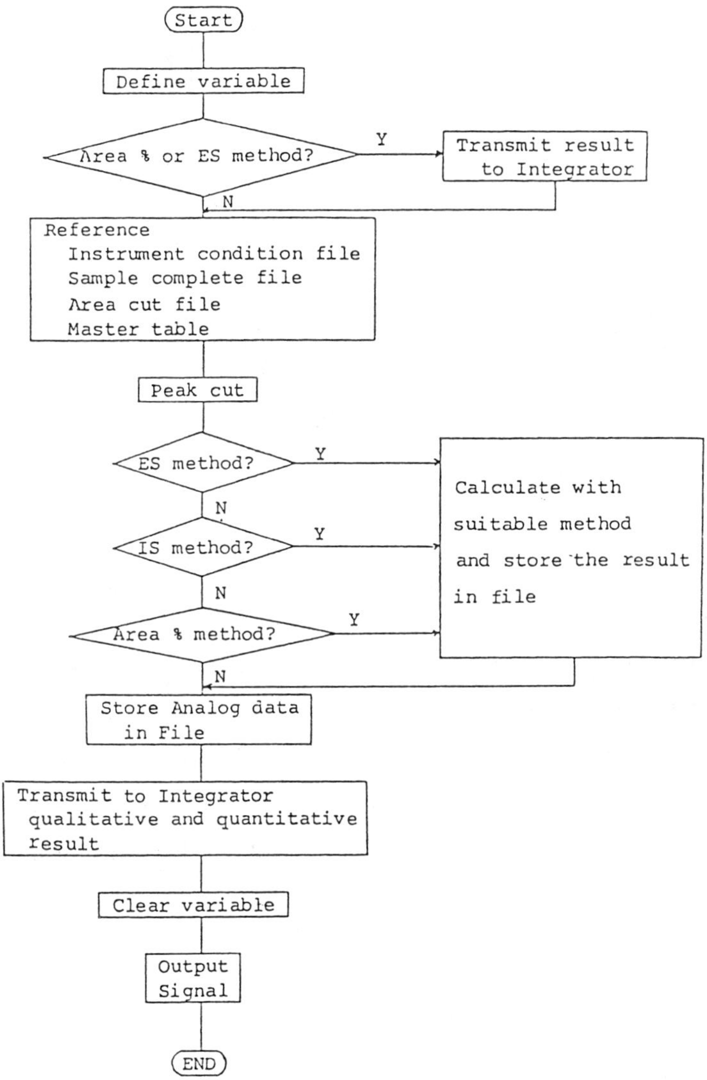

FIGURE 6. Flow chart of data processing program for integrator. ES: External Standard; IS: Internal Standard; Area %: Area Percentage.

For purity test and quantitative determination methods, programs are prepared in three commonly employed versions - the area percentage method, internal standard method, and external standard method.

This system is automated in the entire analysis of chromatograph in combination with automation of experiment operations such as automatic weighing of samples by the Zymate System,[11] pretreatment, and injection operation into HPLC.

To take in the data, the data can be collected by operating two integrators simultaneously. The analog data is converted into digital information at one-second intervals by the program incorporated into the integrator and is realtime transferred to the C-280. As a result, all items of data relating to chromatogram in long-term stability test can be taken in.

RESULTS AND DISCUSSION

Sample Weight Measurement

The sample weight measurement and sample information are printed by the printer connected to the PC-8201 instantly upon end of weighing. A print example relating to weighing of sample measured for HPLC is shown in Table 4.

The storage condition and term read from the adhered label are recorded in the comment column.

TABLE 4. Example of the Output of Sample Weight.

INS#-SEQ#	(LC 2 - 166)
DATE	Dec. 12, 1984
OPERATOR	T. SAEKI
SAMPLE NAME	ABCDE
MODE	SA
LOT NO.	403100
COMMENT	SUS-C-BROWN 20D
REPETITION	1
Sample + Tare	19.281 mg
Tare	0.112 mg
Sampled weight	19.169 mg

TABLE 5. Example of the Output for Loss on Drying.

INS#-SEQ#	(BALI - 3)
DATE	Dec. 17, 1984
OPERATOR	SHIBASAKI
SAMPLE NAME $	ABCDE
TESTING ITEM	DRY
COMMENT	SUN-C-WHITE 60D
LOT NO.	403080
REPETITION	1

Weight of weighing bottle	15.4437 g
Weight of weighing bottle + Sample	16.4585 g
Weight sample	1.0148 g
Weight of weighing bottle + Sample after drying	16.4583 g
Loss on drying	0.0002 g
Percent loss on drying	0.02 %

Loss on Drying and Residue on Ignition

In succession to the information relating to measurement and sample, the acquired data and result of processing are printed out. A print example of loss on drying to the printer is shown in Table 5.

Absorption Spectrum

The result of measurement on storage stability sample is shown in Figure 7 as an output example. The output is delivered by the XY-plotter,[9] and the analog data of three lots of stability sample are displayed in different colors so as to be distinguished easily. In data processing, the maximum wavelength in the wavelength region longer than 220 nm, the minimum wavelength between respective maximum, wavelengths, and absorbance between respective wavelengths are determined and compiled in a list.

Potentiometric Titration

The result of measurement on storage stability sample is shown in Figure 8 as an output example. Similar to Figure 7, the analog and blank test results of three lots of test are displayed in different colors, and the equivalent point is indicated by x-mark. On the result of data processing, the blank test value is subtracted from the equivalent point of titration, and the measured content (mg) and assay percent are printed out according to the calculation formula determined for each item.

HPLC Method

Figures 9 and 10 show the result of measurement of standard solution and stability test solution by employing the internal standard method.

The result of measurement of standard solution (STD RUN) shown in Figure 9 comprises the outputs of apparatus number, experiment number, date of measurement, operator's name, master table number, and sample name,

FIG.NO. 06009-1 U.V.SPECTRA

85/08/23 T.SAEKI

TA-2711 : 25°-75%RH 9 months (1)

NO. LOT NO.	(mg)	MAX.(nm)	ABS.	MIN.(nm)	ABS.	MAX.(nm)	ABS.
1.403100	32.297	279.9	0.516	276.5	0.419	271.2	0.555
2.403110	30.545	279.9	0.490	276.5	0.399	271.2	0.527
3.403120	29.733	279.9	0.476	276.5	0.388	271.2	0.512

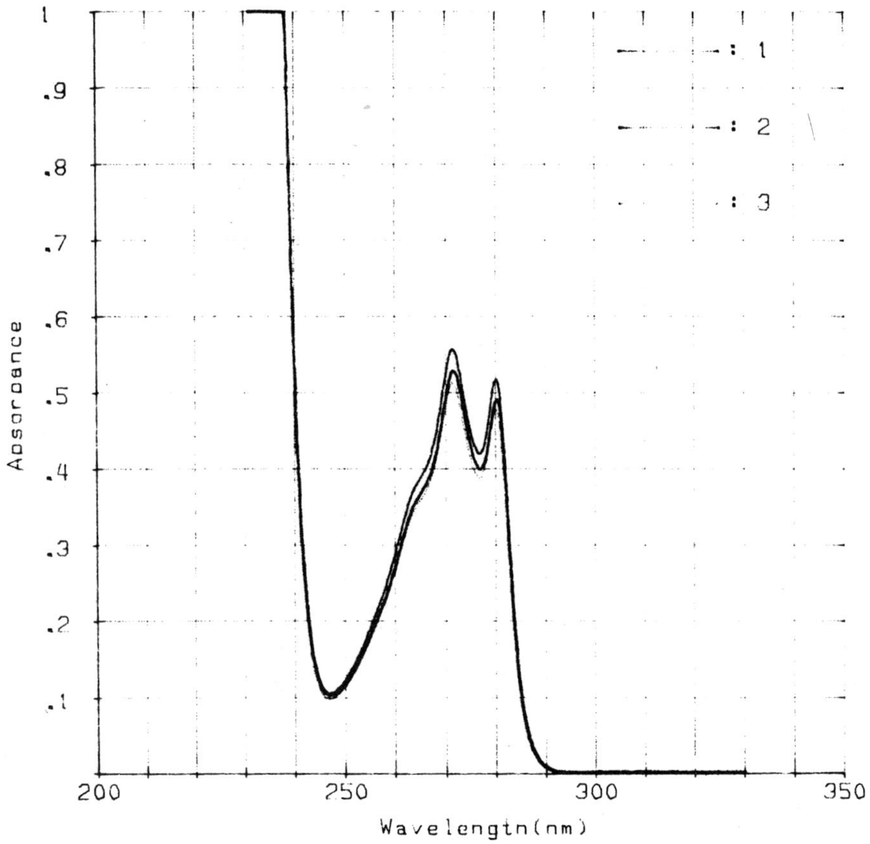

FIGURE 7. Example of the output for spectrophotometry.
(mg): Sample weight; MAX. (nm): Maximum wavelength; ABS.: Absorbance at maximum wavelength.

FIG.NO. 00000-1 TITRATION CURVES

84/12/12 Y.SHIMAOKA

TA-4708 : 25°-75%RH 3 months (1)

FACTOR = 1.010

NO.	LOT NO.	(mg)	E.P.(ml)	FOUND(mg)	ASSAY(%)
1.	EE100	591.31	15.252	590.19	99.81
2.	EE11527	592.27	15.263	590.61	99.72
3.	YA6727	594.38	15.290	591.66	99.54
4.	BLANK		0.014		

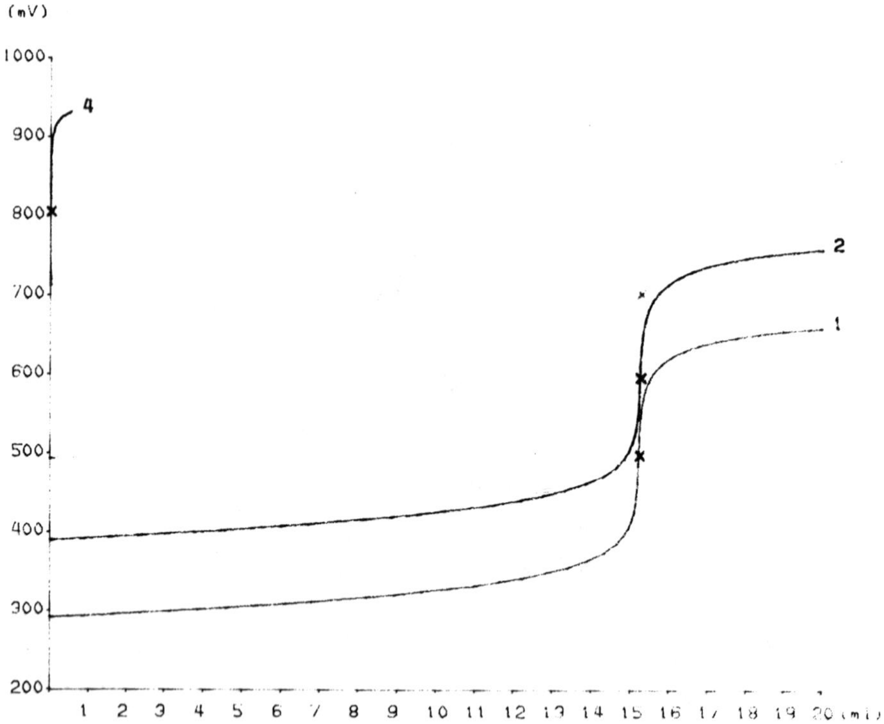

FIGURE 8. Example of the output for potentiometric titration. (mg): Sample weight; E.P. (mL): Endpoint; FOUND (mg): Actual measurement value.

being followed by sequential printouts of measured data of peak number, retention time, peak status, peak area, area ratio, weighed value, coefficient for determination, component name and figure number. On the other hand, the result of measurement of sample solution (SAMP RUN) comprises, besides the items shown in Figure 9, the lot number, storage condition, term, repetition, and weighed value, being followed by sequential printouts of measured content (mg) and assay percent calculated by using the coefficient for determination obtained in the measurement of standard solution. Also, in the area percentage method and external standard method, the results are printed out in respective formats. It is possible, moreover, to deliver the resolution factor showing the column performance of the obtained chromatogram, tailing factor, and theoretical plate number of stages whenever necessary.

SUMMARY

To evaluate the stability of pharmaceutical materials, it is necessary to collect, process, classify and compile a colossal volume of experimental data for a very long period, which requires a huge load of labor.

Recently, by connecting plural analytical apparatus with personal computers, the authors developed an on-line system[15,16,17] capable of taking in and collecting all analytical data relating to the routine stability tests, and arranging and delivering the results of processing together with the analog data through the printer, XY-plotter, and integrator.

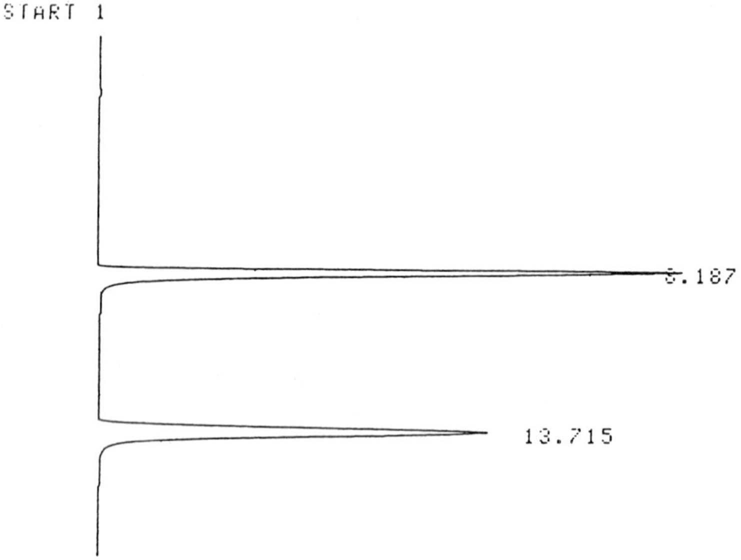


```
SEQ.NO.   LC2-052          84/12/12   T.SAEKI
TAB.NO.   001              SAMP RUN
SAMPLE    ABCDE            LOT.NO.    403100
COMMENT   RT-O-WHITE    3M -1
SAMP WT   20.365
```

NO	RT	MK	AREA	RATIO	FOUND	PERCENT	NAME
1	8.19		742636	1.0466	20.301	99.69	ABCDE
2	13.72		709568	1.0000	*******	*******	IS

Fig. No. 0055 LIQUID CHROMATOGRAPH (ASSAY TEST)

FIGURE 9. Example of the output of standard solution for HPLC. NO: Peak No. RT: Retention time; MK: Peak status; AREA: Peak area; RATIO: Ratio of peak area (ABCDE/IS); Weight: Sample weight (mg) ; COEFF: Coefficient for determination; NAME: Component name.

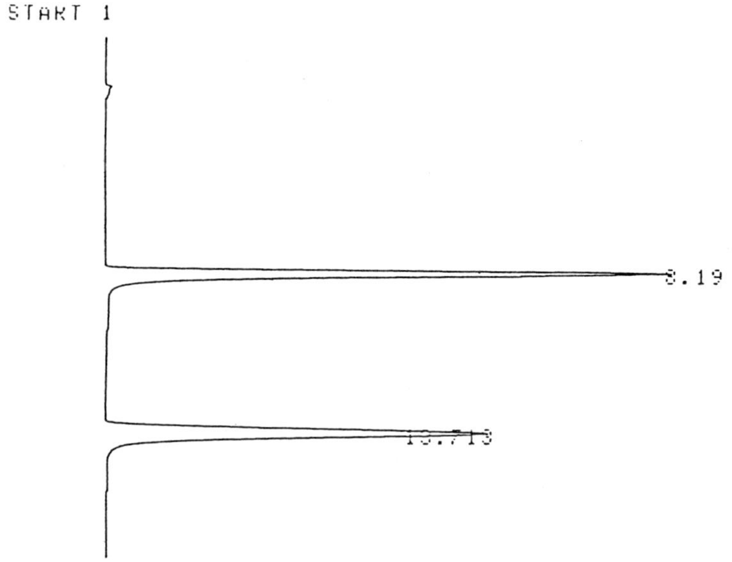

SEQ.NO. LC2-051 84/12/12 T.SAEKI
TAB.NO. 001 STD RUN
SAMPLE ABCDE

NO	RT	MK	AREA	RATIO	WEIGHT	COEFF	NAME
1	8.19		723715	1.0284	19.948	19.397	ABCDE
2	13.71		703733	1.0000	*******	*******	IS

Fig. No. 0054 LIQUID CHROMATOGRAPH (ASSAY TEST)

FIGURE 10. Example of the output of sample solution for HPLC. FOUND: Actual measurement value (mg); PERCENT: Assay percent.

By the operation of this system, labor can be saved and the efficiency may be enhanced in a series of conventional manual operations for tests and analyses, while possible errors in the process of calculation or data compilation can be avoided.

The authors are also planning, in the near future, to connect a host computer to this on-line system so as to automate comprehensively tabulation and retrieval of acquired and stored measured values, compilation of test results, and listing of stability research results.

REFERENCES

1. User's Manual of Personal Computer PC-8201, N_{82}-BASIC Reference Manual, Nippon Electric Co., Ltd.

2. Instruction Manual of Personal Computer C-280 (1983), Pana-Facom Corporation.

3. Fujiwara, K. et al., Interconnecting Microcomputers, Jitsugyo Nippon Publishing Co. (1982).

4. Sugasaki, et al.: Standard Interface, Use of Bus. Interface 69, 214, CQ Publishing. (1983).

5. RS-232C Interface Instruction Manual for ME 30, AE 163 (1982), Mettler.

6. Instruction Manual of RS-232C Multiplexer M-100A (1983), Techno Park Mine Co.

7. Instruction Manual of Communication Interface for UV-365, Shimadzu Corporation.

8. Titroprocessor 636 Instructions for Use, Metrohm AcT CH-9100 Herisau, Switzerland.

9. Digital Plotter VP-6802A Instructions, Matsushita Communication Industrial Co., Ltd.

10. Chromatopack C-R2AX (X), Telecommunication Interface Manual, Shimadzu Corporation.

11. Zymark System Instruction Manual, Zymark Corporation, (1982).

12. PC-8246 Bar Code Reader User's Manual, Nippon Electric Co., Ltd.

13. Bar Code Book 1, Paper Soft Bar Code Program Collection, Mechano Systems Ltd.

14. Kanji Printer FP-80K Instruction Book, Epson.

15. Fujiwara, et al., Analysis, 4, 238 (1981).

16. Minami, Chemical Industry, 30(9), 913 (1980).

17. Nishio, K., Attempts in Laboratory Automation, DECUS Proceedings, 35 (1983).

APPENDIX A: BENCH LAYOUTS

Schematic diagrams and photos of bench layouts are listed by application and page number:

APPENDIX B

Abstracts of the video-poster sessions presented at the Third International Symposium on Laboratory Robotics, Boston 1985.

COLORIMETRIC DETERMINATION OF URANIUM IN AQEOUS SOLUTIONS USING ROBOTICS

Richard P. Ringle
Nuclear Fuel Manufacturing Department
General Electric Co.
Wilmington, NC

ABSTRACT

A laboratory robotic system is used at General Electric Company's Nuclear Fuel Manufacturing Department in Wilmington, North Carolina in a procedure for the colorimetric determination of uranium in aqueous solutions.

In this procedure, aliquots of previously dissolved uranium dioxide samples or standards are manually pipeted into test tubes located in a rack within the working area of the robotic arm. Using a general purpose hand, the test tubes are sequentially picked up and moved to a reagent dispensing station where measured volumes of hydrogen peroxide and sodium hydroxide solutions are added. After being transferred to a vortex station where the solution is thoroughly mixed, the test tube is returned to the rack.

Once solution in all of the test tubes have been prepared and a necessary method color development time has passed, the robot hand picks up a sipper tube and sequentially inserts the tube into the test tubes. A peristaltic pump is activated and draws solution from a test tube into a flow-thru cell in a spectrophotometer. After the cell is flushed with several volumes of solution, the absorbance of the yellow colored complex that has been formed in the solution is measured at 425 nm and printed out using an A/D converter. The software contains provisions for entering the number of standards and samples to be analyzed and timers for the necessary waiting periods.

With only approximately 10% of the system's available time being utilized there is a potential for the significant additional savings and efforts are underway to use it to perform several other analyses. In addition to freeing laboratory personnel from a repetitive monotonous procedure for more challenging and rewarding tasks, personnel are no longer required to repeatedly pipet hydrogen peroxide and caustic sodium hydroxide solutions used in this analysis.

IMMERSION TESTING OF POLYMERIC MATERIALS

S. G. Dunn and G. L. Hagnauer
Army Materials & Mechanics Research Center
Polymer Research Division, Watertown, MA 02172

ABSTRACT

The sorption and diffusion properties of liquids in polymeric materials can be evaluated by measuring the weight changes of specimens as a function of the time they are immersed in the liquids. The measurements also depend upon other variables such as sample thickness, temperature, and operator technique. Testing involves repetitive operations - transferring liquids, removing specimens from sample containers, blotting specimens, weighing specimens, recording weights and immersion times, returning specimens to their respective containers - which are monotonous and can be hazardous, especially if flammable liquids or toxic chemicals are used. In addition, the technician must calculate the weight change and plot % weight change versus immersion time to analyze the data and estimate times for successive measurements. Finally, testing is limited by the large number of specimens and measurements required the normal 8 hour work day.

This poster describes efforts within laboratory to develop a robotics system for immersion testing of polymeric materials (composites, elastomers, and plastics). Our goal is to significantly reduce costs and manpower requirements, while increasing the sample work load and providing for improved precision, accuracy, and safety in measurements. The current system employs a Zymate Controller and Robot with a specially designed robot hand, Power and Event Control Station, Master Laboratory Station with Remote Dispenser, Capping Station, Printer, and infrared sensor, specimen blotter, special holders for samples and containers, a Mettler top loading analytical balance, and a thermostatted water bath. Test specimens have been standardized as either 1 inch diameter circles or 1 inch squares of uniform thickness ranging from 0.02 to 0.2 inch. Each specimen is positioned vertically and immersed in the test liquid inside a specially designed glass container. The containers are capped as a safety measure and to prevent liquid loss. The system is programmed to reproduce, as closely as possible, operations performed by technicians running the test and has been designed to minimize sample handling time and optimize measurements.

The poster describes the immersion test and includes a schematic drawing and photographs of the apparatus. Examples of test results are illustrated and compared with those obtained by a technician. Advantages and limitations of the current test system are listed and questions relating to implementation, anticipated cost savings, and future plans to improve and extend the test will be addressed.

A SIMPLE GAS CHROMATOGRAPHIC INTERFACE FOR A LABORATORY ROBOTICS SYSTEM

W. A. Schmidt, K. M. Stelting,
J. J. Rollheiser and A. T. Chatham
Midwest Research Institute
425 Volker Boulevard, Kansas City, MI 64110

ABSTRACT

MRI has developed laboratory robotics systems for chemistry support for toxicology research. One such system performs chromatographic analyses of animal feed dosed with potentially toxic chemicals. This procedure includes the basic operations of extraction, clarification and dilution, and GC or HPLC analysis, with the robotic system preparing standards and performing all operations. MRI has recently modified a Varian 8000 gas chromatographic autosampler to interface with a Zymate laboratory robotic system. A description of the modifications, as well as experimental results will be included on the poster.

LAB-MADE ROBOTICS STATIONS

K. E. Trumbo, F. W. Brill and S. J. Eitelman
ICI Americas Inc.
Biological Research Center
Goldsboro, NC 27533-0208

ABSTRACT

Zymark markets many useful modules and stations for use with the Zymark Laboratory Automated System, but this was not always so. When robotics was first introduced to pioneering laboratories, some modules and stations were invented by the user in order to expand or improve their robotics system. Most of the time these inventions were put together inexpensively and with simplicity and practicality being the major factors.

There are six such stations that are part of the Zymark Robotics System at ICI Americas Biological Research Center in Goldsboro, North Carolina. Since many of these stations have been put into use, Zymark has marketed similar items. The six stations are: 1) Cap Dispensing Station, 2) Batch Shaker, 3) Balance Door Opener, 4) Heated Vial Rack, 5) Liquid Transfer Station, 6) Robotic Voice Box. All these stations work in conjunction with other Zymark modules in our Robotic System. The purpose of the system is to prepare liquid samples of herbicides and pesticides for gas chromatographic analysis.

APPLICATION OF LABORATORY ROBOTICS IN AGRICULTURAL PLANT TISSUE ANALYSIS

Joseph M. Rao and Michael W. Martin
Allied Corporation, Syracuse Research Laboratory
P.O. Box 6, Syracuse, NY 13209

ABSTRACT

The Crop Sciences Laboratory of Allied Corporation is currently conducting a large research effort in plant genetics. One of their main projects is the development of improved varieties of corn. The amount of total nitrogen and nitrate in the experimental plants is a very important parameter in this study.

The Analytical Sciences Group at Allied's Syracuse Research Laboratory has thus been analyzing approximately 20,000 corn tissue samples per year for total nitrogen and extractable nitrate. The objective of our work is to analyze these samples rapidly, accurately and at minimum cost.

Total nitrogen is currently being performed via an Erba Automatic Nitrogen Analyzer. Extractable nitrate is being performed by the Zymate Laboratory Robotics System and a Wescan Ion Analyzer. The nitrate determination was originally performed by a manual extraction followed by analysis on a Technicon Autoanalyzer.

In the nitrate analysis, corn tissue samples are first dried overnight in an oven. A technician then tightens the caps and places the rack of sample tubes on the robotics table and begins the program. The samples are then uncapped and 120 to 150 mg are weighed using the vibrating hand. The extracting solvent is then added and the sample is vortexed for one minute. The solution is then filtered, diluted, and directly injected into the Wescan.

This automated procedure shows excellent precision and values correlate well to the previous method described above. We can currently run approximately 100 samples per day. Based on our analyst rate per hour, the cost to the Crop Sciences Group is approximately 70% less than with the original analysis scheme.

Future plans involve bar coding the samples and interfacing the system to a microcomputer for complete report writing. We will also investigate automating the sample preparation for the total nitrogen analysis. This involves weighing approximately 20 mg of plant tissue into 5 x 8 mm tin cups.

AUTOMATION OF MICROBIOLOGICAL LARGE PLATE AGAR DIFFUSION ASSAYS EMPLOYING LABORATORY ROBOTICS

Mary Jo Mroczenski-Wildey, and William M. Maiese
Biotechnology Dept., American Cynamid Co.
Medical Research Division, Lederle Laboratories
Pearl River, NY 10965

ABSTRACT

Quantitative microbiological agar diffusion assays are often tedious and labor intensive procedures. The Zymate Laboratory Automation System was used to automate several quantitative diffusion assays to increase their overall efficiency. Software was developed to direct the system to automatically weigh, extract, and dilute samples onto an assay plate containing the microbial indicator organisms. Reference standards are also added to each assay plate at specific locations to aid in subsequent computer-assisted quantitive evaluation of the data. The Zymate System is capable of processing over 325 samples in one batch setting and is flexible so that any of its operating parameters can be quickly changed. Optimum conditions to standardize the automation of the microbiological assays using the Zymate Robotic System and flexible approach to automate quantitive microbiological assays.

AUTOMATION OF FRACTION COLLECTION IN SUPERFUSION STUDIES

Dwight Lubansky
Ciba-Geigy Pharmaceuticals
556 Morris Ave., Summit, NJ 07901

ABSTRACT

A project proposed by one of Ciba-Geigy's researchers required the collection of aqeous effluent of tissue perfusion from 24 samples. The nature of the experiment mandated a time-controlled collection routine with various solutions presented to the tissue at the appropriate times needed to incubate tissue chambers, incubate perfusion solutions, control peristaltic pumps, and collect 240 samples into mini-scintillation vials.

The mechanism designed is capable of moving a tray containing 24 Beckman vial racks. The racks move along under a set of dispensing needles, stopping at each of the 18 positions while the peristaltic pumps are activated to fill the vials with the perfusate. The Beckman racks are used so that the vials can then be transferred to the scintillation counter after being filled with scintillation cocktail and capped.

AUTOMATED ANALYSIS OF AMINO ACIDS IN FERMENTATION BROTH WITH THE ZYMARK LABORATORY ROBOT AND COLUM SWITCHING HPLC METHOD

Peter K. Lai and Tin-Chuen Yeung
G.D. Searle & Co.
4901 Searle Parkway, Skokie, IL 60077

ABSTRACT

A fully automated procedure for the analysis of amino acids in fermentation broth has been developed. This automated procedure includes: robotic technology, HPLC with column-switching technology and on-line data acquisition and reduction.

The Zymate robot system is used in the sample preparation step, which replaces the routine repetitive manual operation of the analyst. In addition, the Zymark Z310 HPLC injector station of the robotic system is interfaced with a Waters 590 solvent delivery system which in turn controls the switching of a Valco 10-port switching valve. The use of column-switching significantly reduces the analysis time from 45 minutes to 8 minutes by back-flushing of late eluting materials off the guard column. This automated procedure provides a more efficient and precise analysis of amino acids in fermentation broth and is especially suitable for high volume of sample analysis.

The design and approaches of interface will be presented.

UTILIZATION OF LABORATORY ROBOTICS TO AUTOMATE THE PREPARATION OF SAMPLES FROM DRUG DISPOSITION STUDIES FOR CHROMOTOGRAPHIC ANALYSIS: DETERMINATION OF CGS10787B

L. A. Brunner and R. C. Luders
Development Dept., Pharmaceuticals Div.
Ciba-Geigy Corp.
Ardsley, NY 10502

ABSTRACT

A Zymate Laboratory Automation System was developed to automate the preparation of plasma and urine samples for the analysis of CGS 107887B (a disease-modifying antirheumatic drug candidate).

The robot aliquots the biological sample, adds internal standard and performs all steps necessary for the liquid-liquid extraction and concentration of the drug candidate. The concentrates are transferred to an autosampler and analyzed using an ion pair reversed-phase high performance liquid chromatographic technique. The robot allows for unattended 24-hr preparation of clinical samples arising from drug disposition studies. The analytical method has both the sensitivity and specificity to quantify CGS 10787B concentrations in the presence of major metabolites and drug related compounds following administration of anticipated doses of the drug.

ROBOTIC PROCEDURE FOR THE DIRECT DETERMINATION OF IVERMECTIN IN ANIMAL PLASMA AT PARTS-PER-BILLION LEVELS

J.V. Pivnichny, A.A. Lawrence and J.D. Stong
Animal Formulation Development
Merck Sharp & Dohme Research Laboratories
P.O. Box 2000, Rahway, NJ 07065

ABSTRACT

A robotic method for determining the veterinary antiparasitic agent, ivermectin at ppb levels in animal palsma is described. The method, which is an adaptation of a previously reported manual procedure (J.Pharm.Sci.,72,1447(1983)) involves twofold extraction by liquid-liquid partitioning, further clean-up by column chromatography on Florisil, and final analysis by HPLC. Removal of solvents by evaporation is required at two points. A modification which significantly reduces processing time for an individual sample is substitution of small (6 mL) disposable Florisil columns (Baker-10) driven by compressed gas for large (25 mL) gravity-fed reusable glass columns. Coupling the Florisil columns to a solvent dispensing station facilitates handling of the large solvent volumes (10-20 mL) prescribed in the original procedure. An additional modification, the substitution of vortex mixing for lateral shaking during liquid-liquid partitioning, eliminates the need for leakproof sealing of screw-cap centrifuge tubes and minimizes the formation of interfacial emulsions. While the overall procedure is quite complex, involving three uncapping/cappings, four vortex mixings, two centrifugations, nine solvent dispensings, two solvet evaporations, and seven transfers by pipet, it exhibits a remarkably low failure rate. It is as accurate as, and more precise than, the manual procedure and; on a weekly basis, allows for sample processing at greater than twice the manual rate.

ROBOTIC SAMPLE PREPARATION AND LIQUID CHROMATOGRAPHIC ASSAY FOR THE ANALYSIS OF PIRMENOL IN HUMAN PLASMA

E. L. Johnson and L. A. Pachla
Pharmacokinetics/Drug Metabolism
Warner-Lambert/Parke-Davis Pharmaceutical Research
Ann Arbor, MI 48105

ABSTRACT

A sensitive, specific and rapid high performance liquid chromatographic procedure, using the Zymate Laboratory Automation System for sample preparation, was developed for the determination of Pirmenol in human plasma. Plasma samples were alkalinized and extracted with cyclohexane. The organic extract was evaporated to dryness, reconstituted with mobile phase, and then chromatographed on a microparticulate spherical trimethylsilane stationary phase with UV detection at 254 nm.

The procedure was linear from 0.125 to 5.0 mcg/mL. The reproducibility of the peak area ratios of the standard curve had relative standard deviations between 4.9 and 9.8% and a relative error between 0 and 3.8% over the linear range. The accuracy for the determination of Pirmenol in human plasma containing 0.5, 2.5, and 4.0 mcg/mL had relative errors of 2.8, 9.2, and 5.5% respectively.

A comparison of the reproducibility of the manual versus robotic procedure will also be discussed.

THE FULLY AUTOMATED LABORATORY: INTEGRATING ROBOTICS AND LIMS

B.J. McGrattan, M.R. Thompson, P. Barrett,
C.G. Fisher, and A.C. Cerino
Laboratory Robotics Dept., Perkin-Elmer
Main Ave., Norwalk, CT 06859-0927

ABSTRACT

The design of an automated laboratory requires consideration not only of how samples are prepared and data are acquired, but what will eventually become of that data. Although a robotics system comprised of a robot and analytical instruments provides a solution for the first two of these considerations, and a Laboratory Information Management System (LIMS) provides a solution to the last problem, the integration of robotics and LIMS offers the possibility of fully automating the laboratory.

There are two conceptual ways to organize this automated laboratory. The first is to consider the robotics system as a completely intelligent instrument capable of preparing samples and acquiring data. The LIMS system interfaces to the robotics system via the robotic controller, as seen below:

```
                          <-----> Robot
                Robotic   <-----> RS-232 Instrument 1
LIMS <-----> Controller <-----> RS-232 Instrument n
                          <-----> Simple Interface 1
                          <-----> Simple Interface 2
```

The alternative method is to interface instruments directly to the LIMS system, and consider the robot as another of these instruments, as illustrated below:

```
Robotic                      <-----> Instrument 1
System    <-----> LIMS <-----> Instrument 2
                             <-----> Instrument n
```

Each of these methods has an advantage. The first approach allows the robotic system to be operated independently of the LIMS, and to acquire and locally store data. The second method considers the robotic system as another instrument on the LIMS, and consequently, the instrument can be operated independently of the robotics system.

The advantage of each of these methods will be discussed in terms of the Perkin-Elmer MasterLab Robotic System and LIMS 2000. Furthermore, methods of downloading programs and the transferring of data between the robot and the LIMS will be presented, as well as applications of the fully automated laboratory.

LABORATORY AUTOMATION AT ELKTON PHARMACEUTICAL LABORATORIES

Cynthia W. Skelley, R. D. Schlabach, S. A. Johnson
D. C. Fischer and R. J. Sturgeon
Merck Sharp & Dohme
Route 340 South, Elkton, VA 22827

ABSTRACT

Semiautomation of dissolution testing at Merck Sharp & Dohme Pharmaceutical Laboratories in Elkton, Virginia was a result of an immediate need to replace existing equipment. Two Merck-made 20 vessel dissolution testers equipped with automatic samplers needed refurbishing or replacement.

Effective 5/1/82 with Supplement 3 of the USP XX, dissolution became a compendial test. The Merck-made equipment met the USP criteria but frequent maintenance intervention and support were necessary.

The approximate cost to refurbish the existing equipment was $30,000. This included a complete overhaul of the electrical and hydraulic systems, replacement of paddle holders and samplers, sandblasting, and painting. The cost of two 12 vessel Van-Kel dissolution testers was $23,500. The cost of four 6 vessel automatic samplers was $28,000. The cost of a Zymate System and accessories for our application was $27,800. The cost of a special table to hold all dissolution equipment was $3700. Use of the robot as an automatic sampler would accomplish the immediate need of sampling as well as adding flexibility for future applications. After reviewing the choices, economically as well as developmentally, it was deemed reasonable to purchase new dissolution testers and a robot system.

At the Elkton site two tablet products are manufactured in the pharmaceutical area. Both of these are antihypertensives and require dissolution testing. One product has one active ingredient that is sampled at 20 minutes. The other has two active ingredients that require sampling at 30 and 60 minutes. As the sampling tasks are labor intensive, automation of that was desired initially. Upon completion of that tasks, programming was next accomplished for the automatic addition of tablets. Validation of the semiautomated dissolution set up was completed satisfactorily, and use of that system is in place.

As previously mentioned acid corrosion to the prior equipment did pose a problem. In order to prevent a recurrance of this to the new equipment an air line was installed at the robot arm and skirt. A steady but slow pressure is applied that reduces the risk of corrosion to the sensitive electronics.

The future objective for the Zymate System includes content uniformity, dissolution, and assay analyses. Purchases of the following instruments will accomplish this:

1. Perkin-Elmer Lambda 4C UV/VIS Spectrophotometer
2. Perkin-Elmer Series 10 HPLC Pump
3. Perkin-Elmer LC 75 Detector
4. Perkin-Elmer LCI 100 Integrator
5. Perkin-Elmer 7500 Computer
6. HPLC Software
7. UV/VIS Software
8. Epson printer
9. Zymark HPLC injector

For dissolution testing the robot will sample the kettles in the same manner as it does now, but instead of dispensing an aliquot to a cup on the turntable, it will introduce an aliquot directly into a flowcell of the spectrophotometer. The computer will then process the absorbance readings and convert them to a percentage of dissolved active ingredient and print a report package of each lot tested. For the product with two active ingredients there will be interference between the actives at one wavelength. Use of the HPLC may be alternative to multiwavelength analyses. Experimentation with the methods is underway.

Content uniformity and assay determinations will be easily accomplished by manually placing samples and standards on the turntable. The sipper hand will then aspirate samples and standards and inject them singly via the HPLC injector. Methods are in place which use the HPLC for such analyses. The peak heights will be converted to percentages and milligrams as required by the procedures, utilizing programs on the computer.

As a long range plan a second robot arm and the concurrent EasyLab software are desired that will allow for preparation of content uniformity and assay samples. Further plans also include installation of a pumping device that will cleanse kettles thoroughly between runs on the dissolution testers. Finally, a device to deliver exact quantities of dissolution medium will be implemented.